A field guide to the
Larger Mammals of South America

Richard Webb and Jeff Blincow

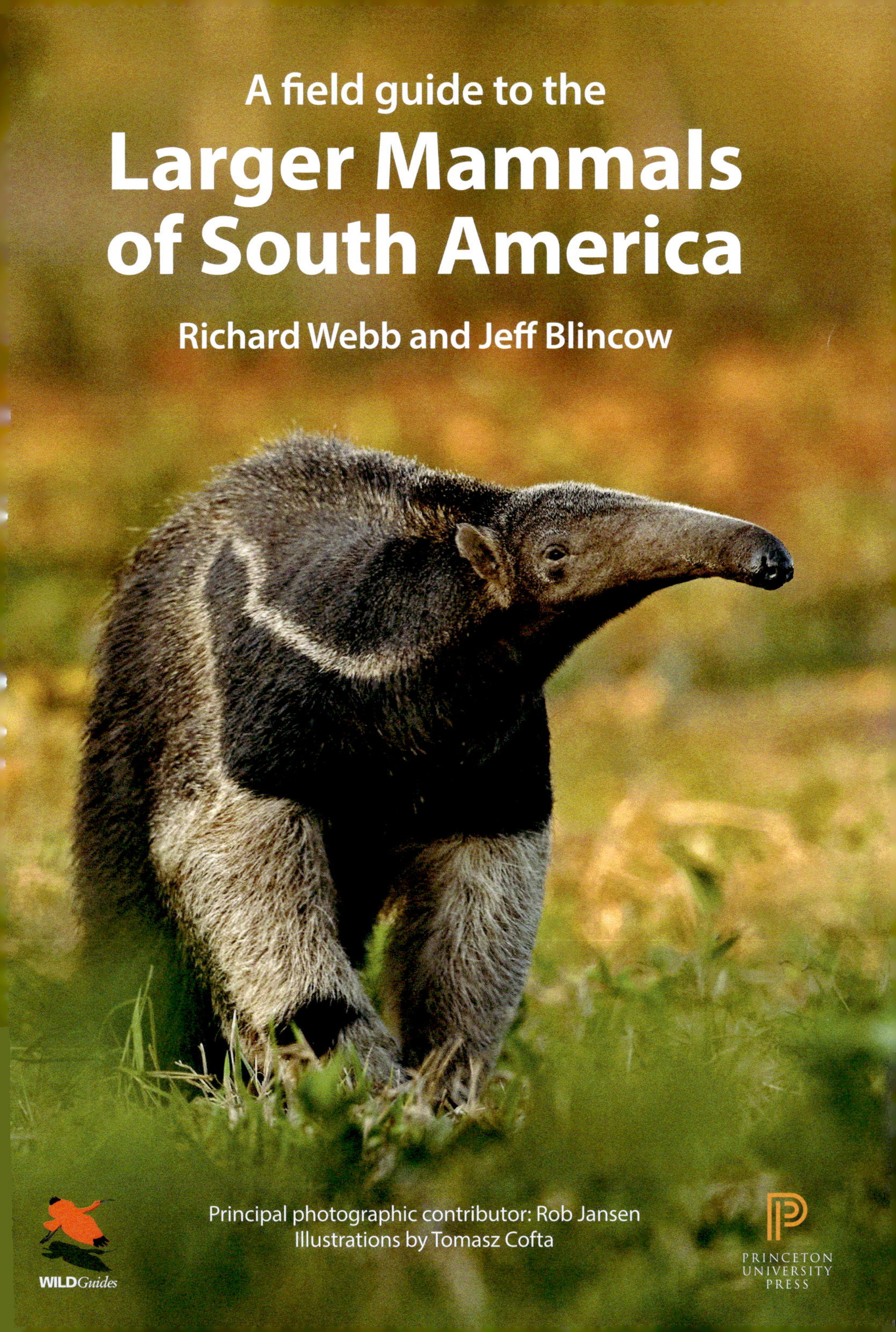

Principal photographic contributor: Rob Jansen
Illustrations by Tomasz Cofta

PRINCETON
UNIVERSITY
PRESS

Published by Princeton University Press,
41 William Street, Princeton, New Jersey 08540
99 Banbury Road, Oxford OX2 6JX
press.princeton.edu

British Library Cataloging-in-Publication Data is available

Library of Congress Control Number: 2023943697
ISBN 978-0-691-17409-9
ISBN (ebook) 978-0-691-25497-5

Editorial: Andy Swash and Megan Mendonça
Cover Design: Rob Still
Production: Ruthie Rosenstock
Publicity: William Pagdatoon and Caitlyn Robson
Copyeditor: David Christie

Cover image: Jaguar by Andrew Griffin
Title page image: Giant Anteater by Ondřej Prosický

MIX
Paper | Supporting
responsible forestry
FSC® C015829
www.fsc.org

Printed in Italy

10 9 8 7 6 5 4 3 2 1

Contents

Dedication

Dedicated to the memory of John Brodie Good, a pioneer of Jaguar and Puma watching tours in South America, and a great supporter of this project prior to his untimely death in 2019.

Puma (rainforest form)

Preface

South America is a tremendously mammal-rich continent, reflecting its diverse range of habitats. These include the vast Amazon rainforest, extensive grasslands, deserts and high mountains, along with the Pantanal, one of the world's great wetlands. Together, these habitats support a wide array of mammals, many of which occur only in South America. Over 420 species of 'larger' mammals (those that are at least the size of a guinea pig) inhabit the continent, including no fewer than 184 species of primates based on current taxonomy.

Having visited South America over 40 times between us since the 1980s, it has been a long-standing frustration of ours that although there are numerous guides to the mammals of this incredibly rich continent there are no field guides that cover the majority of the species likely to be identifiable in the field.

Fortunately, in recent years the interest in mammals in South America has increased exponentially with the greater availability of information on the internet, particularly on websites such as www.mammalwatching.com, fuelling increased interest among mammal watchers, birders, local conservation groups, reserve, land and lodge owners, and more general naturalists alike. The number of publications on South American mammals has also increased rapidly in recent years.

This book is designed to provide a comprehensive overview of the larger mammals of mainland South America for the serious mammal watcher, and a one-stop field guide for those people who have an interest in the mammals of South America but do not want to have to assemble a library of individual guides for each country in the continent.

We concentrate on the larger species that can be more readily identified in the field, rather than smaller species, such as tuco-tucos and smaller rodents, that are more difficult to see well. In part this is due to limitations on space, particularly as taxonomic developments are leading to a rapid increase in the number of species known to occur in South America, and in part due to the availability of photographs.

We hope that this book will provide people with the tools necessary to identify accurately the larger mammals that they encounter in the field. We also hope that it will encourage people to document their sightings and to:

- Submit them to the relevant recording bodies in the countries/areas that they are visiting.
- Document significant range extensions and other interesting information in an appropriate scientific journal to ensure that organizations such as the International Union for Conservation of Nature (IUCN) have access to the latest distribution data, and that the information is available to future researchers.
- Publish them on sites such as www.mammalwatching.com.

Richard Webb and Jeff Blincow
January 2024

Acknowledgements

Thanks are due to all the people who have produced the many papers referred to during the preparation of this book. The speed of advances in the knowledge of Neotropical mammals, and in particular their taxonomy, is truly staggering, and this book would not have been so complete but for their endeavors. The following provided additional important information for use in the book: Julián Baigorria, Mark Bowler, Pablo Cerqueira, Enrique Couve, Ewan Davies, Guillermo D'Elía, Edson Endrigo, Anderson Feijó, Jon Hall, Craig Hilton-Taylor, Jaime Jimenez, Diego J. Lizcano, Laura Marsh, Russell Mittermeier, Regina Ribeiro, Luis Ruedas, Anthony Rylands, Frederico Tavares and Claudio Vidal. Thanks are also due to Martin Jones, David Christie, Nick Wilcox-Brown, Brian Clews and Gill Swash for their contributions to designing, copy editing, image processing and proofing, and especially to Andy Swash at Princeton WILD*Guides* for his contribution and tireless efforts during the latter stages of production, and in seeing the book to fruition.

Thanks are also due to our families, in particular Suzanne Webb, for their patience and support, especially during the latter stages of the project. Finally, particular thanks go to the late John Brodie Good of WildWings for recognizing the opportunities for running mammal tours to South America 20 years ago, giving RW the opportunity to travel widely in South America, and in doing so providing the inspiration to write this book.

Photographers and Artist

One of the key features of the book is the remarkable array of photographs that are featured, the vast majority of which have never previously appeared in print. The high quality of these photos is testament to the skill and dedication of the contributing photographers, to all of whom we extend our grateful thanks. In total, 563 photographs are included, representing the work of 230 photographers. Collating these images proved to be a monumental task that involved contacting hundreds of photographers over the years in which the book has been in preparation. We would, therefore, also like to thank everyone who kindly agreed to allow us to include their images in the final shortlist, even if, in the event, these were not selected.

Every image that is included in the book is listed in the *Photographic Credits* section, which starts on *page 474*, with the name of the photographer. We would, however, like especially to mention Rob Jansen, who was the main contributor with 113 photos (almost one-fifth of the total), and Andy and Gill Swash with 39. Twelve other photographers contributed five or more images: Jon Hall (14), Cheryl Antonucci (13), Chris Collins (13), Pablo Cerqueira (12), Ignacio Yufera (12), Alex Meyer (9), James Adams (7), Steve Davis (6), Roger Ahlman (5), Phillip Edwards (5), Marc Faucher (5) and Kevin Schafer (5). Without the willing contribution of so many photographers it would not have been possible to produce a book of this nature.

Despite our best endeavors, it did not prove possible to obtain photographs of every species included in the book, and in some cases the images that were available were not suitable for reproduction at a large size. We were therefore delighted when the talented artist Tomasz Cofta agreed to illustrate these species, as well as providing the wonderful artwork covering all the species of marmoset and titi monkey. Our grateful thanks are due to Tomasz for his significant contribution.

Introduction

The primary purpose of this book is to enable the reader to identify mammals that they may encounter in South America. The bulk of the book is taken up by the individual species accounts, and the introductory sections have been kept deliberately brief in order to maximize the space available for the species accounts and photographs.

The book covers the larger land mammals that can be more readily identified in the field, rather than smaller species that are more difficult to see well and are consequently often difficult to identify other than in the hand. It includes all species the size of a guinea pig or larger, but excludes tuco-tucos, mouse opossums, mice, rats and bats. Some marine mammals, such as seals and sea lions breeding around the coasts of South America, vagrant seals and fur seals, and river dolphins and manatees, have also been included. Other cetaceans that have been recorded in the seas around South America are listed. For one frequently encountered group of 16 species, guinea pigs and cavies, many of which are almost impossible to identify in the field, a summary of their measurements, habitat preferences and distribution has been provided. In total, 420 species are illustrated.

The species are grouped taxonomically by family, and there is an introduction to each family that includes generic information on aspects of the group's ecology, thereby avoiding the need to repeat such information in each of the individual species accounts. These introductory sections also highlight recent taxonomic changes.

Taxonomy

As with many other animal groups, mammal taxonomy is in a constant state of flux and finding a stable taxonomy at this point has proved to be impossible. The original intention was to follow the taxonomy adopted by the International Union for Conservation of Nature (IUCN) and in most cases the speciation in this book follows the IUCN Red List (www.iucnredlist.org). However, with new species regularly being discovered or described as a consequence of taxonomic reviews, some additional species which have yet to be formally recognized by IUCN but are recognized by other widely followed taxonomic authorities, such as the Mammal Diversity Database (www.mammaldiversity.org), have been included. For primates the latest taxonomy of the IUCN Primate Specialist Group has been followed.

Where new species have been included this is indicated in the introduction to the family accounts, with references being noted. These introductions also detail other proposed taxonomic changes that have not been adopted in the book. Some changes included in the *Illustrated Checklist of the Mammals of the World* (Burgin *et al.,* 2020), *e.g.* several changes to the taxonomy of opossums in the genera *Metachirus* and *Philander*, are noted in the introduction to the appropriate group account but the taxa concerned have not been afforded a full species account.

The taxonomic order of mammals is also in a state of flux and the *Illustrated Checklist of the Mammals of the World* and *All the Mammals of the World* (Lynx Nature Books, 2023), and the Mammal Diversity Database, have totally changed our understanding of the taxonomic order of mammals on the basis of the latest phylogenetic studies. This book follows the taxonomic order in common usage prior to the publication of the *Illustrated Checklist of the Mammals of the World*. In addition, in some cases the sequence of species within a family or order has been adusted so that similar species appear on the same two-page spread to enable direct comparison.

As a consequence of the taxonomic uncertainties alluded to above, certain decisions have had to be made regarding the inclusion of some taxa and the positioning of others in the sequence. It is acknowledged that some of these decisions may not be agreed upon by everyone using the book, but they were based on the best information available at the time of writing.

The species accounts

The species accounts presented in this book follow a consistent format. The following notes provide background information on the various components.

Measurements

The measurements shown are based on a search of published information. Since there is considerable variation in the measurements given in different publications, however, those given here take into account the extremes from a number of different sources. In most cases, size ranges are given, but for some species it has been possible to include only average measurements. Separate measurements are shown for both males and females where such information is available.

Head/Body (HB)	The length of the body from the tip of the nose to the base of the tail when the head is fully extended.
Shoulder (Sh)	The height at the shoulder is provided for some groups, such as ungulates, and is the measurement from the foot to the top of the shoulder blade when the animal is standing upright.
Tail	The length of the tail from its base to the tip.
Weight (Wt)	The range of weights, or average (mean) weights.

IUCN Red List status

For each species the 2023 global conservation status is shown. This is based on the IUCN Red List (www.iucnredlist.org), which uses the categories listed below. The coded icons shown are used throughout the book. The Red List category assigned to a species may be altered periodically as new assessments are undertaken.

EW	**Extinct in the Wild**	Known to survive only in captivity or as a naturalized population (or populations) well outside the former range.
CR	**Critically Endangered**	Considered to be at an extremely high risk of extinction in the wild.
EN	**Endangered**	Considered to be at a very high risk of extinction in the wild.
VU	**Vulnerable**	Considered to be at a high risk of extinction in the wild.
NT	**Near Threatened**	Considered likely to be elevated to one of the higher risk categories in the not-too-distant future.
LC	**Least Concern**	Not considered threatened, as widespread and/or abundant within its range.
DD	**Data Deficient**	Insufficient data available to assign to a category.
NE	**Not Evaluated**	Yet to be assessed and evaluated, normally because the species has only recently been described and/or is yet to be recognized by IUCN.

Description

This section includes a description of the species including, as appropriate, its overall size, body shape, color and other distinctive characteristics, such as particular ways of moving. For readily identifiable species with no obvious confusion species the descriptions are often relatively brief, but for those species that are difficult to identify, or are recently described, greater detail is provided. For species with a number of distinct subspecies, particularly in those instances where subspecies may prove to be separate species, additional information is provided, including details of relevant references.

Similar species

Lists and provides details of the key identification features of potential confusion species within the range of the species concerned. For certain groups, *e.g. Pithecia* saki monkeys, where there remains some doubt regarding the exact ranges of individual species, information is also provided on similar species with adjacent ranges. Other species mentioned are cross-referenced, unless they appear on the opposite page.

Habitat

Covers the habitat(s) in which the species is known to occur in South America. Where details are not available for South America, information from elsewhere is provided.

Distribution

Although the distributions and maps included are based primarily on those shown in the IUCN Red List accounts, or on the website of the relevant IUCN Specialist Group, a number of these now appear to be out of date. In such cases the distributions also take into account:

- information from other published sources, including the first eight volumes of the *Handbook of the Mammals of the World*, the two-volume *Illustrated Checklist of the Mammals of the World*, and *All the Mammals of the World* (Lynx Nature Books, 2023);
- maps available on the Map Of Life website www.mol.org;
- range extensions published in the scientific literature;
- for primates, Noel Rowe's website www.alltheworldsprimates.org.

Where monographs have been published for individual groups, such as felids, canids, primates and squirrels, the distribution information given follows the most recently published details.

The only codes included in the text are those for compass directions when referring to countries: N (north), S (south), E (east) and W (west). References the left and right banks of rivers relate to the direction of flow (*i.e.* when the viewer is facing downstream).

Maps

The maps are based on the same criteria as outlined in the *Distribution* section and show the species' range in South America, including inshore islands such as Santa Catarina, in south-east Brazil. For many species, including all those that are widespread, the map shows the whole of South America, but for those species with a restricted range the distribution is shown at a larger scale so that it is more clearly defined.

In instances where a species occurs as one or more isolated populations away from its main range, which may not be obvious owing to the scale of the map, small red arrows have been added.

Former ranges, where the species is now believed to be extinct, have been excluded, although they may be referenced in the Distribution section. Given the scale of the maps, it is generally not possible to show the precise distribution of the species concerned, and within the range shown it will be found only in suitable habitat. In addition, habitat degradation is, in most cases, reducing the range of species and consequently the effectiveness of a distribution map at representing precisely where a species might be found.

Photographs and Illustrations

Thanks to the recent advances in digital photography, the decision was made to use photographs rather than artwork for this book in order to show the animals as they appear in life. Wherever possible, the photographs that are included were taken in the wild. In some cases, however, it has been necessary to include photographs from camera traps, and in other instances, where photographs taken in the wild were unavailable, the photographs are of captive individuals. In a few cases, the background of images of captive animals has been replaced for aesthetic purposes; this is indicated in the *Photographic Credits* section on *page 474*.

For the relatively few species for which it has not been possible to obtain suitable photographs, artwork has been included. The illustrations were prepared with reference to published papers and images online, based on the structure of similar species within the genus.

Anatomy of a Mammal

The annotated pictures shown here indicate the main parts of the body referred to in the species accounts. Other terms are covered in the *Glossary* on *page 462*.

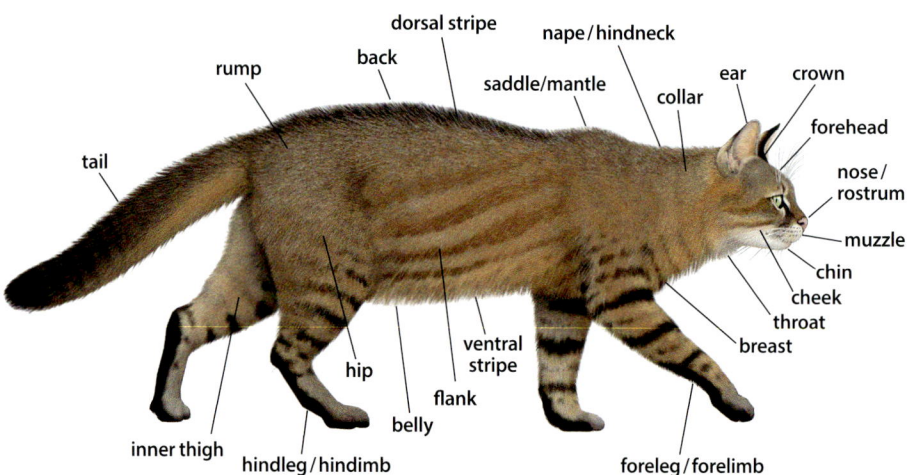

A number of anatomical terms are specific to primates, as indicated on the annotated pictures shown here. Other terms are covered in the *Glossary* on *page 462*.

sagittal cap/crest

supraocular

crown

arch/arc

subocular

facial fringe

sideburn

supraocular spot/frontal tuft/blaze

ear tuft

mane

forehead

malar stripe

cheek

chin

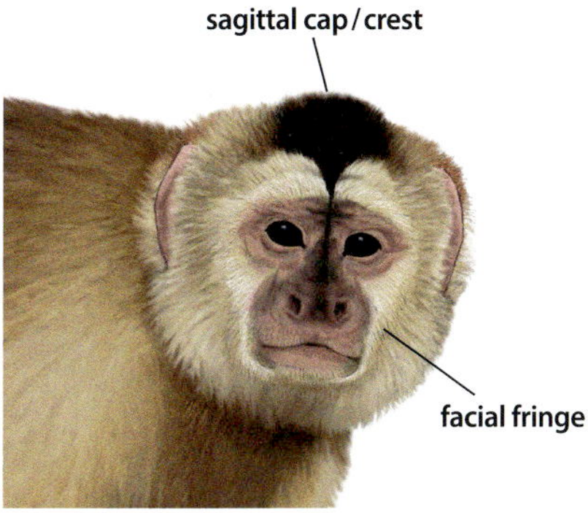

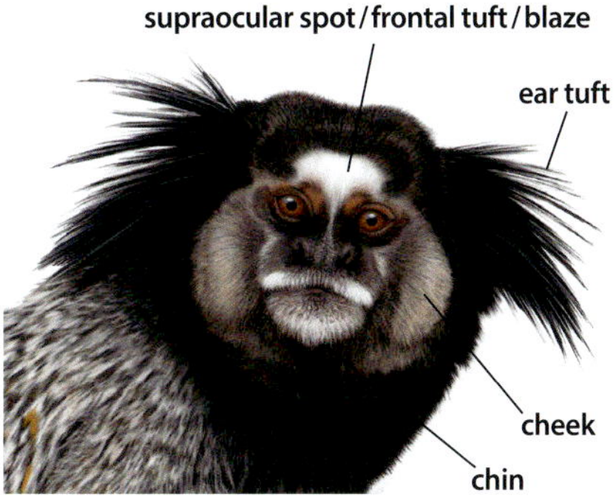

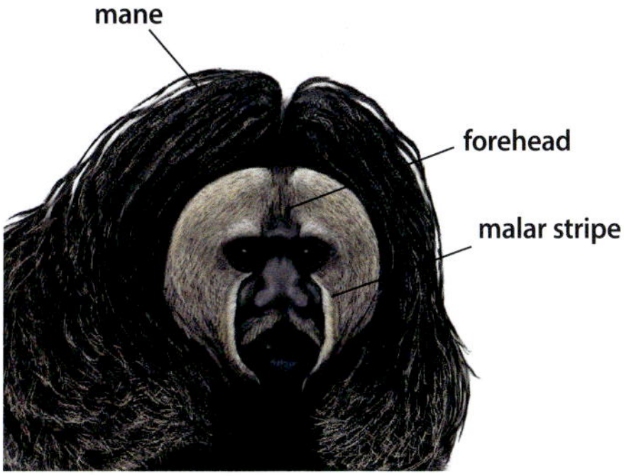

Anatomical terms that are specific to cetaceans are indicated on the annotated picture shown here. Other terms are covered in the *Glossary* on *page 462*.

dorsal fin

melon

beak/rostrum

fluke

flipper/pectoral fin

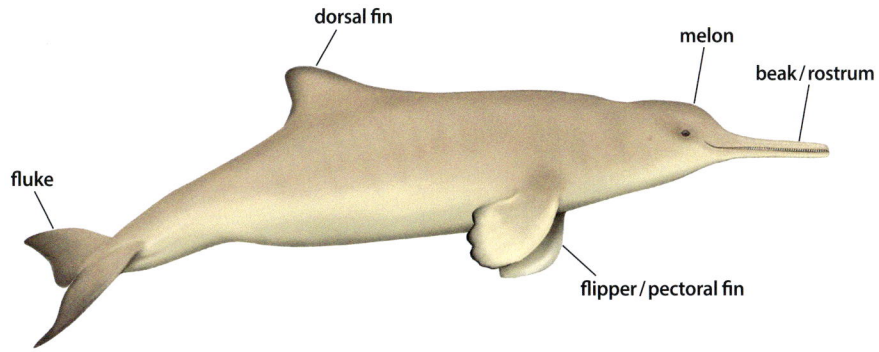

The Geography of South America

This book covers the whole of South America, including Trinidad. The geopolitical areas referred to in the text are indicated on the map on *page 14*, and the main rivers are shown on the map on *page 15*.

South America can be divided broadly into 17 terrestrial ecoregions, the approximate distributions of which are shown on the map *opposite*. These ecoregions reflect the topography and key geographical features of the continent, many of which are evident on the satellite image *below*.

Source: www.worldmap1.com/south-america-map.asp

Terrestrial ecoregions of South America

This map shows the 17 key terrestrial ecoregions specific to South America. These influence the distribution of species, and are referred to regularly in the text.

Geopolitical areas of South America

This map shows the approximate location of the main geopolitical areas (*i.e.* countries and states/departments/provinces/regions) in South America that are referenced in the species accounts.

Main rivers of South America

This map includes the names of the main rivers that are referenced in the species accounts, and may be particularly useful in understanding the distributions of certain species, notably the primates.

1. Sinú
2. San Jorge
3. Cesar
4. Catatumbo
5. Atrato
6. San Juan
7. Raposo
8. Guayabero
9. Esmeraldas–Guayllabamba
10. Guamués
11. Apaporis
12. Uatuma
13. Nhamundá
14. Chinchipe
15. Tapiche
16. Perene
17. Urubamba
18. Manu
19. Ribeirão Carmelita
20. Peixoto de Azevedo

The Species Accounts

The species accounts that form the bulk of the book are structured in a consistent manner. See *pages 8–10* for an explanation of each of the components, and the annotated maps on *pages 14–15* for locations of geopolitical areas and main rivers.

Monito del Monte

Monitos del Monte

Small nocturnal and arboreal marsupials which construct spherical nests of water-repellent bamboo leaves lined with moss or grass. They feed on fruits, larvae and insect pupae and will hibernate in winter when food becomes scarce. D'Elía *et al.* (2016) proposed that Monito del Monte should be split into three species:

Southern Monito del Monte *Dromiciops gliroides*, restricted to the southern parts of the 'wider' species' former range in both Chile and adjacent W Argentina, south of the range of *D. mondaca* and including the island of Chiloe.

Pancho's Monito del Monte *D. bozinovici*, found in the Bío Bío, Araucania and possibly southern Maule regions of Chile and in adjacent areas of Neuquén in W Argentina.

Mondaca's Monito del Monte *D. mondaca*, known from two localities in the Coastal (Mahuidanche) Cordillera of the Región de Los Ríos, Chile.

IUCN have yet to accept the proposed split, so the original taxonomy is followed here, with only one species being recognized.

NT Monito del Monte *Dromiciops gliroides*

Description: Resembles a large mouse, with silky, short, dense fur that is fawn-gray on the upperparts and dirty yellowish-white underneath. The face is pale gray with a distinct black ring around each eye, and the crown and adjoining nape are shaded rufous. The ears are short and rounded, and have short hair. The prehensile tail is thick at the base and tapers to the tip. The tail has thick, fawn-colored fur at the base and dark brown fur towards the tip. Females are, on average, significantly longer and heavier than males and have a distinct abdominal pouch.

HB:	8.3–13.0 cm
Tail:	9.0–13.2 cm
Wt:	16–42 g

Similar species: None in range.

Habitat: Cold temperate rainforests, where it inhabits thickets of Chilean Bamboo and other native forest species from sea level to 1,500 m.

Distribution: S Chile from south of Concepción south to the island of Chiloé and adjacent Argentina from Neuquén to Chubut Provinces.

The number of medium-sized to large opossums currently occurring in South America is the subject of considerable debate. IUCN recognizes 21 species. In addition, there are a large number of mouse opossums, fat-tailed opossums and shrew-opossums in South America, but these have been excluded from this book owing to their small size and the difficulty of observing and identifying them in the field.

Burgin *et al.* (2020) recognize five additional species of opossum:

Brown Four-eyed Opossum (*p. 28*) split into **Guianan Brown Four-eyed Opossum** *Metachirus nudicaudatus* and **Common Brown Four-eyed Opossum** *M. myosuros*.

Pebas Four-eyed Opossum *Philander pebas*, **Common Four-eyed Opossum** *P. canus*, **Dark Four-eyed Opossum** *P. melanurus* and **Northern Four-eyed Opossum** *P. vossi*, all split from Gray Four-eyed Opossum *P. opossum* (*p. 27*).

Conversely, Burgin *et al.* (2020) lump Mondolfi's Four-eyed Opossum *P. mondolfi* and Olrog's Four-eyed Opossum *P. olrogi* with Common Four-eyed Opossum *P. canus*. Since this proposed taxonomic treatment has yet to be adopted by IUCN, these changes are not reflected here.

Opossums are long-tailed marsupials and most are largely terrestrial, although many also climb and swim well. Black-shouldered Opossum, Bushy-tailed Opossum and the three woolly opossums are almost entirely arboreal.

Opossums are generally solitary, crepuscular, nocturnal and omnivorous, and feed on small mammals, lizards, frogs, fish, invertebrates, birds' eggs, fruit, flowers, nectar and seeds. They will also feed on carrion and some species visit garbage dumps and chicken runs in urban areas. Larger species, such as Big Lutrine Opossum, will also eat birds and mammals up to the size of cavies and squirrels.

Northern Black-eared Opossum

LC **Big Lutrine Opossum**

Lutreolina crassicaudata

Description: Previously considered conspecific with Massoia's Lutrine Opossum (*p. 20*), but now recognized as a distinct species based on genetic and morphological characteristics. Larger than Massoia's Lutrine Opossum and with a more extensively furred tail. Uniformly colored, ranging from buff to dark brown above, with slightly paler (sometimes with a reddish tinge) fur below. The short head is relatively plain and unmarked, and the small rounded ears are brown. The short, stout legs and elongated body conveys an otter-like or weasel-like appearance. Does not overlap in range with Massoia's Lutrine Opossum. Two geographically separated subspecies are recognized (see *Distribution*).

Similar species: None in range.

Habitat: A wide variety of habitats, often close to water, including wet grasslands, dry pastures, savanna grasslands and woodlands, and humid, gallery and dry Chaco forests.

Distribution: ssp. *crassicaudata* occurs in NE and central E Argentina, Uruguay, E Paraguay,

HB:	24.3–40.0 cm
Tail:	24.5–35.8 cm
Wt:	200–910 g

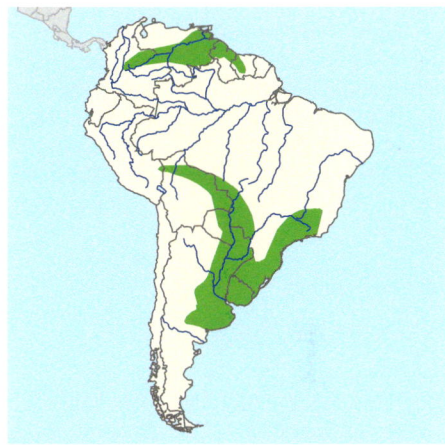

SE Brazil and parts of central N Bolivia, and extreme SE Peru.

ssp. *turneri*, which may prove to be a distinct species, occurs in central E Colombia, central Venezuela, N Guyana.

LC **Massoia's Lutrine Opossum** *Lutreolina massoia*

HB:	18.8–23.9 cm
Tail:	20.9–25.5 cm
Wt:	(mean) 284 g

Description: Previously considered conspecific with Big Lutrine Opossum (*p. 19*), but now recognized as a distinct species based on genetic and morphological characteristics. The upperparts are uniformly brownish-olive. The head is relatively short, and is sooty-black from just below the eyes to the nose. The ears are short, brown and rounded. The cheeks, underparts and insides of the legs are orange-cinnamon. The front of each foreleg is dark olive-brown. The first third of the tail is thickly furred, while the remainder is lightly furred and black, apart from the final 15 mm which are naked and yellowish-orange.

Similar species: None in range.

Habitat: Usually associated with dense ground cover and watercourses, including marshes and swamps, within the montane Yungas forests, at 450–2,000 m.

Distribution: Montane forests in the Yungas of NW Argentina (Jujuy, Salta and Tucumán Provinces) and southern Bolivia (Tarija and Chuquisaca Departments).

Anderson's Four-eyed Opossum *Philander andersoni*

LC

Description: Dark gray with short, dense fur, and a well-marked wide black dorsal stripe which extends from the neck to the furred basal portion of the tail, and which contrasts with the gray sides of the body. The underparts are pale gray with a creamy-yellow wash or mottling, or grayish-buff and tinged with orange on some individuals. The face is dark gray with creamy cheeks and a creamy spot above each eye. The ears are large and pale brown, with a cream spot at the base. The feet are black (pink or brown in many other *Philander* opossums). The basal 15–20% of the tail is furred. The rest is naked and brown with a whitish tip which extends up to 30% of the total tail length.

Similar species: **Gray Four-eyed Opossum** (*p. 27*) and **McIlhenny's Four-eyed Opossum** (*p. 24*) both lack a contrasting wide black dorsal streak. **Mondolfi's Four-eyed Opossum** (*p. 25*) is smaller, pale gray and lacks an obvious dorsal stripe.

Habitat: Mature and disturbed lowland rainforest in the Amazon Basin and Andean foothills.

HB:	22·3–30·7 cm
Tail:	25·5–33·2 cm
Wt:	230–600 g

Distribution: Up to 500 m in S Venezuela, SE Colombia, E Ecuador, and south into N Peru (Ucayali Department) and W Brazil (Amazonas State).

LC Southern Four-eyed Opossum *Philander quica*

Description: Previously known as *P. frenatus* (Burgin *et al.*, 2020). The upperparts, including the flanks, are dark gray with no dorsal stripe. The fur is short and smooth. The head is dark gray with a small creamy spot above each eye. The ears are large and pink with black borders and with white fur at their base. The underparts are creamy-gray to white, being slightly grayer on the throat. The tail is roughly the same length as the head and body, is furred for the first 15–20% of its length, naked and dark brown for 50% of its length, and naked and whitish for the remaining 30–35%.

Similar species: None in range.

Habitat: Lowland evergreen forests, and appears to be largely restricted to Atlantic Forest and absent from humid Chaco forests to the west.

Distribution: Restricted to NE Argentina, E Paraguay, and SE Brazil as far north as Bahia.

HB:	26.5–32.7 cm
Tail:	25.3–32.6 cm
Wt:	220–910 g

LC **Orinoco** (Deltaic) **Four-eyed Opossum** *Philander deltae*

Description: The smallest *Philander* opossum and the only one within its range. Dark brown above with short velvety fur, and with a broad dorsal stripe that runs to the base of the tail. The tail is the same length as the body, and the first 20% is sparsely furred, the middle section is unfurred and dark brown, and the final 25% is unfurred and completely unpigmented. The flanks are slightly mottled. The throat, chest and belly and the backs of the limbs are uniformly cream-colored. The front of each limb is brown. The face is brown, with a very small, poorly defined pale spot above each eye, and the cheeks and chin are cream-colored. The small ears are beige with faint black edges.

HB:	approx. 21 cm
Tail:	approx. 30 cm
Wt:	148–350 g

Similar species: None in range.

Habitat: Evergreen, permanently flooded swamp forest, or seasonally flooded marsh forest.

Distribution: Restricted to the Orinoco River delta and around the Gulf of Paria, in NE Venezuela.

LC **McIlhenny's Four-eyed Opossum** *Philander mcilhennyi*

Description: Large and almost entirely black or blackish, with longer fur than most *Philander* opossums, giving it a slightly shaggy appearance. The sides are grayer and the hairs are tipped silver. The underparts are dark gray, although some individuals have a paler chin and throat. The head is black or blackish with a large pale spot above each eye, and with pale cheeks. The tail has black fur at its base for about 25% of its length, the naked middle section is black, and the outer section is pale. The feet are blackish.

Similar species: Anderson's Four-eyed Opossum (*p. 21*) has a distinctive wide black dorsal stripe and has a cream spot at the base of each ear. **Gray Four-eyed Opossum** (*p. 27*) is shorter-furred and has a paler overall appearance.

Habitat: Dry tropical forest, inundated (várzea) and upland (terra firme) forests, secondary growth and disturbed habitats, including gardens. Where known to overlap in range with Gray Four-eyed Opossum, McIlhenny's Four-eyed Opossum occurs in terra firme forests, while Gray Four-eyed Opossum occurs in várzea forests.

Distribution: The Amazon Basin of eastern-central Peru (Ucayali, Loreto and Huánuco Departments), and the states of Acre and Amazonas in W Brazil.

HB:	28.7–30.7 cm
Tail:	26.5–37.7 cm
Wt:	396–640 g

LC Mondolfi's Four-eyed Opossum *Philander mondolfii*

HB:	24.3 cm
Tail:	26.5–29.0 cm
Wt:	260 g

Description: The upperparts are pale gray with short, woolly fur; some individuals have brown fur on the back and around the base of the tail. The throat and flanks are olivaceous. The chin, the throat and the backs of the legs are slightly paler than the upperparts and contrast with the creamy abdomen. The face and the back of the head are dark brown, with a large cream-colored spot above each eye. There is a dark ring around each eye and the cheeks are cream-colored. The ears are quite large and cream-colored with a wide black border and yellowish hairs. The tail is slightly longer than the body, the first 20% being densely covered by short hairs, the middle section is dark brown, and the outer part, up to 35% of the tail, is sparsely or totally unpigmented.

Similar species: Gray Four-eyed Opossum (*p. 27*) is larger and darker, with shorter fur, and a significantly shorter tail. **Anderson's Four-eyed Opossum** (*p. 21*) is larger and blackish (not pale gray) and has an obvious dorsal stripe.

Habitat: Occurs mainly in pre- and submontane forests at 50–800 m, but also in Venezuelan lowlands dominated by tree savannas, semi-deciduous riparian forests and tall, partially flooded evergreen forests.

Distribution: Two main populations: one in the north-eastern and north-western sectors of the Piedmont Hill systems of central Venezuela; and another in the foothills of the eastern slope of the Cordillera Oriental of Colombia, and of the northern and southern slopes of Cordillera de Mérida in Venezuela. There is also a single specimen from the state of Amazonas in S Venezuela.

DD Olrog's Four-eyed Opossum *Philander olrogi*

Description: The upperparts are dark gray or grayish-black, with a darker dorsal line and slightly paler flanks. The underparts, chin, throat and chest are ochre-buff. The front legs and the front of the hind legs are creamy-buff, while the back of the hind legs is of the same color as the upperparts. The head is blackish-brown, with cream cheeks and a well-defined cream spot above each eye. The ears are rounded and bicolored, with broad dark margins. The tail is furred for 20% of its length, the middle 75% is blackish-brown, and the outer 5% is creamy-buff.

Similar species: The color of the upperparts, tail and ears is similar to that of the Peruvian and Bolivian populations of **Gray Four-eyed Opossum**. The tail of Olrog's Four-eyed Opossum, however, is covered by short hairs and the ears lack yellow hairs at the posterior base, whereas Gray Four-eyed Opossum has the tail covered by long hairs and there are numerous yellow hairs on the proximal base of each ear.

Habitat: Humid savannas dominated by palm trees in Bolivia, where it has been found alongside Gray Four-eyed Opossum. A report from humid secondary lowland forest in Loreto, in north-east Peru, has been disputed and the identification of this individual is unclear.

HB:	23.5–29.1 cm
Tail:	26.0–30.7 cm
Wt:	284–550 g

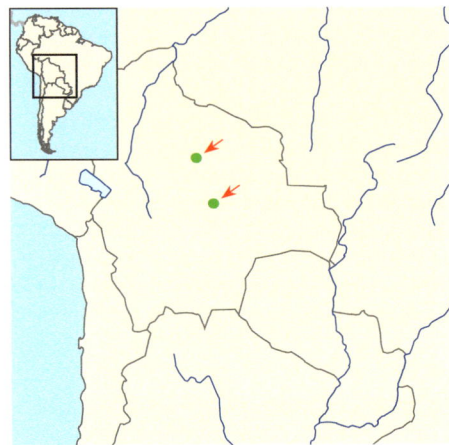

Distribution: Two localities in Beni and Santa Cruz Departments in E Bolivia, at 150–250 m. (See also the reference to a Peruvian record under *Habitat*.)

LC Gray Four-eyed Opossum

Philander opossum

Description: Includes **Pebas Four-eyed Opossum** *P. pebas,* **Common Four-eyed Opossum** *P. canus,* **Dark Four-eyed Opossum** *P. melanurus* and **Northern Four-eyed Opossum** *P. vossi* (Burgin *et al.,* 2020). A short-furred, dark gray-brown to blackish-gray opossum finely grizzled with white hairs. The underparts, including the top of the feet, are often creamy or yellow but can be brownish-gray or even dark gray. The head is blackish, with cream cheeks and a cream spot above each eye and at the base of each ear. The ears are black and naked. The tail is slightly longer than the body, and the first 20% is furred and of the same color as the upperparts; the remainder is almost naked and blackish for two-thirds of its length and white towards the tip, except in northern Ecuador, where the tail is completely dark gray. When disturbed will readily climb trees.

Similar species: Very similar to **Anderson's Four-eyed Opossum** (*p. 21*) and best separated by the lack of a wide black mid-dorsal streak. **McIlhenny's Four-eyed Opossum** (*p. 24*) is black with long fur and a dark belly. **Mondolfi's Four-eyed Opossum** (*p. 25*) is smaller and paler, with shorter fur and a significantly shorter tail. **Guianan Brown Four-eyed Opossum** (*p. 28*) is brown and lacks the tricolored tail. (See **Olrog's Four-eyed Opossum** for differences from Peruvian and Bolivian populations.)

HB:	25.0–31.5 cm
Tail:	25.3–32.9 cm
Wt:	200 g–1.4 kg

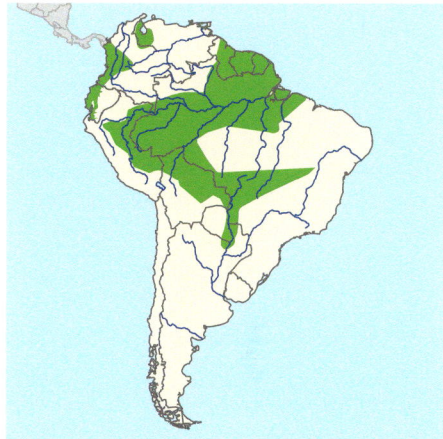

Habitat: Locally common in humid, primary, secondary and gallery forests, often along watercourses. Most commonly in moist/humid areas, but also in drier habitats such as Chaco forests. Often on the ground in dense undergrowth along river banks.

Distribution: Two populations: W Venezuela, N and W Colombia and W Ecuador; and E Peru, E Bolivia, Guyana, Suriname, French Guiana, W & central Brazil, central Paraguay and south into N Argentina. Range extends through Central America and into Mexico.

LC Brown Four-eyed Opossum   *Metachirus nudicaudatus*

Description: This species is also known as **Guianan Brown Four-eyed Opossum**, and includes **Common Brown Four-eyed Opossum** *Metachirus myosuros* (Burgin *et al.*, 2020). Medium-sized and long-legged with reddish-brown to gray-brown upperparts, darkest along the midline and rump and becoming tan-colored on the sides and below the dark brown, naked ears. The underparts are pale yellow. The face is dark brown with a narrow dark stripe running from the muzzle to the nape, and with buff-colored cheeks and a spot above each eye. The eyes are dark, each surrounded by a dark brown ring. The tail appears naked but is sparsely haired for its entire length, and is largely brown above and white below, with a white tip. Normally silent, but will run off clicking its teeth when disturbed.

Similar species: *Lutreolina* opossums (*pp. 19–20*) are shorter-legged and have less conspicuous facial markings. **Gray Four-eyed Opossum** (*p. 27*) is gray with white grizzling on the fur. All *Philander* opossums (*pp. 21–27*) have a distinctly tricolored tail.

Habitat: Widespread in mature evergreen forests in lowlands and foothills and in deciduous or dense secondary forest. Appears to favor mature forests with little undergrowth.

HB:	21.7–29.9 cm
Tail:	28.0–40.5 cm
Wt:	254–619 g

Distribution: Widespread at up to 1,500 m from N Colombia to Paraguay and NE Argentina; apparently absent from a large part of N Venezuela, W Peru and parts of E Brazil. Range extends through Central America as far as Nicaragua.

LC Water Opossum

Chironectes minimus

HB:	22.1–40.0 cm
Tail:	28.1–43.0 cm
Wt:	510–790 g

Description: Uniquely patterned: the upperparts are gray with a narrow dark dorsal line and four broad black bands across the back. The front and back bands extend from the shoulders and rump on to the front and hind legs respectively, and the middle two bands extend on to the sides of the body. A blackish mask extends from the nose to above the eyes and then back behind the ears, where it joins the black crown. There is a narrow gray band above the eyes between the mask and the crown. The throat and the chest are white. The rounded ears are short and blackish. The feet are reddish-brown, and the hind feet are webbed. The tail is furred at its base for 3–4 cm, naked and black for 80% of its length, and white-tipped. This species is an excellent swimmer and usually catches its prey in water. When swimming, the eyes and the top of the head show just above the water's surface.

Similar species: None in range.

Habitat: Semi-aquatic, most records coming from fast-flowing rivers, streams and lakes in hilly areas at up to 1,860 m, although occurs also in mangroves in Central America. May be rare or absent from silt-laden lowland rivers and streams. Carnivorous, feeding mainly on small fish, crabs, crustaceans, insects and frogs.

Distribution: Patchily distributed with two populations: one in Bolivia, Peru, Ecuador, Colombia, Venezuela, Guyana, French Guiana, Suriname and the mouth of the Amazon River in Brazil; and another in SE & S Brazil, E Paraguay, extreme NE Argentina and N Uruguay. Range extends through Central America and into Mexico.

LC Brazilian White-eared Opossum *Didelphis albiventris*

HB:	30.0–44.2 cm
Tail:	29–45 cm
Wt:	500 g–2.5 kg

Description: Upperparts gray to whitish-gray, or occasionally blackish, with long, coarse, gray- or white-tipped guard hairs. The underparts, including the throat, are whitish. The legs are blackish, on the forelimbs the dark coloration extending up on to the shoulders and nape to create an almost collared appearance. The face is whitish with white cheeks, a black patch through each eye, and a narrow black patch on the forehead which tapers to a point between the eyes. The large ears are predominantly white. The tail is similar in length to the body and is furred at the base; the remainder of the tail is naked, the half nearest the body being black and the remainder white.

Similar species: Southern Black-eared Opossum (*p. 33*) lacks black facial markings, and has black (not white) ears and less fur on the tail.

Habitat: A wide range of habitats: marshes, grasslands, agricultural areas and open forest, including dry forests such as Cerrado, Caatinga and Chaco. Tolerates human disturbance and occurs in some large cities.

Distribution: NE & central Brazil west into Bolivia and central & S Paraguay, and south into Uruguay and Argentina as far south as Río Negro Province and as far west as the Monte Desert.

LC Guianan White-eared Opossum

Didelphis imperfecta

Description: The head is pale gray, whitish or yellowish, with a less contrasting dark brown mask, and a black stripe down the forehead. The ears are large and black with a white tip. The limbs are black. The tail is of similar length to the body and furred at the base; the rest of the tail is naked, the half nearest the body being black and the remainder yellowish-white.

Similar species: Brazilian White-eared Opossum is larger, has black and gray color forms and lacks a collared appearance. **Northern Black-eared Opossum** (*p. 34*) is larger, lacks distinct facial markings, and has a shorter furred portion of the tail.

Habitat: Tropical, subtropical and temperate evergreen forests and forest edges up to 2,000 m, and in flooded fields and savanna.

Distribution: Venezuela, Guyana, Suriname, French Guiana and adjacent regions of Brazil. Recently discovered in Colombia (Gonzalez *et al.*, 2020).

HB:	31.7–39.0 cm
Tail:	30–41 cm
Wt:	600 g–1.2 kg

LC **Andean White-eared Opossum** *Didelphis pernigra*

HB:	34–44 cm
Tail:	32.0–41.2 cm
Wt:	720 g–2 kg

Description: Generally a dark opossum with long shiny black fur on the upperside. The underparts are slightly paler than the upperparts, and the throat is rusty-buff. The head is whitish with light rusty cheeks, and with a black patch running from the nose to just behind each eye, rather than to the base of the ears as on some other *Didelphis* opossums. It has a broad black line on the forehead which tapers to a point between the eyes. The ears are pinkish-white. The tail is furred at the base, black for the next 40–60% and white towards the tip.

Similar species: Northern Black-eared Opossum (*p. 34*) is larger and paler, and has a largely plain face and black ears.

Habitat: A wide range of habitats, including wet and dry forested habitats and páramo in the Andes, riparian habitats in western Peru, and also secondary forests, cultivated areas and even around towns and cities. Known to visit hummingbird feeders and compost heaps around lodges in Ecuador.

Distribution: The Andes from NW Venezuela through Colombia, Ecuador and Peru and into W Bolivia. At 2,000–3,700 m in Ecuador.

LC **Southern Black-eared** (Big-eared) **Opossum** *Didelphis aurita*

Description: A dark gray opossum with long gray guard hairs but with yellowish-brown underfur visible through these hairs, particularly on the flanks and underside, the throat and chest being yellowish. The head is yellowish-brown with indistinct brown eye patches, yellowish or whitish cheeks, and a blackish line down the forehead. The legs are darker and the feet black. The large, naked ears are blackish. The tail is of similar length to the body and is furred at the base, with the remainder of the tail naked, the first half being black and the remainder white.

Similar species: Brazilian White-eared Opossum (*p. 30*) generally has more clearly defined black facial markings, white (not black) ears, and more fur on the tail.

Habitat: Primary and secondary Atlantic Forest and monkey-puzzle (*Araucaria*) forest, including disturbed areas and forest fragments. Unlike other *Didelphis* opossums, it appears to be equally at home on the ground and up to 20 m above it.

HB:	31–39 cm
Tail:	31–37 cm
Wt:	700 g–1.5 kg

Distribution: E Brazil from Bahia to Rio Grande do Sul, into Paraguay east of the lower Paraguay River, and NE Argentina.

LC **Northern Black-eared Opossum** *Didelphis marsupialis*

Description: A large shaggy-looking opossum with a mix of long black and white guard hairs. Upperparts can appear blackish, gray or even whitish. The underparts are cream, yellow or orange and the legs are black. The face is pale with an indistinct narrow black ring around each eye, rather than the masked appearance of other *Didelphis* opossums, and some individuals have an indistinct blackish line on the forehead. The cheeks are cream or yellowish-orange. The ears are naked and entirely black. The tail is slightly longer than the body and has a short-furred section at the base; the remainder of the tail is naked and normally a third black and two-thirds white, but it can be all black or half black and half white. Although often considered to be terrestrial, it is frequently found feeding higher up in trees.

HB:	26.3–46.5 cm
Tail:	29.5–46.5 cm
Wt:	570 g–2.40 kg

Similar species: Andean White-eared Opossum (*p. 32*) is smaller and darker, and has a more well-marked face, and white ears.

Habitat: A wide variety of habitats, including tropical and subtropical forest, secondary and gallery forest, but absent from altitudes above 2,200 m and from extremely arid habitats. Found also near human settlements.

Distribution: Most of northern South America as far south as N Peru, N Bolivia and N Brazil, including most of the Amazon Basin. Also Trinidad. Range extends through Central America and into Mexico.

Additional photo p. 18

LC Derby's Woolly Opossum

Caluromys derbianus

Description: Color varies from pale gray to orange, often with orange patches on the neck and shoulders, center of the back and rump, and a gray patch between the shoulders. The underparts are creamy white. The fur is long and woolly. The diagnostic tail is long and furred above for half of its length; the remainder is naked and mottled dark brown and white, apart from the tip, which is white. The only opossum with half of the tail unfurred. The ears are large and pale pink, and the face is grayish with a dark brown stripe down the center, and a brown ring around each eye. The eyes are large and brown. When caught in a spotlight, it will often freeze with the tail hanging down.

Similar species: None in range.

Habitat: A relatively common species in forested areas, including mature and disturbed evergreen rainforest, dry forest, gardens and plantations. Frequently found at the forest edge, and may be encountered sitting motionless on telephone wires. Primarily arboreal.

HB:	22.5–32.1 cm
Tail:	38.4–45.7 cm
Wt:	245–370 g

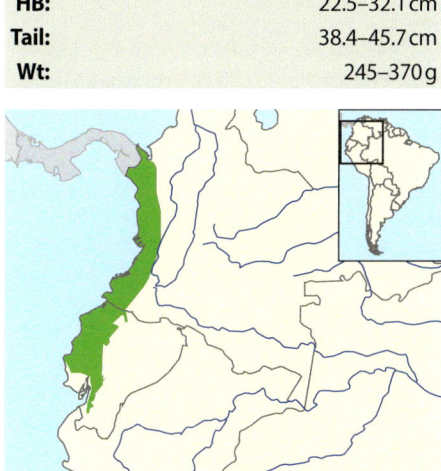

Distribution: W Ecuador and the Cauca Valley of W Colombia, from sea level to 2,600 m. Range extends through Central America and into Mexico.

LC Brown-eared Woolly Opossum *Caluromys lanatus*

Description: The upperparts are reddish-brown to pale brown, with the shoulders, rump and legs generally brighter than the rest of the upperparts. The fur is long, dense and woolly. Sometimes has a pale gray dorsal stripe between the shoulders. The underparts are orange or yellowish-white, with a grayer middle section. The head is gray with a dark stripe down the center of the face, and with a reddish-brown ring around each eye and extending down towards the nose. The eyes are dark brown and the ears are naked and pinkish. The upper tail is furred for 40–70% of its length and the underside of the tail for 20–35%. The naked section of the tail is whitish, with brown mottling near its base.

Similar species: None in range.

Habitat: Widespread in a variety of lowland humid forests, including primary, secondary and gallery forests, and in plantations. Found also in dry xerophytic forests and dense savanna. Normally below 500 m, but has occasionally been recorded to 2,600 m.

HB:	20.1–31.9 cm
Tail:	33.0–44.6 cm
Wt:	300–520 g

Distribution: East of the Andes from W & S Venezuela, central Colombia, E Ecuador, Peru, Bolivia and Paraguay to W, central & S Brazil and NE Argentina.

LC **Bare-tailed Woolly Opossum** *Caluromys philander*

Description: The smallest *Caluromys* opossum, this species has uniform pale brown to warm reddish-brown upperparts, this coloration extending on to the sides of the body. The underparts are orange, sometimes with a grayish center. The head is gray, and the face has an indistinct dark brown stripe down the center and a brownish ring around each eye. The eyes are reddish-brown. The ears are pale brown with yellowish spots at the corners. The tail is 50% longer than the body. Only 10% of the tail is furred, the remainder is naked and usually dark gray-brown, sometimes mottled with cream. The feet are pale gray or white. Although mainly solitary, several individuals may be seen feeding in the same fruiting tree.

Similar species: Most likely to be confused with **four-eyed opossums** *Philander* spp. (*pp. 21–27*), which have pale spots over the eyes, or woolly mouse opossums *Marmosa* spp. [not illustrated], which have no central facial stripes.

Habitat: Rainforest, subtropical forest and marginal forest, as well as semi-natural areas

HB:	16.0–27.9 cm
Tail:	25.0–40.5 cm
Wt:	140–390 g

such as plantations, secondary vegetation and abandoned human settlement areas. Appears to prefer thick, closed vegetation in the upper levels of trees, but found also in the open upper canopy. Also in dry forests, but more closely associated with moist habitats.

Distribution: Up to 1,800 m from N Venezuela through Guyana, Suriname and French Guiana to NE & south-central Brazil and extreme E Bolivia. Also Trinidad.

LC **Black-shouldered Opossum** *Caluromysiops irrupta*

Description: Distinctive, with frosted grayish-brown upperparts and large diagnostic black patches that extend from the wrists and forearms across the shoulders to meet on the back, and then run parallel to each other along the back, fading to brown on the rump. The sides of the body and the underparts are paler buff-gray or white, some individuals having orange on the throat. The head is gray-brown with buff-white cheeks. There is an indistinct dark line between the eyes, but no eye-rings or facial stripes. The ears are rounded, and bright yellow inside. The tapered tail is 20% longer than the body and the upper tail is almost completely furred, being frosted grayish-brown like the rest of the upperparts but with a dirty white tip. The underside of the tail is furred for 25% of its length.

HB:	25–33 cm
Tail:	31–34 cm
Wt:	300–500 g

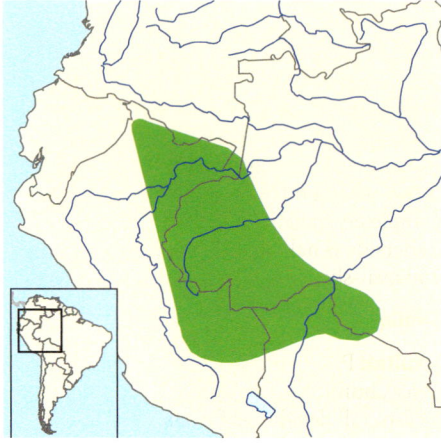

Similar species: None in range.

Habitat: Mature tropical humid forests.

Distribution: Known from five scattered localities at altitudes of up to 700 m in E Peru, extreme SE Colombia and W Brazil (the upper Juruá River). Thought likely to occur also in Bolivia.

LC Bushy-tailed Opossum

Glironia venusta

Description: Distinctive, having a striking head pattern with a pale gray or creamy stripe from the nape down the center of the face to the pink nose. The cheeks are separated from the center of the face by two broad dark brown or black stripes, each extending from the nose, through the eye and up in front of the ear onto the nape, and creating a masked appearance. The upperparts are fawn or cinnamon-brown, with a dark dorsal stripe on some individuals. The underparts are gray, brownish or buff-white. The upper tail is completely furred virtually to the tip, but the underside is naked. The ears are dark, naked and oval-shaped.

Similar species: None in range.

Habitat: Poorly known, but recorded in dense, humid tropical and semi-deciduous forests, tall deciduous forests, dry woodlands and dwarf evergreen forests. Has been encountered in disturbed habitats and foraging at the forest edge.

HB:	17–21 cm
Tail:	19.5–22.5 cm
Wt:	130 g

Distribution: Up to 500 m in the Amazonian regions of Bolivia, Brazil, Ecuador, Peru and S Colombia.

Armadillos

The IUCN Anteater, Sloth and Armadillo Specialist Group recognizes 22 species of armadillo, all of which occur in South America. Feijó *et al.* (2018) proposed treating Southern Long-nosed Armadillo *Dasypus hybridus* as a subspecies of Seven-banded Armadillo *D. septemcinctus*, and Feijó *et al.* (2021) proposed that subspecies *squamicaudis* of Southern Naked-tailed Armadillo should be treated as a full species, **Cerrado Naked-tailed Armadillo** *Cabassous squamicaudis*, but neither proposal had been adopted by IUCN at the time of writing and consequently this treatment is not followed here.

Armadillos are covered by a protective, bony shell, or carapace, which has movable bands in the middle to enable the animal to bend. The underparts are soft and either naked or lightly furred. The feet have long, broad claws for digging. They are terrestrial and largely solitary animals. Many are predominantly nocturnal, although some species are active also by day. They feed primarily on insects, particularly ants and termites, but will occasionally feed also on vertebrates and plants.

***Dasypus* armadillos (9 species)** have a long, slender nose that is normally more than half the length of the head, plus long naked ears, and a long tail more than half the combined length of the head and body. The final two-thirds of the tail are encased by narrow rings of two or more rows of scales, and the tail tapers to a slender tip. The carapace is generally smooth, the exception being Hairy Long-nosed Armadillo, on which it is covered by a dense coat of hairs. The carapace consists of scapular and pelvic shields separated by between 6 and 11 movable bands. The front feet have four long claws and the hind feet five claws.

***Chaetophractus* armadillos (2 species)** have prominent hairs on the upperparts which range in color from tan to buff and black. They can resemble Six-banded (Yellow) Armadillo *Euphractus sexcinctus*, but have a proportionately broader head shield, shorter ears, and a separate movable band on the front edge of the scapular shield.

***Cabassous* armadillos (4 species)** are medium-sized armadillos with dorsal plates arranged in transverse rows across the length of the carapace. They have a short, broad snout, small eyes and moderately large, funnel-shaped ears. The distinctive naked tail is slender and lacks the armor of other armadillos, but it may have small, thin, widely spaced plates.

***Tolypeutes* (2 species)** are small to medium-sized armadillos with a distinctive rounded carapace and normally three movable bands. They have an elongated flat head shield. They also have conspicuous hair below the carapace. The only armadillos that roll into a ball.

The remaining five species are from the monotypic genera ***Calyptothractus, Chlamyphorus, Euphractus, Zaedyus*** and ***Priodontes***.

Six-banded Armadillo (LEFT); Southern Three-banded Armadillo (RIGHT)

NT Southern Long-nosed Armadillo *Dasypus hybridus*

Description: One of the smallest of the long-nosed armadillos, this species has a narrow snout and long ears. The upper surfaces of the head, body and tail are covered by a dark gray carapace of bony scales, with very little hair. The central part of the carapace is divided into 6–8 movable bands. The scales on the scapular and pelvic shields are hexagonal, while those on the bands are rectangular and marked with a 'V'-shaped groove that divides them into three triangular sections. The scales on the head are variable in shape. The tail is ringed for about two-thirds of its length and has a slender tip. The underparts are dark brown and sparsely haired. There are four toes on each front foot and five on each hind foot.

Similar species: Nine-banded Armadillo (*p. 45*) is much larger, has fewer movable bands, a proportionately longer tail and much longer ears. **Seven-banded Armadillo** (*p. 49*) is smaller and darker, with a proportionately longer tail and much longer ears.

Habitat: Usually found from sea level up to 2,300 m in grassland habitats and Pampas, and less commonly in forested habitats. Occurs also in some degraded habitats (arable land, pastures and plantations). Normally burrows in sandy soils.

HB:	26–31 cm
Tail:	15.0–19.1 cm
Wt:	1.09–2.40 kg

Distribution: S Brazil, Uruguay and south to the province of Buenos Aires, in Argentina. There are two records from Paraguay. Not so widespread as previously thought, as reports from farther west, close to the Andes, are now known to be the result of misidentifications.

NE Kappler's Long-nosed Armadillo
Dasypus kappleri

Description: Like Pastaza and Beni (*p. 44*) Long-nosed Armadillos (with neither of which it overlaps in range), distinguished from other *Dasypus* armadillos by the enlarged downward-projecting scales at the knee, its wide tail base, and the lighter skin color on the part of the head that is not covered by the head shield. Appears elongated, with a long, narrow and conical head. The ears are long, funnel-shaped and close together, almost joining at the base. The long, totally armored tail is broad-based and more than 80% of the length of the combined head and body; the scales on the proximal tail rings are keeled. The carapace is divided into a scapular shield, a pelvic shield with a unique pattern of smooth, uniformly sized scales, and 7–8 movable bands between the two shields. The face is pink and the upperparts dark gray-brown. The underparts are paler, and the belly has no protective armor. The species has characteristic skull features that can be observed on museum specimens but not on live animals.

HB:	51–58 cm
Tail:	30.5–45.6 cm
Wt:	8.5–11.8 kg

Similar species: Nine-banded Armadillo (*p. 45*) is much smaller, with a proportionately longer tail.

Habitat: The Orinoco and Amazon River Basins, where restricted to tropical moist lowland forests. In savanna areas generally confined to forest patches.

Distribution: Found in E Venezuela (south of the Orinoco River), Guyana, Suriname, French Guiana, and Brazil east of the Negro–Branco Rivers and north of the lower Amazon River.

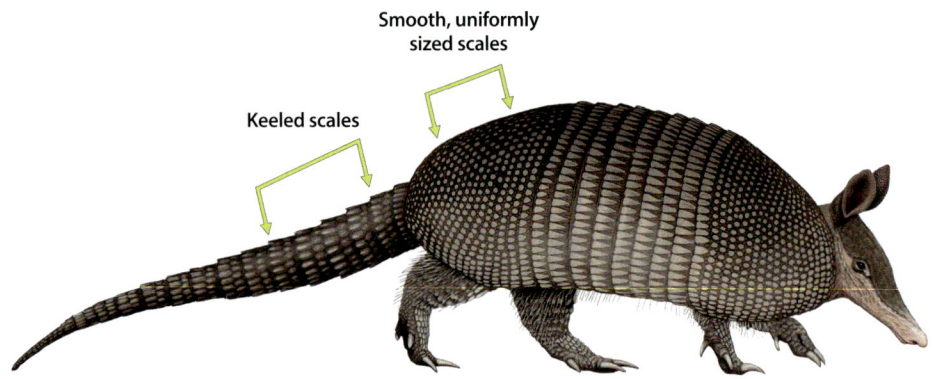

Smooth, uniformly sized scales

Keeled scales

NE Pastaza Long-nosed Armadillo

Dasypus pastasae

Description: Like Kappler's and Beni (*p. 44*) Long-nosed Armadillos (with neither of which it overlaps in range), distinguished from other *Dasypus* armadillos by the enlarged downward-projecting scales at the knee, its wide tail base, and the lighter skin color on the part of the head that is not covered by the head shield. Appears elongated, with a long, narrow and conical head. The ears are long, funnel-shaped and close together, almost joining at the base. The long, totally armored tail is broad-based and more than 80% of the length of the combined head and body; the scales on the proximal tail rings are flattened. The carapace is divided into a scapular shield, a pelvic shield with rough-textured, irregularly sized scales, and 7–8 movable bands between the two shields. The face is pink and the upperparts dark gray-brown. The underparts are paler, and the belly has no protective armor. The species has characteristic skull features that can be observed on museum specimens but not on live animals.

Similar species: Nine-banded Armadillo (*p. 45*) is much smaller and has a proportionately longer tail.

Habitat: Found exclusively in lowland tropical rainforest.

HB:	51–58 cm
Tail:	33–48 cm
Wt:	8.5–10.5 kg

Distribution: Occurs in the foothills of the eastern Andes in Colombia, Ecuador, Peru and Venezuela south of the Orinoco River, and extends into W Brazil between the Madeira and Branco Rivers. It appears to be sympatric with Kappler's Long-nosed Armadillo in E Venezuela.

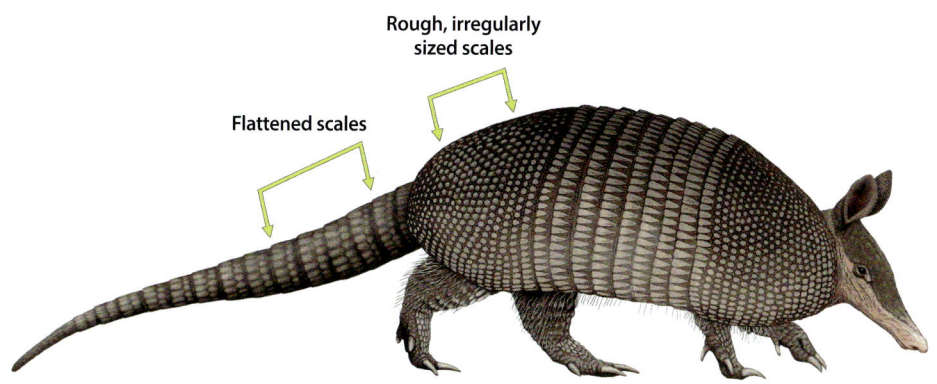

Rough, irregularly sized scales

Flattened scales

43

NE Beni Long-nosed Armadillo *Dasypus beniensis*

Description: Like Kappler's (*p.42*) and Pastaza (*p.43*) Long-nosed Armadillos (with neither of which it overlaps in range), distinguished from other *Dasypus* armadillos by the enlarged downward-projecting scales at the knee, its wide tail base, and the lighter skin color on the part of the head that is not covered by the head shield. Appears elongated, with a long, narrow and conical head. The ears are long, funnel-shaped and close together, almost joining at the base. The long, totally armored tail is broad-based and more than 80% of the length of the combined head and body; the scales on the proximal tail rings are flattened. The carapace is divided into a scapular shield, a pelvic shield with rough-textured, irregularly sized scales, and 7–8 movable bands between the two shields. The face is pink and the upperparts dark gray-brown. The underparts are paler, and the belly has no protective armor. The species has characteristic skull features that can be observed on museum specimens but not on live animals.

Similar species: Nine-banded Armadillo is smaller, with a proportionately longer tail.

Habitat: Restricted to tropical lowland rainforest.

HB:	51–58 cm
Tail:	33–48 cm
Wt:	8.5–10.5 kg

Distribution: Occurs to the south of the lower Amazon and Madeira Rivers in Brazil and south of the Madre de Dios River in Bolivia. The southern limit of its range is to the north of the dry Chaco forests and savannas in Bolivia and the Caatinga and Cerrado in Brazil.

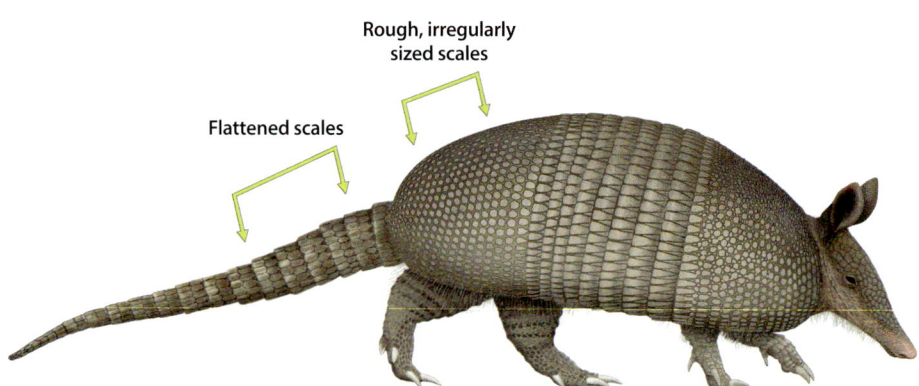

Rough, irregularly sized scales

Flattened scales

LC Nine-banded Armadillo

Dasypus novemcinctus

Description: A medium-sized elongated-looking armadillo with a long, narrow and conical head. The ears are long, funnel-shaped and close together. The long, totally armored tail is approximately two-thirds the length of the head and body combined. The carapace is divided into a scapular shield, a pelvic shield, and 8–10 movable bands between these two. The small plates which compose the scapular and pelvic shields are polygonal in shape and smaller than the rectangular plates on the mobile bands. The upperparts are dark gray-brown, while the flanks are yellowish. The belly is yellowish-pink and sparsely covered with white hairs. Easily separated from other *Dasypus* armadillos by the 8–10 movable bands and the long tail and ears; can be distinguished also by the blackish coloration with yellowish-white triangular scales on the posterior edge of the movable bands. Has proportionately the longest ears (just under half of head length) and tail (equal to or greater than body length) of all the *Dasypus* armadillos.

Similar species: Most closely resembles **Seven-banded Armadillo** (*p. 49*), but that species is much smaller and lacks pale edges to the 6–7 movable bands. The **naked-tailed armadillos** (*pp. 60–61*) lack armor on the tail and have more widely spaced ears.

HB:	36–57 cm
Tail:	26–45 cm
Wt:	3–7 kg

Habitat: A wide variety of habitats at up to 2,600 m (normally below 1,500 m), including degraded habitats such as arable land, pastures, rural gardens, urban areas, plantations, and heavily degraded subtropical and tropical forests.

Distribution: Widespread across South America, including Trinidad, as far south as central Argentina. Range extends through Central America and Mexico, and into S USA.

DD Hairy Long-nosed Armadillo

Dasypus pilosus

Description: Poorly known but distinctive, with long, thick, yellow to light brown hair growing out of and covering the carapace. The same type of hair is present on the cheeks and the proximal portion of the limbs, and the underparts are covered by shorter, sparser hair. The long and slender snout is more than half the length of the head. The eyes are narrow slits, the ears are long, conical and hairless, and the slender tail tapers to a point. The front part of the tail is protected by rings of scales. The front feet have four strong claws and the hind feet have five.

Similar species: None in range.

Habitat: Subtropical montane deciduous and evergreen forests, including sub-páramo habitats, from 2,600 m to 3,000 m. In areas with dense or shady cover and limestone formations.

Distribution: Restricted to the Peruvian Andes, where it is known from the departments of San Martín, La Libertad, Pasco, Huánuco, Junín and Amazonas.

HB:	32–44 cm
Tail:	23–31 cm
Wt:	1.0–1.5 kg

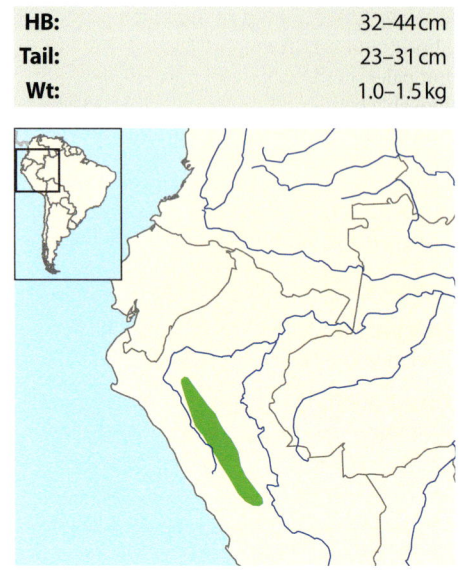

NT **Northern** (Llanos) **Long-nosed Armadillo** *Dasypus sabanicola*

Description: Long-eared, the ears being almost joined at the base. The carapace is dark, but paler on the sides, and normally has 8–9 movable bands with triangular scales. The underparts are yellowish-pink and are sparsely covered with pale hairs. The tail is covered with armored rings. The legs are short, and the four toes on the forefeet and five on the hind feet are long and have strong claws.

Similar species: None in range.

Habitat: Mainly in flooded savannas, open lowlands and riparian forests at up to 500 m.

Distribution: The Llanos of Venezuela and Colombia.

HB:	25–31 cm
Tail:	17–21 cm
Wt:	1–2 kg

DD Yungas Lesser Long-nosed Armadillo *Dasypus mazzai*

Description: Formerly known as **Yepes's Long-nosed Armadillo** *D. yepesi*. The carapace has 7–9 (usually 8) movable bands, the ears are roughly half the length of the skull, and it has a relatively long tail more than half the length of the head and body. The sides of the carapace are lighter-colored than those of other *Dasypus* armadillos.

Similar species: None in range.

Habitat: At 450–1,800 m in a variety of environments ranging from xeric habitats to humid montane forest.

Distribution: Known from nine locations in Jujuy, Salta and Santa Fe Provinces in Argentina, but thought likely to occur also in SW Bolivia and possibly Paraguay.

HB:	approx. 31 cm
Tail:	18–23 cm
Wt:	2.0–2.5 kg

LC Seven-banded Armadillo

Dasypus septemcinctus

Description: The smallest of the *Dasypus* armadillos and with a long, narrow, conical head and long ears that are relatively close together. The carapace is grayish-brown to dark brown, the sides being slightly darker than the dorsum, and with polygonal-shaped plates covering the pelvic and scapular regions: these are smaller than the rectangular plates found on the 6–7 movable bands. The tail is relatively long, being slightly more than half the length of the head and body, and the proximal half of the tail has 9–13 distinct rings each with two rows of scales.

Similar species: Nine-banded Armadillo (*p. 45*) is much larger and lighter-colored, with a longer tail, and has 8–10 rather than 6–7 movable bands.

Habitat: Generally a grassland and savanna species, but occurs also in gallery forest in SE Brazil. Appears to be adaptable to human disturbance and secondary habitats.

Distribution: Lower Amazon Basin of Brazil to the Gran Chaco of Bolivia, Paraguay, and

HB:	24.0–30.5 cm
Tail:	12.5–17.0 cm
Wt:	1.0–1.8 kg

possibly N Argentina, although its southern limit is unclear.

 **LC Screaming Hairy Armadillo** *Chaetophractus vellerosus*

Description: A smallish armadillo, similar in size to Pichi (*p. 52*). The upperparts are grayish with pinkish or yellowish edges to the scutes, and both the upperparts and the underparts are extensively haired. The long hair on the upperparts is tan-colored and the hair on the cheeks, throat, limbs and belly is whitish. The nose is pinkish. The large head shield is convex and curves over the eyes. The scapular shield is relatively small. The carapace is rounded, with 7–8 movable bands, and is covered with light-colored elongated bristly hairs, while the underside of the body is covered with close hairs. **Andean Hairy Armadillo** *C. nationi* is no longer considered a valid species and is now considered synonymous with Screaming Hairy Armadillo.

Similar species: Pichi (*p. 52*) and **Large Hairy Armadillo** also have gray ears, but these are proportionately shorter and do not project above the head. Pichi also has a distinctive, sharply pointed serrated edge to the carapace.

Habitat: Occurs at up to 4,600 m primarily in xeric environments in lowland and upland areas, as well as in high-altitude grasslands with loose sandy soils. Found also in pasture and agricultural areas. In Buenos Aires

HB:	20–30 cm
Tail:	8.0–13.8 cm
Wt:	640 g–1.33 kg

Province found on sandy-calcareous soils in grasslands with low vegetation and high vegetation cover.

Distribution: Chaco region of Bolivia, Paraguay and Argentina, with a disjunct population in eastern Buenos Aires Province, Argentina. The Andean form is found at high altitudes in Chile, Bolivia and Argentina.

LC Large Hairy Armadillo

Chaetophractus villosus

Description: The largest *Chaetophractus* armadillo. The head shield is broad, covering the dorsal part of the head and extending almost to the end of the thick snout. The carapace has 18 transverse bands, the central 7–8 of which are movable. The central portion of the carapace is divided by bands of skin that provide flexibility to the otherwise rigid upperparts. Hairier than most armadillo species, the underparts of this species are densely covered with whitish or light reddish-brown hairs, while long, coarse brown hairs project from the plates covering the upperparts.

Similar species: Six-banded Armadillo (*p. 53*) is usually larger and often more yellowish, with pale guard hairs on the carapace, but it can also appear more reddish-tan with dark brownish hairs. **Pichi** (*p. 52*) has a distinctive, sharply pointed serrated edge to the carapace. **Screaming Hairy Armadillo** has proportionately longer ears that project well above the head.

Habitat: Occurs at up to 1,500 m in a wide variety of grasslands (including Pampas and Chaco), savanna, and forest habitats. Found also in cultivated landscapes and some degraded habitats, such as arable land, pastures, rural gardens, and plantations. In Paraguay restricted to xerophytic areas of the

HB:	26.1–40.0 cm
Tail:	11.2–15.6 cm
Wt:	2–5 kg

Chaco, where it occurs in matorral, edges of Chaco woodland, ranch land and agricultural areas.

Distribution: Found in the Gran Chaco of Bolivia, Paraguay and Argentina, and as far south as Santa Cruz, in Argentina, and Magallanes, in Chile. Also introduced to Tierra del Fuego Province, in Argentina.

NT Pichi

Zaedyus pichiy

Description: A rather small armadillo. The carapace has seven movable bands ranging in color from light yellow to almost black with a white dorsal line, but is most frequently dark brown, and yellower along the lateral edges of the body. Fine blackish hairs and long yellow, brown and white bristles protrude between the individual plates on the back, while coarse yellowish hairs cover the soft skin on the underside of the body. The bare skin on the face and limbs is almost black.

Similar species: Screaming Hairy Armadillo (*p. 50*) has proportionately longer ears that project well above the head, and it lacks the distinctive sharply pointed serrated edge of the carapace. **Large Hairy Armadillo** (*p. 51*) lacks the distinctive sharply pointed serrated edge to the carapace.

Habitat: Found up to 2,500 m in areas with sandy soils, including xeric grasslands and shrublands, as well as Patagonian steppe habitats.

HB:	22–31 cm
Tail:	8–13 cm
Wt:	700 g–1.5 kg

Distribution: Central & S Argentina and Chile, as far south as the Strait of Magellan.

LC **Six-banded** (Yellow) **Armadillo**   *Euphractus sexcinctus*

Description: A medium-sized armadillo with a large relatively flattened convex carapace and a triangular-shaped head with small ears and a row of plates on the top of the head behind the head shield. A row of scutes extends along the back of the neck. It lacks movable bands at the front edge of the scapular shield. The carapace is narrower towards the head and has 6–7 movable bands. It is pale yellow to reddish-brown and scantily covered by buff to white bristle-like hairs, whereas the hairy armadillos are covered by dense hairs. The tail is scaled. The front feet have five toes with moderately developed claws.

Similar species: Small individuals are most likely to be confused with larger individuals of the **Large Hairy Armadillo** (*p. 51*), which are generally darker, hairier and more reddish, especially ventrally, with conspicuous tufts of hair on the cheeks, legs and throat.

HB:	40–50 cm
Tail:	20–25 cm
Wt:	3–7 kg

Habitat: Inhabits open areas, savannas, Cerrado, shrubland and dry, semi-deciduous forest. Also in secondary forests, and may occur in primary Amazonian Forest.

Distribution: S Suriname and adjacent areas of Brazil south to Bolivia, Paraguay, Uruguay and N Argentina.

Additional photo *p. 40*

DD Greater (Chacoan) **Fairy Armadillo**   *Calyptophractus retusus*

Description: An unmistakable small armadillo with a pinkish to yellowish-brown carapace. The carapace is thin and flexible and attached along the sides of the body, with a separate rounded pinkish plate over the rump. The lateral edge of the carapace is wavy, and it has 24 dorsal bands which are fused to the pelvis and the spine. The tail is pinkish, short, lightly armored, and flattened with a rounded tip. The underparts, including the legs and the sides of the head and the body, are covered with long, dense, fine white hair, and there are also sparse white hairs on the upperparts. The head shield is broad and the eyes and ears are small. Spends most of its time below ground: the front feet are scoop-shaped with large, curved claws for burrowing, and the hind feet have sharp claws.

Similar species: None in range.

Habitat: Patchily distributed in areas with loose, sandy soils and absent from areas with clay soils. Occurs in disturbed habitats and may be encountered close to villages.

HB:	12.0–17.5 cm
Tail:	3–4 cm
Wt:	90–130 g

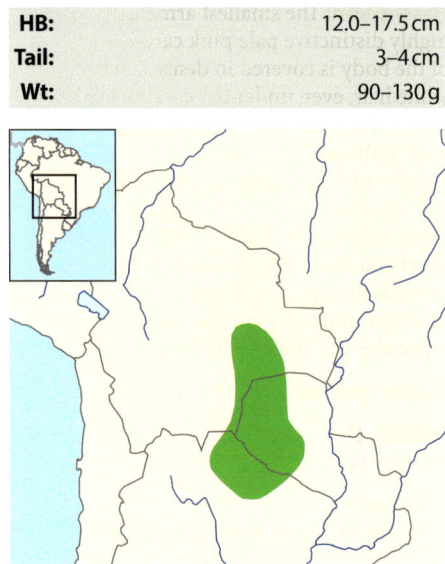

Distribution: Restricted to the Gran Chaco region of central & SE Bolivia, W Paraguay and extreme N Argentina.

DD Pink Fairy Armadillo

Chlamyphorus truncatus

Description: The smallest armadillo, with a highly distinctive pale pink carapace. Much of the body is covered in dense, smooth fine white hair, even under the carapace, which has 24 dorsal bands and is attached to the body only along the spine. The head shield is broad and the eyes and ears are tiny; the head has also a unique pair of tiny horns on both sides of the frontal bone. The tail is pink, short and rigid, becoming flat and pointed towards the tip. Spends most of its time in underground burrows, the front claws being especially well developed for digging.

Similar species: None in range.

Habitat: Dry grassland and sandy plains with shrubby vegetation at up to 1,500 m.

Distribution: Restricted to the provinces of southern Buenos Aires, Catamarca, Córdoba, La Pampa, La Rioja, Mendoza, Río Negro, San Juan and San Luis, in central Argentina.

HB:	11–15 cm
Tail:	2.5–3.5 cm
Wt:	approx. 100 g

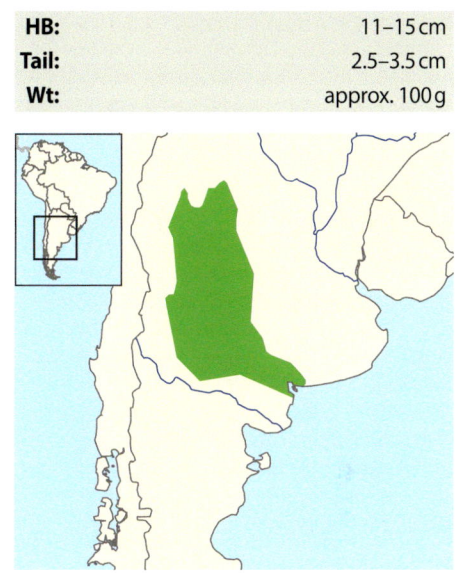

NT Southern Three-banded Armadillo   *Tolypeutes matacus*

Description: A small armadillo with a steeply rounded carapace, which is able to roll itself into a tight ball when threatened. The head shield is triangular, and the armor-plating that covers the body is divided into two domed shells, with three armored movable bands in between. Generally sandy-colored, but blackish individuals occur in some parts of the range. Long, thick, pale sandy-colored bristles protrude along the lower edge of the carapace. The large ears are flattened and have a rough edge. The belly is quite heavily furred with dark brownish hair. Bare skin on the sides of the face is similarly dark brown, but the tip of the snout and the nose are pinkish. The legs are short and strong, armored but with a covering of thick brownish hair. Has sharp powerful claws on the front feet and blunter hoof-like claws on the hind feet.

Similar species: None in range.

Habitat: Found mostly at up to 800 m in areas of thorny forest and scrub in the dry Chaco, but present also in palm savanna and gallery forest in the humid Chaco. Occurs also in Cerrado in Brazil.

HB:	20–25 cm
Tail:	5–8 cm
Wt:	800 g–2.2 kg

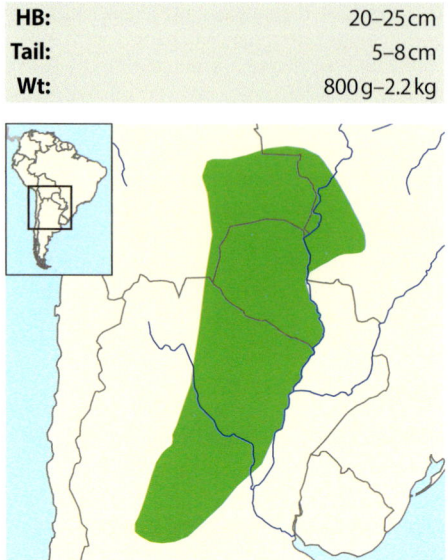

Distribution: E Bolivia and SW Brazil, south through the Gran Chaco of Paraguay to N & central Argentina as far south as San Luis Province.

Additional photo p. 40

EN Brazilian Three-banded Armadillo

Tolypeutes tricinctus

Description: A small armadillo with a steeply rounded carapace, which is able to roll itself into a tight ball when threatened. The carapace is hard and convex with three transverse armored movable bands. The head is triangular in shape, is flattened dorsally and is covered by a shield of approximately 30 plates. The ears are rounded, and can be folded in half when the animal curls up into a ball. The tail is short and completely covered by small round plates. The carapace is dark yellow and the abdomen has long yellow and white hairs.

Similar species: None in range.

Habitat: Caatinga in NE Brazil and in the eastern parts of the Cerrado in central Brazil.

Distribution: NE Brazil, throughout the highlands around the middle course of the São Francisco River and on the eastern slope of the Parnaíba River valley, as well as on the western slope of the Borborema Plateau.

HB:	23–25 cm
Tail:	5–8 cm
Wt:	1.0–1.8 kg

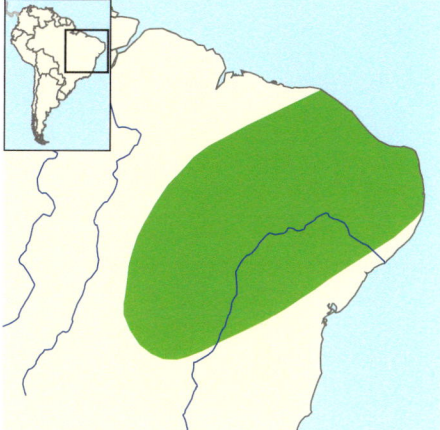

LC # Greater Naked-tailed Armadillo   *Cabassous tatouay*

Description: The largest of the naked-tailed armadillos, with a brownish-yellow carapace. The muzzle is rounded and robust. The upper surface of the head is covered by a scaled shield, with smaller scales on the ears and on the cheek below each eye, and there are three bands of small plates on the back of the neck. The carapace has solid shields over the shoulders and hips, with 10–13 movable bands between these. The ears are large and funnel-shaped, and placed laterally on the head. The tail is conical, with a few small isolated scales on the underside.

HB:	36–49 cm
Tail:	15–20 cm
Wt:	3.4–6.4 kg

Similar species: Chacoan Naked-tailed Armadillo is as much as 50% smaller, and has much shorter ears that do not extend above the top of the head. **Southern Naked-tailed Armadillo** (*p. 61*) is reliably distinguished from Greater Naked-tailed Armadillo only on the basis of measurements and scale counts (particularly of the head shield).

Habitat: Tropical lowland and sub-montane forest, possibly including secondary forest. Occurs also in more open areas, including agricultural habitats.

Distribution: E & S Brazil, Uruguay, NE Argentina and SE Paraguay.

NT Chacoan Naked-tailed Armadillo

Cabassous chacoensis

Description: The smallest of the naked-tailed armadillos. The upper surface of the head, body and limbs is armored with thick bony plates, separated by flexible bands of skin. The head is wide, tapering to a short, broad snout, and it has small ears with distinctive fleshy borders. The upperparts are brown or blackish, while the underparts are a dull yellowish-gray. The claws of the forefeet are long and powerful, particularly the middle claw, which is especially large and shaped like a sickle. The tail is poorly armored with a sparse covering of small, thin scales.

Similar species: Greater Naked-tailed Armadillo is up to 50% larger, with much longer ears extending well above the top of the head.

HB:	30–35 cm
Tail:	9.0–9.6 cm
Wt:	1.0–1.5 kg

Habitat: Restricted to Chaco thorn forests and not found in cultivated areas.

Distribution: Restricted to the Gran Chaco of W Paraguay and north-central Argentina.

DD **Northern Naked-tailed Armadillo** *Cabassous centralis*

Description: A broad, relatively flattened armadillo, this species has a short (less than half of the head-and-body length), pinkish-gray tail that appears naked. The carapace is dark gray-brown with a clear pinkish band around its lower part. The scales are large and square-shaped, and it has 10–13 inconspicuous, movable bands on the back. The head is broad, with widely spaced and moderately large, funnel-shaped ears. The eyes are extremely small.

Similar species: Southern Naked-tailed Armadillo is larger. **Nine-banded Armadillo** (*p. 45*) has ears set closer together and also has an armored tail.

Habitat: Lives at up to 3,000 m in dry to moderately moist deciduous and semi-deciduous forests, at forest edges in rocky terrain, and in open habitats such as dry savanna. Found also in secondary forest habitat including a mix of forest and agricultural land. Occurs in tropical moist montane forests and in the sub-páramo of the central Andean highlands in Colombia.

HB:	30–38 cm
Tail:	13.0–18.3 cm
Wt:	2.0–3.5 kg

Distribution: West of the Andes in NW Ecuador, W & N Colombia and NW Venezuela. Range extends through Central America and into SE Mexico.

LC Southern Naked-tailed Armadillo *Cabassous unicinctus*

Description: Medium-sized armadillo, but with considerable variation in size. The carapace is dark gray, normally with a bright yellow border to the lower edge, and it extends over the base of the legs. It also extends up between the ears and has dorsal plates arranged as bands over its entire length, including 10–13 movable bands. The scales are small, inconspicuous and square. The underparts are tan-colored and unscaled. The head is broad, with a blunt nose, and the large rounded, funnel-like ears are set widely apart. The head shield extends between the ears. The gray tail is long and narrow, and lacks protective scales; it can be pale-tipped.

Similar species: Northern Naked-tailed Armadillo is smaller and does not have a uniform pale-tipped tail. **Greater Naked-tailed Armadillo** is reliably distinguished from Southern Naked-tailed Armadillo only on the basis of measurements and scale counts (particularly of the head shield), but it tends to be more brownish-yellow rather than dark gray.

Habitat: Tropical lowland and sub-montane forest, possibly including secondary forest, gallery forest and Cerrado savannas.

HB:	29.0–44.5 cm
Tail:	8.7–20.0 cm
Wt:	1.6–3.6 kg

Distribution: East of the Andes from N Colombia, Ecuador, Peru, and N & E Bolivia through to Venezuela, Guyana, French Guiana and Suriname in the north and to Mato Grosso do Sul (Brazil) in the south. Recently discovered in Paraguay.

NT Giant Armadillo

Priodontes maximus

Description: Unmistakable: the largest armadillo, having enormous legs and feet with a greatly enlarged claw on the front feet. The head is covered by protective scales and is relatively small and conical, and the small ears are set widely apart at the back of the head. The carapace can be gray or reddish to dark brown, with a wide yellow stripe along each side. There are 11–13 movable bands on the back and a further 3–4 on the neck. The carapace does not cover the lower sides or legs. The tail is long and tapered, and covered with small scales. The underparts are unscaled and pinkish.

Similar species: None in range.

Habitat: Occurs at up to 500 m, often close to water within undisturbed primary rainforest habitats, but also in more open Cerrado habitats in parts of Brazil.

Distribution: Widespread east of the Andes in Colombia, N Venezuela, Guyana, Suriname, French Guiana, Ecuador, Peru, Bolivia, Paraguay, Brazil and N Argentina.

HB:	75–100 cm
Tail:	40–50 cm
Wt:	20–60 kg (exceptionally to 80 kg)

IUCN recognizes ten species of anteater, following the proposal by Miranda *et al*. (2017) that Silky Anteater should be treated as seven distinct species. Three of these seven are newly described species, while three are reinstated species.

All anteaters are largely solitary. Giant Anteater is terrestrial and also a good swimmer, and the two tamanduas are both terrestrial and arboreal. Silky anteaters are primarily arboreal and nocturnal, while Giant Anteater and tamanduas are both diurnal and nocturnal. They feed primarily on ants and/or termites, but tamanduas have been observed feeding also on bees, honey and palm fruit. Silky anteaters appear to feed exclusively on ants.

Southern Tamandua (ᴛᴏᴘ); **Giant Anteater (with infant on back)** (ʙᴏᴛᴛᴏᴍ)

LC **Northern Tamandua** *Tamandua mexicana*

Description: A medium-sized anteater with a long, tapered nose and a blotchy, prehensile tail that is furred for a third of its length. The most common coloration is golden-brown, with a distinctive patch of black fur ('vest') over the flanks, back and shoulders, but it can also be cream-colored. Very similar to Southern Tamandua but appears not to overlap in range.

Similar species: None in range.

HB:	52–77 cm
Tail:	40.0–67.5 cm
Wt:	3–6 kg

Habitat: Tropical and subtropical dry and moist forest, including mixed deciduous and evergreen forests. Also in mangroves, in grasslands with scattered trees, and in secondary forest and disturbed habitats.

Distribution: West of the Andes from Colombia to W Ecuador, NW Peru and NW Venezuela. Range extends through Central America and into S & E Mexico.

LC Southern Tamandua

Tamandua tetradactyla

Description: Often very similar in appearance to Northern Tamandua and most 'vested' individuals cannot be safely separated in the field, although the two species appear not to overlap in range. Southern Tamandua can be highly variable in appearance, those in N Brazil, S Venezuela and the Guianas often being totally blond and without a 'vest'. Animals from western Amazonia may be 'vested', partially 'vested', blond or completely black. The dark parts of the body may be black, sooty-gray or brown, and the pale areas may range from blond to golden-brown.

Similar species: None in range.

Habitat: Lives at up to 2,000 m in a variety of habitats, including gallery forest adjacent to savanna, lowland and montane moist tropical rainforest and mangroves.

Distribution: East of the Andes in Colombia, Venezuela, Trinidad and the Guianas, south through E Ecuador, Peru, Bolivia, Paraguay and Brazil to N Uruguay and N Argentina.

HB:	47–77 cm
Tail:	40–67 cm
Wt:	3.5–8.4 kg

Additional photo *p. 63*

VU Giant Anteater

Myrmecophaga tridactyla

Description: A distinctive tricolored anteater with a long, tapered nose and a long bushy tail, creating a unique elongated appearance. The overall coloration is brownish, with a black stripe banded by white or cream running from the throat and across each shoulder on to the middle of the back. The forelegs are cream, each with a black band on the wrist. There are four huge claws on the front feet and five on the hind feet. Walks on the knuckles, with the claws folded up into the palms for protection. Females are often observed to carry young on their back.

Similar species: None.

Habitat: Tropical moist forest, including upland forest in the Brazilian Amazon, dry forest, savanna habitats and open grasslands, and in the Gran Chaco and timber plantations. Easily located at long range in open grasslands, and can often be approached closely so long as the observer remains downwind.

Distribution: East of the Andes and south as far as N Argentina. Now thought to be extinct in Uruguay and in several states in

HB:	100–140 cm
Tail:	60–90 cm
Wt:	22–45 kg

S Brazil. Reported in Ecuador west of the Andes, but this requires confirmation. Range extends discontinuously through Central America as fas as Honduras.

Additional photo *p. 63*

LC **Common Silky Anteater**

Cyclopes didactylus

Description: A tiny woolly anteater. The nose is relatively blunt compared with that of other anteaters, and this species has a long, tapered, furry prehensile tail, although the underside of the final quarter and the entire tip are naked. It has brownish-yellow upperparts and gray underparts, legs, rump and tail. This species is the only silky anteater with clearly marked dark dorsal and ventral stripes. Some populations of Common Silky Anteater may have an indistinct ventral stripe or lack it entirely. Silky Anteaters could be confused with a small squirrel, but squirrels do not have a furry prehensile tail.

Similar species: Xingu Silky Anteater (*p. 72*) is generally gray, lacking yellowish tones. **Rio Negro Silky Anteater** (*p. 68*) lacks distinct dorsal and ventral stripes. **Red Silky Anteater** (*p. 73*) has a reddish appearance and lacks destinct stripes.

Habitat: Primary and secondary forests, including semi-deciduous and evergreen tropical moist lowland forest, gallery forest and mangrove forest. Most often encountered among small stems, vines and lianas.

Distribution: Occurs on Trinidad and from E Colombia, through E & S Venezuela and the Guianas, into N and NE Brazil,

HB:	20.1–20.5 cm
Tail:	16.5–29.5 cm
Wt:	approx. 300 g

as far south as the São Francisco River. Has a disjunct distribution in Brazil: one population occurs to the north-east of the Amazon River in the states of Pará and Amapá, towards Maranhão and Piauí; the other is found in the north-eastern Atlantic Forest, including the states of Rio Grande do Norte, Paraíba, Pernambuco, and Alagoas.

NE Rio Negro Silky Anteater *Cyclopes ida*

Description: The upperparts, legs and tail are usually gray, sometimes with an indistinct dorsal stripe. The underparts are light yellow and lack a ventral stripe. The front of the head has a slightly concave profile.

Similar species: Red Silky Anteater (*p. 73*) is distinguished by its brighter reddish coloration and lack of dorsal and ventral stripes. **Thomas's Silky Anteater** (*p. 71*) tends to be yellower, with gray legs and tail. **Common Silky Anteater** (*p. 67*) has a distinct dorsal stripe, and **Central American Silky Anteater** is yellowish without any gray tones.

Habitat: Amazonian Forest, particularly in areas of seasonally flooded black-water forest, and along small tributaries and canals.

Distribution: Found in Brazil west and south of the Negro River, with its southern limit being the Amazon River. Also occurs in N Peru and E Ecuador. There is an additional record from the eastern Andean forests of Colombia, but the species' northern limit is unknown.

HB:	20.1–20.5 cm
Tail:	16.5–29.5 cm
Wt:	approx. 300 g

NE Central American Silky Anteater — *Cyclopes dorsalis*

Description: Distinctly yellow throughout, with no grayish parts. The dorsal stripe is irregular but distinctive. A ventral stripe is generally lacking, but, when present, it is weakly marked. The front of the head has a straight rather than slightly concave profile.

Similar species: Rio Negro Silky Anteater is grayish with an indistinct dorsal stripe.

Habitat: Tropical rainforest and mangrove swamps.

Distribution: Pacific coast of Ecuador and N & NW Colombia, and also in the inter-Andean valleys of Colombia. Range extends through Central America and into SE Mexico.

HB:	18.5–21.5 cm
Tail:	17.0–22.5 cm
Wt:	155–275 g

NE **Amboro** (Yungas) **Silky Anteater** *Cyclopes catellus*

Description: Overall brownish-yellow, the tail and limbs appearing more yellow than the rest of the body. Lacks a dark dorsal stripe but has an extensive well-developed ventral stripe. The front of the head has a straight profile.

Similar species: None in range.

Habitat: Andean-slope forests and Yungas forests.

Distribution: Forests on the Andean slopes of central Bolivia and in Bahuaje-Sonene National Park in Peru.

HB:	20.1–20.5 cm
Tail:	16.5–29.5 cm
Wt:	approx. 300 g

NE Thomas's Silky Anteater

Cyclopes thomasi

Description: Orange to reddish-brown, with gray limbs and tail. Has a faint, poorly developed ventral stripe but no dorsal stripe. The front of the head has a straight rather than concave profile.

Similar species: Red Silky Anteater (*p. 73*) is reddish, but with the tail and limbs more yellowish-red. **Rio Negro Silky Anteater** (*p. 68*) is usually gray above with a faint dorsal stripe.

Habitat: None in range.

Distribution: W Brazil and east-central Peru, from the Juruá River in the north, south-west to the area of the Ucayali River in the Peruvian provinces of Pasco and Ucayali. The precise western limits of its range are uncertain.

HB:	20.1–20.5 cm
Tail:	16.5–29.5 cm
Wt:	approx. 300 g

NE Xingu Silky Anteater *Cyclopes xinguensis*

Description: Gray above, with yellow on the rump. The underparts are pale yellowish and the tail and limbs are gray. There is a distinct dorsal stripe, but the ventral stripe is indistinct and irregular. The front of the head has a straight profile.

Similar species: Common Silky Anteater (*p. 67*) has clearly marked dorsal and ventral stripes and has yellow tones on the upperparts. **Red Silky Anteater** is reddish above, with yellowish-brown limbs and tail, and lacks a dorsal stripe.

Habitat: Amazonian rainforest.

Distribution: Brazil south of the Amazon River, west of the Xingu River and east of the Madeira River. The southern limits of its range are unclear.

HB:	20.1–20.5 cm
Tail:	16.5–29.5 cm
Wt:	approx. 300 g

NE Red Silky Anteater *Cyclopes rufus*

Description: The upperparts are reddish, with the tail and limbs more yellowish-red. Lacks both dorsal and ventral stripes. The reddish coloration and lack of dorsal and ventral stripes distinguish it from the other silky anteaters. The front of the head has a straight profile.

Similar species: Rio Negro Silky Anteater (*p. 68*) is mainly gray, with yellowish underparts. **Thomas's Silky Anteater** (*p. 71*) is reddish-brown, but has gray limbs and tail and a faint ventral stripe. **Common Silky Anteater** (*p. 67*) has clearly marked dorsal and ventral stripes, as well as yellow tones to the upperparts. **Xingu Silky Anteater** is grayish with a yellowish rump and underparts, and has a distinct dorsal stripe and indistinct ventral stripe.

Habitat: Amazonian rainforest.

Distribution: Rondônia in W Brazil, in the interfluvium of the Madeira and Aripuanã Rivers. The northern limit is thought to be the Amazon River, while the southern limit is probably the Guaporé River.

HB:	20.1–20.5 cm
Tail:	16.5–29.5 cm
Wt:	approx. 300 g

Five species of sloth occur in South America. Miranda *et al.* (2022) proposed splitting **Southern Maned Three-toed Sloth** *Bradypus crinitus* as a separate species from (Northern) Maned Three-toed Sloth *B. torquatus*. This treatment has yet to be adopted by IUCN and is not followed here. Sloths are largely solitary and arboreal, periodically descending from trees to urinate and defecate and in order to move to other forest patches. They are good swimmers, holding their body high in the water. Sloths are primarily nocturnal feeders, mainly eating leaves, but some species will also eat fruit and lichens. During the day they tend to sleep high in the canopy.

LC Linnaeus's (Southern) Two-toed Sloth · *Choloepus didactylus*

Description: The fur is generally brown with long cream-colored tips, creating a variegated appearance, and the legs are often darker brown than the body. The throat is the same color as the upper chest, and the face is often the same color as the body.

Similar species: Hoffmann's Two-toed and **Pale-throated Three-toed** (*p. 77*) **Sloths** have a contrasting pale throat. **Brown-**

HB:	54–88 cm
Tail:	1.0–3.3 cm
Wt:	4–11 kg

throated Three-toed (*p. 78*) and **Pale-throated Three-toed Sloths** both have three claws on the front feet, pale patches on the back, a more discernible tail and paler limbs.

Habitat: Up to 2,400 m in tropical moist lowland and montane forest, including secondary growth.

Distribution: S Colombia and Venezuela (the Orinoco Delta and south of the Orinoco River), throughout the Guianas, and E Brazil south into Maranhão State on the Atlantic coast and west along the Amazon/Solimões Rivers into the upper Amazon Basin of Ecuador and Peru.

LC Hoffmann's Two-toed Sloth

Choloepus hoffmanni

Description: Usually dull creamy brown with darker brown legs, although the color can range from brown, through tan to almost whitish. The fur on the head is pale gray, often with a greenish cast on the crown. The face is pale, with short hair and with a brown bulbous nose, and the area around the face is white. The throat is pale and contrasts with the chest. The eyes are brown. The feet are long and narrow, with two claws on the front feet and three on the hind feet. Young animals are dark brown with short woolly fur.

Similar species: **Linnaeus's Two-toed Sloth** does not have a contrastingly pale throat. **Brown-throated Three-toed** (*p. 78*) and **Pale-throated Three-toed** (*p. 77*) **Sloths** have three claws on the front feet.

Habitat: Up to 3,300 m, in the canopy of deciduous and mixed-deciduous lowland and montane tropical forest. Mainly in areas of continuous forest, but can be found also in forest fragments.

Distribution: Two disjunct populations: one from W Venezuela through N & W Colombia

HB:	50–70 cm
Tail:	1.4–3.0 cm
Wt:	2.7–10.0 kg

to W Ecuador, and the other from E Ecuador through E Peru to N Bolivia and W Brazil, although the full extent of the range in Brazil is unclear. Range extends through Central America as far as Honduras.

VU Maned Three-toed Sloth

Bradypus torquatus

Description: Although structurally similar to Brown-throated Three-toed (*p. 78*) and Pale-throated Three-toed Sloths, this is a distinctive species with a uniformly pale brown to tan-colored head and upperparts, and with two conspicuous patches of long black hairs extending from the base of the neck, over each shoulder to about halfway down the back. The black mane starts to appear on juveniles and is already conspicuous on sub-adults. The face and chin are brown. The species lacks the orange patch and dark stripe between the shoulders of the other three-toed sloths. Two forms are recognized, Northern *torquatus* and Southern *crinitus* (see *page 74*), but these are indistinguishable in the field.

Similar species: None in range.

Habitat: Northern *torquatus* – from sea level to 1,000 m in humid and seasonal forests of the Atlantic Forest, in transitional areas with the Cerrado, and in restinga. Southern *crinitus* – from sea level to 1,290 m in humid lowland and montane forests of the Atlantic Forest, and also in restinga.

Distribution: Brazil. Northern *torquatus* – in the states of Sergipe and Bahia from Itaporanga d'Ajuda on the southern bank

HB:	59.0–75.2 cm
Tail:	4.0–5.4 cm
Wt:	4.6–10.1 kg

of the of Vaza-Barris River in Sergipe to the northern bank of Mucuri River in Bahia. Southern *crinitus* – in the states of Espírito Santo and Rio de Janeiro from the southern bank of the Doce River in Espírito Santo to the municipality of Arraial do Cabo, in Rio de Janeiro. Occurs west to Nova Friburgo in Rio de Janeiro State.

LC Pale-throated Three-toed Sloth   *Bradypus tridactylus*

Description: The overall coloration is usually grizzled smoky gray or chocolate-brown, with large cream or orange-tinged patches on the lower back and rump. Males have a patch of short orange fur with a brown central stripe between the shoulders. The top of the head and the forequarters are more uniform and darker tan or brown than the hindquarters. The head is small and rounded, with no visible ears, and the face has a small dark nose, a dark mask through the eyes, and a pale orange, yellow or whitish forehead the color of which extends on to each cheek and the throat. The legs are stocky, each foot having three claws, and the tail is short and stubby.

Similar species: Brown-throated Three-toed Sloth (*p. 78*) has a brown rather than pale throat. **Hoffmann's** (*p. 75*) and **Linnaeus's** (*p. 74*) **Two-toed Sloths** have two (not three) claws on the front feet, lack pale patches on the back and have no discernible tail. They also have darker limbs.

Habitat: A rarely seen sloth found in lowland and montane tropical moist forest, but appears to be most common in dense secondary forests and disturbed forests with a good supply of young leaves.

HB:	44.5–75.5 cm
Tail:	2.2–11.0 cm
Wt:	3.3–6.5 kg

Distribution: E Venezuela south of the Orinoco River into Guyana, Suriname, French Guiana and N Brazil south to both banks of the Amazon River.

LC **Brown-throated Three-toed Sloth** *Bradypus variegatus*

Description: Usually pale gray with long, coarse, shaggy hair, giving it grizzled appearance, and with large white patches on the lower back and hind legs. Males have a patch of short orange fur with a brown central stripe between the shoulders. The head is small and rounded, with no visible ears, and the face has a small dark nose, a dark mask through the eyes and a whitish forehead. The throat is brown. The legs are stocky, with three claws on each foot, and the front legs are longer than the hind legs. The tail is short and stubby.

Similar species: Pale-throated Three-toed Sloth (*p. 77*), as its name indicates, has a pale rather than brown throat. **Hoffmann's** (*p. 75*) and **Linnaeus's** (*p. 74*) **Two-toed Sloths** have two (not three) claws on the front feet, lack pale patches on the back, and have no discernible tail. They also lack the grizzled appearance of Brown-throated Three-toed Sloth and have darker limbs.

Habitat: Up to 1,100 m in a variety of forest types, including seasonal mesic tropical forest, semi-deciduous forest (inland Atlantic Forest), cloud forest and lowland tropical forest. In Brazil found in forested areas of the Amazon, the Atlantic Forest, and possibly in the contact zones between these biomes and the Cerrado.

HB:	40–75 cm
Tail:	4.7–9.0 cm
Wt:	3.7–6.0 kg

Distribution: Colombia into W & S Venezuela and south into Ecuador, E Peru and Bolivia, Brazil, and historically N Argentina. It may now be extinct in Argentina and there are no records from Paraguay. The distribution overlaps that of Pale-throated Three-toed Sloth on both banks of the Amazon River east of the Negro River. Range extends through Central America as far as Honduras.

Until recently it was generally accepted that five species of lagomorph, including the introduced European Rabbit (*p. 452*) and European Hare (*p. 452*), occur in South America. However, the taxonomy of the *Sylvilagus* rabbits is in a state of flux.

Wilson *et al.* (2016) and Diersing & Wilson (2017) split Forest Rabbit (Tapeti) *Sylvilagus brasiliensis* into two and three species respectively. The three forms appear not to be identifiable in the field, although Tirira (2017) states that Andean Rabbit *S. andinus* is smaller and has shorter ears than *S. brasiliensis*.

Ruedas *et al.* (2017) described a further split of Santa Marta Cottontail *S. sanctamartae*, and the IUCN Lagomorph Specialist Group now recognizes **Tapeti** *S. brasiliensis*, **Andean Rabbit** *S. andinus*, **Santa Marta Cottontail** *S. sanctamartae* and **Rio de Janeiro Dwarf Cottontail** *S. tapetillus* as distinct species.

Ruedas *et al.* (2017) also described **Suriname Tapeti** *S. palentus* as a new species, but this has yet to be accepted by IUCN.

Ruedas *et al.* (2019) published a further review of the taxonomy of the Forest Rabbit *S. brasiliensis* complex, in which they suggest that there are seven valid species in South America: **Forest Rabbit** *S. brasiliensis*, **Fulvous Tapeti** *S. fulvescens*, **Bogotá Tapeti** *S. apollinaris*, **Santa Marta Cottontail** *S. sanctamartae*, **Western Tapeti** *S. surdastur*, **Ecuadorian Tapeti** *S. daulensis* and **Nicefor's Tapeti** *S. nicefori*, although they also consider *S. tapetillus* to be part of *S. brasiliensis*.

Finally, the American Society of Mammalogists' Mammal Diversity Database (www. mammaldiversity.org) recognizes **Colombian Tapeti** *S. salentus* as a distinct species, but this treatment is not currently followed by IUCN.

Given the general uncertainty over this complex and the identification of the proposed taxa, all of these forms are treated as part of Forest Rabbit in this book. The ranges of the new species recognized by IUCN, however, are shown within the account for Forest Rabbit (*p. 82*).

Lagomorphs are herbivores and feed on grasses and other herbaceous flowering plants. They are terrestrial and largely crepuscular or nocturnal. They are generally solitary or found in small groups, although European Rabbit is highly gregarious and often seen in large groups.

Forest Rabbit

DD # Venezuelan Lowland Rabbit   *Sylvilagus varynaensis*

Description: A poorly known species. Buffy above, with a reddish patch on the nape, and whitish below, with a reddish-cinnamon throat. The front of each leg is reddish-cinnamon, mixed with light cream-and-white hairs towards the back of the leg. The inner thigh is reddish-cream. The tail is reddish-cinnamon above and reddish-buff below. The face is tawny-cinnamon with buff cheeks, and with a white ring around each eye and a pale buff outer surface of each ear.

Similar species: May overlap in range with **Forest Rabbit** (*p. 82*) and **Eastern Cottontail**, but both are smaller and paler, with a narrower and shorter skull and a less distinctive fur pattern. Eastern Cottontail has a conspicuous white underside of the tail.

Habitat: Most frequent in low shrubby-herbaceous savanna adjoining tropical dry forest, and feeds primarily on low shrubs.

Distribution: Restricted to the states of Barinas, Portuguesa and Guárico, in lowland Venezuela.

HB:	42–45 cm
Tail:	2.1–2.7 cm
Wt:	1.5–1.8 kg

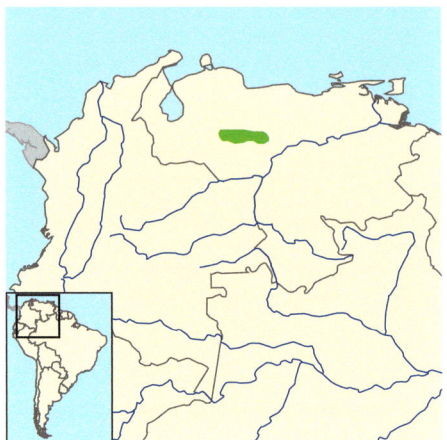

LC Eastern Cottontail

Sylvilagus floridanus

Description: A medium-sized rabbit with a conspicuous white under-tail. The upperparts are buff, grizzled with black, with paler flanks. The nape and legs are orange, the feet white. The throat is buff-colored and the belly is white. The ears and legs are moderately long. The tail is brown above and white below.

Similar species: Forest Rabbit (*p. 81*) and **Venezuelan Lowland Rabbit** do not have a conspicuous white under-tail. Forest Rabbit also is shorter-tailed, appearing almost tailless in the field, has slightly shorter ears, and is normally darker than Eastern Cottontail. Venezuelan Lowland Rabbit has a reddish-cinnamon throat and upper tail.

Habitat: Largely restricted to upland arid and semi-arid areas in South America.

Distribution: Colombia and Venezuela. Occurs also from Costa Rica to SW & E USA, and into Canada, with isolated pockets in NW USA and SW Canada.

HB:	33.7–48.0 cm
Tail:	4.7–6.6 cm
Wt:	630 g–1.5 kg

NE Forest Rabbit (Tapeti) complex   *Sylvilagus brasiliensis*

Description: A small, short-tailed and short-eared rabbit. The upperparts range from yellowish-brown grizzled with black, reddish or darker brown to gray and even blackish with a russet nape patch. The flanks and upper tail are slightly paler, the throat is dark, and the chest and belly are whitish. The legs and feet are orangey-brown and the underside of the tail is dull buff.

Similar species: Eastern Cottontail (*p. 81*), which is medium-sized and has a conspicuous white under-tail. May overlap in range also with **Venezuelan Lowland Rabbit** (*p. 80*), which is generally paler and has a reddish-cinnamon throat patch and upper tail.

Habitat: A wide range of disturbed and undisturbed habitats, including tropical rainforests, deciduous forests and second-growth forests. Found also in agricultural areas bordering forests.

Distribution: Found at up to 4,800 m from Colombia south through Ecuador, Peru, Bolivia and Paraguay and into N Argentina. Present in S & central Brazil, but its distribution in the Amazon region is unclear. Occurs also in French Guiana, Suriname, Guyana and NE Venezuela.

See *page 79* for an explanation of the taxonomic treatment of this species complex, combined here under the name *S. brasiliensis*. The following species are recognised by IUCN:

HB:	29–42 cm
Tail:	1.3–3.5 cm
Wt:	500 g–1.25 kg

This map shows the broad distribution of this species complex, and represents the traditional understanding of its range.

Santa Marta Cottontail *S. sanctaemartae* from the Sierra Nevada de Santa Marta, Colombia.

Andean Rabbit *S. andinus* occurs along the high-altitude treeless páramo zone from S Venezuela to Peru.

Rio de Janeiro Dwarf Cottontail *S. tapetillus* is known from only three specimens, all collected at a single locality in the Vale do Paraíba, in Rio de Janeiro State, Brazil, prior to 1892. It may already be extinct as a result of habitat loss in the area.

Additional photo *p. 79*

Squirrels

FAMILY | **Sciuridae**

Seventeen species of squirrel occur in South America. Thorington *et al.* (2012) and Wilson *et al.* (2016) provide detailed information on the ecology of individual species and on geographical variation of subspecies. Tirira (2017) recognizes three additional species, but these have been excluded pending adoption by IUCN:

Sabanilla Dwarf Squirrel *Microsciurus sabanillae* from the eastern foothills of Ecuador and Peru at 1,000–1,700 m.

Simons's Dwarf Squirrel *M. simonsi* from the western lowlands and foothills of Ecuador at up to 1,900 m.

White-naped Squirrel *Simonsciurus nebouxii*, a taxonomic split from Guayaquil Squirrel, occurs in south-western Ecuador and Peru at up to 2,272 m.

de Abreu Jr *et al.* (2020) suggested that there may be eight genera and 29 species of squirrel within the Neotropical region, but this treatment is yet to be adopted by IUCN and has not been followed here given the lack of more widespread acceptance at the time of writing.

All South American species are diurnal and solitary. Most are both terrestrial and arboreal. They feed on seeds, fruits, leaves, flowers, fungi, sap and insects. Some species will also take birds' eggs.

LC Guianan Squirrel

Sciurus aestuans

Description: Medium-sized and generally olive-gray grizzled with brown and yellow, although, with nine subspecies currently recognized, this species varies in appearance across its range. The upper tail is of the same color as the rest of the upperparts, but the under-tail is rufous. The vent can be white, cream, buff or yellowish. It has a pale yellowish-brown eye-ring and can have a buff-yellow spot behind each ear.

HB:	16.0–18.6 cm
Tail:	16.3–25.0 cm
Wt:	160–380 g

SSP. *ingrami*

The nine subspecies are:

SSP. *aestuans*, occurring north of the Amazon, has white spots behind the ears and a reddish wash on the body and feet.

Continued on next page...

Guianan Squirrel (*continued*)

ssp. *alphonsei,* from the northern coast of Brazil, has pale yellowish upperparts and grayish underparts.

ssp. *garbei,* from E Brazil (Bahia and Espírito Santo), has chestnut-ochraceous upperparts with more orange underparts, and a paler throat.

ssp. *henseli,* from S Brazil and NE Argentina, has ash-colored sides and white underparts.

ssp. *ingrami,* from the eastern and southern coasts of Brazil, has olivaceous upperparts with white to buff underparts.

ssp. *macconnelli*, from mountain regions of S Venezuela, Guyana and probably N Brazil, is brown-olivaceous in appearance.

ssp. *quelchii,* from S Guyana and north-central Brazil, is olivaceous, with a yellow belly.

ssp. *sebastiani* is restricted to San Sebastian Island in SE Brazil; it is darker brown than mainland forms and has a reddish-brown rather than gray-brown tail.

ssp. *venustus*, from Mount Duida in Venezuela, is similar in coloration to ssp. *aestuans*, but is much smaller.

Similar species: Yellow-throated Squirrel has orange-ochraceous underparts and shows a distinct contrast between the throat and the chest.

Habitat: A range of forest types, including tropical rainforests, swamp and other wet forests, gallery forests, secondary forests, plantations, gardens and urban parks. Although found at all forest levels, it is most common at 5–12 m.

Distribution: Throughout French Guiana, Suriname and Guyana, in S Venezuela, SE Colombia and widespread in Brazil, although absent from W Brazil. An isolated population in Colombia is thought to be potentially a separate species.

ssp. *ingrami*

DD Yellow-throated Squirrel

Sciurus gilvigularis

Description: Short-furred, the upperparts grizzled with ochraceous-buff and black. The underparts are dark ochraceous-orange, palest on the throat and lower abdomen and darkest on the chest and upper abdomen. It has a narrow pale buff eye-ring but lacks a spot behind the ears. The tail, which can appear to be faintly banded, is grizzled with buff and black. Individuals from Venezuela, Guyana and the northern part of the species' range in Brazil have a fulvous wash on the tail, while those from the southern part of its Brazilian range have a white wash on the tail.

Similar species: Guianan Squirrel (*p. 83*) is slightly larger and lacks both the orange-ochraceous underparts and the contrast between the throat and the chest.

Habitat: Coastal and evergreen forests, often with palms, lianas and rattans. Appears to be absent from logged and secondary forests.

Distribution: Three disjunct populations: in S Venezuela, Guyana and N Brazil.

HB:	15.5–17.7 cm
Tail:	16.5–19.5 cm
Wt:	150–165 g

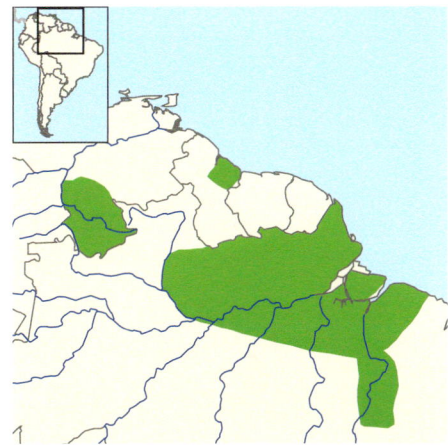

LC Red-tailed Squirrel

Sciurus granatensis

Description: A highly variable species with 28 subspecies recognized in South America. The upperparts are most frequently deep red, although they can range from grizzled black and yellow to charcoal with a yellowish tinge. Often shows a black longitudinal stripe down the center of the back. The underparts range in color from white to orange-red. The upper tail is red to rusty-colored, frequently grizzled with black, and/or with a black tip. The under-tail is yellowish-brown to black, often with reddish frosting. Those found east of the Andes are most likely to be white-bellied, populations in coastal Ecuador and Colombia have more blackish forms with a black-tipped red tail, and those in Venezuela have more brownish variants. Subspecies are highly variable, but form five distinct groups:

- Reddish with white underparts (6 sspp.)
- Dark olivaceous and brown with yellow, orange or red underparts (10 sspp.)
- Reddish with a black-tipped tail (4 sspp.)
- Reddish with a black medial dorsal stripe (4 sspp.)
- Brownish with a black-tipped tail (4 sspp.)

Similar species: None in range.

Habitat: A wide range of forested habitats, including secondary forest and disturbed areas

HB:	20.0–28.5 cm
Tail:	14–28 cm
Wt:	212–520 g

(even picnic areas and parks). Up to 3,000 m in the Andes. Feeds at all levels from the ground to the upper canopy, but most frequently in the mid- to upper canopy of trees.

Distribution: W Ecuador, W Colombia, N Venezuela and Trinidad. Range extends through Panama and into Costa Rica.

LC Bolivian Squirrel

Sciurus ignitus

Description: A medium-sized squirrel with gray-brown upperparts, conspicuous protruding ears, and a buff patch behind each ear. The underparts are whitish-buff to pale orange. The tail is often frosted yellow to pale orange. Five subspecies are recognized.

SSP. *ignitus,* from east of the Andes in Bolivia and adjacent Peru, has ochraceous-buff underparts.

SSP. *argentinius*, from Bolivia and Argentina, has yellowish underparts and reddish-tinged ears.

SSP. *boliviensis,* from much of Bolivia, is larger than other forms with white/whitish underparts.

SSP. *cabrerai*, from Brazil, has dark chestnut upperparts with reddish hair tips, fulvous ears with orange edges, and a distinct tuft of fulvous fur behind the ears. The upper surface of each foot is reddish.

SSP. *irroratus*, from Peru east of the Andes, has yellow underparts.

This species feeds mainly on the ground and in the lower and middle levels of trees.

Similar species: None in range.

HB:	14–22 cm
Tail:	15–23 cm
Wt:	183–242 g

Habitat: Found in evergreen lowland tropical forests transitional between humid and dry tropical forests, at 600–2,700 m.

Distribution: Found in Peru, extreme W Brazil, Bolivia and extreme NW Argentina.

SSP. *ignitus*

SSP. *ignitus*

DD **Andean Squirrel**

Sciurus pucheranii

Description: A small squirrel with thick, soft, dark reddish-brown upperparts and gray or yellow underparts. The upper tail is black, the individual hairs having a white tip, and the under-tail is gray to black with white tips. Some individuals have a black crown. Three subspecies are recognized.

SSP. *pucheranii*, from the eastern Andes of Colombia, has brownish-gray underparts with a buff tinge to the pectoral region.

SSP. *caucensis*, from the western Andes of Colombia, has brownish underparts with an ochraceous-buff wash.

SSP. *medellinensis*, from near Medellin, Colombia, usually has a blackish longitudinal midline, lacking on the other two subspecies, and its underparts are washed with white.

Mainly terrestrial, although has been reported foraging in the canopy.

Similar species: None in range.

Habitat: A poorly known species in montane Andean and sub-Andean forests at 2,200–3,500 m.

HB:	14.0–18.4 cm
Tail:	11.9–16.0 cm
Wt:	100–150 g

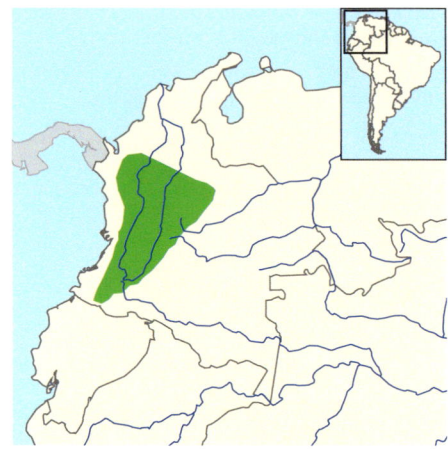

Distribution: Restricted to the Andes of Colombia.

SSP. *medellinensis*

^{DD} Sanborn's Squirrel

Sciurus sanborni

Description: A small squirrel with uniform olive-brown upperparts including the tail, which sometimes shows black banding. The eye-ring, the mouth and the patch behind each ear are contrastingly pale buff. The underparts are white to yellowish-orange, with yellowish feet. Forages on the ground and at higher levels.

Similar species: None in range.

HB:	15.2–17.5 cm
Tail:	16.1–18.4 cm
Wt:	No info. available

Habitat: Lowland Amazonian rainforest at 300–580 m.

Distribution: Restricted to the department of Madre de Dios in SE Peru and may extend into Puno Department.

LC Guayaquil Squirrel

Sciurus stramineus

HB:	18–32 cm
Tail:	25–33 cm
Wt:	460–495 g

Description: Two distinct but variable forms occur. The dark form has a black head, ears and feet, with grizzled black and white upperparts. The hips, rump and base of the tail are grizzled with orange, and the underparts range from cream through tan to faint rusty. A paler form has charcoal to black upperparts heavily frosted with white, giving it a pale gray overall appearance. The hips, rump and base of the tail are tinged with buff to orange, and it has a distinct white to buff patch on the nape. The underparts are grayish and the ears and feet are black. Both forms have a black tail with white frosting.

Similar species: None in range.

Habitat: Wet evergreen and dry tropical forests, including secondary forests. Coffee plantations and urban areas.

Distribution: Restricted to SW Ecuador and NW Peru. Introduced to Lima, in W Peru.

PALE FORM

DD Fiery Squirrel

Sciurus flammifer

Description: A large squirrel with black to brown upperparts suffused with yellow, although melanism and partial albinism have been reported. There is a reddish tinge to the head, ears and rump, and the cheeks and chin have a yellow to orange wash. The underparts are white to cream and are separated from the upperparts by an orange line along each flank. The base of the tail is black with bright orange-red frosting, the remainder of the tail being orange-red. Largely arboreal, this species forages high in the dense canopy of large trees, particularly in low humid marshy palm groves.

HB:	27.2–30.3 cm
Tail:	24.2–31.0 cm
Wt:	550–630 g

Similar species: None in range.

Habitat: Found in semi-deciduous and evergreen forests in the tropical lowlands, but normally absent from disturbed or secondary forests.

Distribution: Restricted to S Venezuela south of the Orinoco River, from the Colombian border east to Ciudad Bolívar.

LC Northern Amazon Red Squirrel

Sciurus igniventris

Description: A relatively large squirrel with dark chestnut-red or rusty-orange upperparts grizzled with black. Large thinly haired ears, sometimes with indistinct yellowish patches behind them, extend above the often black crown. The underparts are pale orange, red or white and contrast strongly with the upperparts, from which they are sometimes separated by a dark lateral line. The bushy tail is orange or rusty-colored, with a black base. The feet are bright red or orange. Melanistic individuals have been reported. The eastern form ssp. *igniventris* has ochraceous-reddish upperparts and ferruginous underparts, while the western form ssp. *cocalis* has a dark blackish central stripe from the head to the tail, and pale ochraceous-buff underparts.

Similar species: Very similar to **Southern Amazon Red Squirrel** and often indistinguishable in the field (see the account for that species for details of differences).

Habitat: Common in both mature and disturbed rainforests. Spends about 10% of its time in foraging on the ground, and the remainder of the time in traveling or feeding at all levels of the canopy.

HB:	24.0–29.5 cm
Tail:	24.0–30.5 cm
Wt:	500–900 g

Distribution: Lowland areas of E Peru, E Colombia, S Venezuela and W Brazil (ssp. *igniventris*) and E Ecuador and N Peru east of the Andes (ssp. *cocalis*).

ssp. *igniventris*

LC Southern Amazon Red Squirrel

Sciurus spadiceus

Description: Closely resembles Northern Amazon Red Squirrel and likewise has dark chestnut-red or rusty-orange upperparts grizzled with black. Large thinly haired ears extend above the crown. The underparts are pale orange, white or yellowish and contrast strongly with the upperparts. The bushy tail is orange or rusty-colored, with a black base. The feet are dark red mixed with black, or solid black. Melanism has been reported. Three subspecies are recognized.

ssp. *spadiceus*, from Brazil, has grizzled pale yellow and dusky upperparts and ochraceous-buff underparts, with a reddish-orange wash on the cheeks and head.

ssp. *steinbachi*, from Bolivia is larger and paler, with a yellowish wash on the cheeks and head.

HB:	24–29 cm
Tail:	23.5–34.0 cm
Wt:	570–660 g

ssp. *tricolor,* from Ecuador and Peru, has dark brown to blackish upperparts with an ochraceous wash, and pale yellow underparts.

Similar species: Almost indistinguishable from **Northern Amazon Red Squirrel** in the field but that species is more coarsely grizzled on the back; Southern Amazon Red Squirrel tends to be finely grizzled. Northern Amazon Red Squirrel has uniform red or orange-red feet (other than in melanistic individuals), whereas Southern Amazon Red Squirrel has black hairs mixed in with the red hairs.

Habitat: Common in mature and disturbed lowland rainforests in the western Amazon Basin and in the foothills of the Andes. Normally forages on the ground, in undergrowth, and in palm trees.

Distribution: S Colombia, E Ecuador, E Peru, N Bolivia and W Brazil.

SSP. *tricolor*

DD Junín Red Squirrel

Sciurus pyrrhinus

Description: A large squirrel with dark red upperparts grizzled with black, and sometimes flecked with white. The head is generally darker than the upperparts. The sides of the muzzle and the underside of the lower jaw are lighter than the rest of the head, and the ears are edged in black. The upperparts contrast with the underparts, which range from white to cream or orange, sometimes with patches of the other colors. The tail is chestnut to black towards the base and orange-red towards the tip. Forages on the ground and at higher levels and also visits clay licks.

HB:	24–28 cm
Tail:	20.8–24.0 cm
Wt:	No info. available

Similar species: None in range.

Habitat: Found primarily in subtropical and montane forests at 600–2,500 m, but also in lowland rainforest.

Distribution: Known only from the eastern Andean slopes of central Peru, but thought possibly to occur also in Bolivia and Ecuador.

LC **Central American Dwarf Squirrel** *Microsciurus alfari*

Description: A small but relatively long-limbed squirrel with dull olivaceous-brown to olivaceous-black upperparts tinged with red in some individuals, and often with white spots at the base of the ears. The underparts can be buff, gray or pale orange-buff. The tail is dark olivaceous-brown with orange frosting.

Similar species: Overlaps in range with **Western Dwarf Squirrel** (*p. 97*) in extreme NW Colombia: the two are almost indistinguishable in the field although that species generally has pale to reddish-orange underparts.

Habitat: Found in dense evergreen forests, including cloud forest, at up to 2,600 m. Occurs individually and in pairs, and feeds on the ground and in the canopy.

Distribution: NW Colombia, near the border with Panama. Range extends through Central America to S Nicaragua.

HB:	10.8–14.6 cm
Tail:	8–13 cm
Wt:	72–105 g

LC Amazon Dwarf Squirrel

Microsciurus flaviventer

Description: Upperparts dark brown with reddish to olivaceous tones, and a pale patch behind the ears. The underparts range from grayish with an orange wash, to pale or deep orange. The tail is grizzled brown to black with slight steel-gray frosting. Eight subspecies are recognized.

SSP. *otinus*, from W & NW Colombia, has white-tipped ears and a tail frosted with white.

SSP. *similis,* from S Colombia, has orange-rufous underparts and lacks pale patches on and behind each ear.

SSP. *napi*, from east of the Andes in Ecuador and S Colombia, and SSP. *peruanus*, from NW Peru, have yellowish-rufous underparts, with a large white patch behind each ear.

SSP. *simonsi,* from central Ecuador, is dark, with a yellow eye-ring and fulvous underparts.

SSP. *sabanillae,* from S Ecuador, has ochraceous underparts and lacks patches behind the ears.

HB:	12–16 cm
Tail:	9.6–15.0 cm
Wt:	60–128 g

SSP. *flaviventer*, from W Brazil, is blackish-olive above with a darker central dorsal stripe, the underparts are reddish-yellow, and the tail is washed with yellow.

SSP. *rubrirostris,* from central Peru, has orange-ochraceous underparts and a long tail with yellow frosting.

Tirira (2017) treats **Sabanilla Dwarf Squirrel** *M. saballinae* and **Simons's Dwarf Squirrel** *M. simonsi* as separate species.

Similar species: Neotropical Pygmy Squirrel (*p. 99*) is much smaller. **Western Dwarf Squirrel** has grizzled brown upperparts suffused with pale yellow or orange-buff and some individuals have a black median line.

Habitat: Evergreen tropical rainforests at up to 2,000 m. Common in terra firme forests but does not occur in flooded forests. Forages on the ground and up to 5 m in trees and known to accompany mixed feeding flocks of insectivorous birds.

Distribution: Widespread in W Amazon Basin of Colombia, Ecuador, Peru, W Brazil and possibly just into Bolivia. Occurs also in W & NW Colombia.

SSP. *napi*

LC **Western Dwarf Squirrel**

Microsciurus mimulus

Description: A small squirrel with grizzled brown upperparts tinged with pale yellow or orange-buff. The underparts are pale to reddish-orange. The tail is grizzled black-brown with slight yellow or gray frosting. Three subspecies are recognized in South America, but this is likely to be subject to revision following recent molecular studies.

ssp. *mimulus,* from NW Ecuador and SW Colombia, shows a black central dorsal line on some individuals.

ssp. *boquetensis*, from NW Colombia, has paler upperparts and ochraceous-rufous underparts.

ssp. *isthmus*, from W Colombia, has more yellowish upperparts with a reddish tinge to the rump.

Generally solitary, but occasionally seen in pairs foraging on the ground or in the lower levels of trees.

HB:	13.5–14.8 cm
Tail:	9.4–11.6 cm
Wt:	120 g

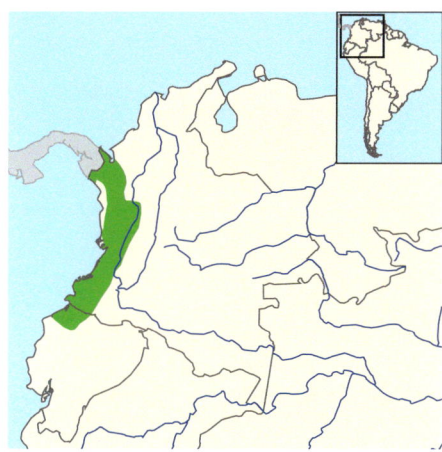

Similar species: Overlaps in range with **Central American Dwarf Squirrel** (*p. 95*): the two are almost indistinguishable in the field although that species generally has grayish or yellowish underparts. **Amazon Dwarf Squirrel** has dark brown upperparts with reddish to olivaceous tones.

Habitat: Evergreen forests and highland forests. It occurs in higher-elevation forests, where the range overlaps with that of Central American Dwarf Squirrel.

Distribution: NW Colombia to NW Ecuador. Range extends into Panama.

ssp. *mimulus*

ssp. *isthmus*

DD Santander Dwarf Squirrel

Microsciurus santanderensis

Description: Formerly treated as a subspecies of Neotropical Pygmy Squirrel, this small squirrel has dark brownish upperparts and grayish underparts. It has a black central dorsal stripe and has pale ochraceous patches behind the ears. The tail is grizzled black-brown with gray frosting.

Similar species: None in range.

Habitat: Low to mid elevations in humid forests, and reportedly found also in marshlands and montane forests at elevations of 100–1,000 m and 2,700–3,800 m.

Distribution: Colombia, occurring on the western slopes of the eastern Cordillera, and also on the right bank of the middle Magdalena River, in the vicinity of Barrancabermeja and San Vicente de Chucurí, in the department of Santander.

HB:	13.6–15.6 cm
Tail:	13.6–15.2 cm
Wt:	No info. available

LC Neotropical Pygmy Squirrel

Sciurillus pusillus

Description: A tiny squirrel with gray upperparts and tail suffused with yellow. The head is cinnamon to reddish with conspicuous white or buff patches behind the ears, and the ears are black-tipped. The underparts are buff to gray. Three subspecies are currently recognized, although their respective geographic distributions are unclear and there may prove to be two or more distinct species involved.

SSP. *pusillus*, from Guyana, Suriname and French Guiana, as described above.

SSP. *glaucinus*, from lower Amazonian Brazil, is generally paler.

SSP. *kuhlii*, from the upper Amazon of Peru, typically lacks pale patches behind the ears, and is more buff to brown.

Similar species: Amazon Dwarf Squirrel (*p. 96*) is much larger.

Habitat: Mature lowland evergreen Amazon forests. Individuals and family groups feed on sap at all levels within the forest canopy, although most common around 10 m above the ground.

HB:	8.9–11.5 cm
Tail:	8.9–12.0 cm
Wt:	33–45 g

Distribution: Four disjunct populations: in NE Peru, S Colombia (subspecies unknown), lower Amazonian Brazil, and NE Brazil into French Guiana, Suriname and Guyana.

SSP. *pusillus*

New World Porcupines   FAMILY | **Erethizontidae**

New World porcupines occur from North America south to southern Brazil. There are 18 species, 16 of which occur in South America. These include Amazonian Long-tailed Porcupine *Coendou longicaudatus*, split from Yellow Quill-tipped Porcupine *C. prehensilis* (Menezes *et al.*, 2021). Although yet to be confirmed by IUCN, this treatment is adopted here. Voss *et al.* (2013) provide an important overview of the taxonomy of New World porcupines.

New World porcupines are small to medium-sized rodents covered with sharp quills and have a prehensile tail. Largely nocturnal and arboreal, they feed mainly on fruit, young seeds, leaves, buds and sometimes bark. Generally slow-moving but can move quickly when threatened.

LC **Bicolor-spined Porcupine**   *Coendou bicolor*

Description: A medium-large porcupine with bicolored quills which are white at the base, then brown or black, and may have short pale tips. The quills extend beyond the fur and are noticeably longer on the nape, shoulders and upper back than they are on the lower back and rump, creating the impression that the upper back is darker (blackish) than the yellowish lower back and rump. The face, the sides of the body, the tail and the legs normally appear speckled with yellow, but can be dark brown or black. The underparts are pale grayish-brown. Lowland animals are normally longer-tailed than those from montane regions, the tail being close to or slightly longer than the length of the head and body.

HB:	37.8–50.0 cm
Tail:	33.5–54.0 cm
Wt:	3.4–4.7 kg

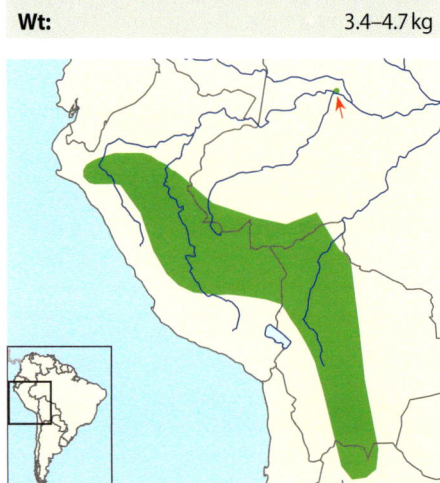

Similar species: None in range.

Habitat: Mature and disturbed lowland and montane rainforest.

Distribution: Eastern Andean foothills and adjacent Amazonian lowland forests from San Martín, in NE Peru, south to NE Bolivia in the lowlands of Beni and up to 2,500 m in the Yungas of Cochabamba. Reported also from NW Argentina, and in relict montane forests on the western side of the Andes in N Peru. Recently recorded from the state of Acre, in W Brazil within 50 km of the Bolivian border, and from Amazonas State in Brazil.

DD Quichua Porcupine

Coendou quichua

Description: Previously considered a subspecies of Bicolor-spined Porcupine. Medium-sized and with a speckled and spiny appearance owing to the fur being shorter than the quills. The back is blackish and appears speckled as a result of the bicolored (white with black tips) quills and the longer tricolored bristle-quills with white bases and tips and a black middle section. The underparts are less speckled than the rest of the body. The tail is black or speckled with black and white above and black below, and is highly variable in length, ranging from 55% to 90% of the length of the head and body. There is a considerable amount of geographical variation, some individuals from Ecuador appearing blackish without pale tips to the bristle-quills, although most have whitish, buff or reddish-brown tips.

Similar species: **Amazonian Long-tailed Porcupine** (*p. 104*) is much larger and generally longer-tailed. **Blackish Hairy Dwarf Porcupine** (*p. 113*) has long fur which covers the quills, and appears hairy rather than spiny.

Habitat: Inter-Andean valleys and tropical, subtropical, temperate and high Andean forests at up to 3,600 m.

HB:	33.2–44.0 cm
Tail:	26.0–41.3 cm
Wt:	2–3 kg

Distribution: NW Ecuador and trans-Andean Colombia. Range extends through Panama and into Costa Rica.

DD Eastern Amazonian Dwarf Porcupine   *Coendou nycthemera*

HB:	29–38 cm
Tail:	26–35 cm
Wt:	approx. 950 g

Description: A medium-sized blackish porcupine lacking emergent dorsal fur. The quills are bicolored, the base being white or yellowish for the first third and the remaining two-thirds being black. Where the base of the quills is visible, for example around the head, shoulders and forearms, it can appear yellowish or whitish. Some individuals appear speckled owing to the presence of tricolored defensive quills with white or pale reddish-brown tips. The rump appears black owing to the short, upright quills with heavily worn tips. The underparts are gray-brown or dark brown with soft hair-like bristles. The long prehensile tail is roughly 90% of the length of the head and body and is paler at the base, becoming darker for the remaining two-thirds.

Similar species: None in range.

Habitat: A lowland tropical-rainforest species occurring in primary forest.

Distribution: Brazil, occurring in Pará and Amazonas States south of the main Amazon River, from the Madeira River east to Ilha de Marajó.

NE **Yellow Quill-tipped Porcupine** *Coendou prehensilis*

Description: A small to medium-sized porcupine with tricolored quills along the dorsal crest and bicolored quills on the rump. The fur is brown or sometimes tricolored but never covers the quills on adults, and the general color pattern is light yellowish dashed with dark brown. This pattern is created by the predominance of tricolored quills, the longest quills being light yellowish.

Similar species: None in range.

Habitat: Atlantic Forest.

Distribution: Restricted to a small area of the north Atlantic Forest of Brazil, where its range extends as far south as the north bank of the São Francisco River. Treated by some authorities as conspecific with Amazonian Long-tailed Porcupine (*p. 104*).

HB:	29–48 cm
Tail:	31–43 cm
Wt:	2.35–2.90 kg

This is the only photograph of this species that was available at the time of publication: it shows a dead animal on a railway line.

LC Amazonian Long-tailed Porcupine   *Coendou longicaudatus*

Description: A large, heavily grizzled black and white porcupine with a long tail and considerable geographical variation. Lacks emerging fur and appears very spiny. The head is rounded and has large, bulbous pink lips and nose and a white face. The small ears are inconspicuous but the whiskers are long and broad, reaching back to the shoulders. Two subspecies are recognized. ssp. *longicaudatus* appears dark brown with short white dashes caused by the presence of bicolored quills among the tricolored quills of the dorsal crest and short whitish tips of the tricolored quills. The underparts are covered by short, soft, pale gray or whitish quills, and the feet are broad and appear gray-brown above. The long, robust tail is as long as the head and body. The tail is whitish at the base, covered in short quills for two-thirds of its length, and the lower third of the upper tail is naked. ssp. *boliviensis* has a mixed brown and white color pattern. Sub-adults have light brownish fur mixed in with the dorsal quills, this fur being particularly conspicuous on the rump.

Similar species: None in range.

Habitat: Lowland rainforest, dry deciduous forests, and riparian woodlands in savanna areas. During the day rests up in hollow trees and shady places in the sub-canopy.

HB:	45–57 cm
Tail:	46–60 cm
Wt:	2.33–5.57 kg

Distribution: Widespread across the Amazon and Cerrado biomes, with two confirmed records for the Bolivian Chaco; also on Trinidad. ssp. *longicaudatus* is found in the Amazon forest east of the Andes, although examined specimens are limited to the west bank of the Xingu River. ssp. *boliviensis* is associated with forested areas in open, dry parts of the Bolivian Chaco and Brazilian Cerrado.

DD Baturité Porcupine

Coendou baturitensis

Description: A large porcupine with a dorsal crest comprised of a mixture of long tricolored and short bicolored quills, giving the species an overall appearance of being brown with whitish dashes. The underparts have a mixture of brownish and grayish tones, although some individuals exhibit strong rust-brownish tones. The quills on the sides of the body are shorter than those on the back and appear darker. The quills are not covered by hair. The nose is large and bulbous. Juveniles have long soft hairs spread over the back among the quills. The dark gray tail becomes more densely quilled with age.

Similar species: None in range.

Habitat: Little known, but apparently found in the moist forest within the Caatinga biome.

Distribution: Occurs in the eastern Amazonian states of Pará and Maranhão, and in montane forested enclaves (Brejos de altitude) of Ceará State, in NE Brazil. Menezes *et al.* (2021) extended the known range of

HB:	46.0–54.9 cm
Tail:	32.5–47.0 cm
Wt:	3.42–3.50 kg

the species 1,460 km west from its previously known range in the Baturité Mountains.

LC **Stump-tailed Porcupine** *Coendou rufescens*

Description: A medium-sized, mid-brown to black porcupine, the upperparts being largely covered by tricolored quills with brownish or reddish tips. The face may appear whitish with a pinkish-gray muzzle. The relatively short tail is only 40% of the length of the head and body, and the sharp quills on the rump and at the base of the tail are bicolored, having a yellow base and darker tip. The remainder of the tail is largely naked. The underparts are pale brown and spiny.

Similar species: None in range.

Habitat: Usually at 1,500–3,000 m, but has been recorded from 800 m and to 3,500 m. Most records are from wet montane cloud forests, but this species has been found also in seasonally dry forest in NW Peru.

HB:	34–41 cm
Tail:	13.6–16.4 cm
Wt:	No info. available

Distribution: Western, central and eastern cordilleras of Colombia, on both slopes of the Ecuadorean Andes and in N Peru, with a further isolated population in Bolivia (Cochabamba).

DD Western Amazonian Dwarf Porcupine *Coendou ichillus*

Description: A small blackish-yellow porcupine. The blackish hairs are sparse and hidden among the quills and bristles. The short defensive quills are bicolored, each having a yellowish base and a dark brown or blackish tip. On the head, some of the quills are tricolored (with ivory-white base and tip separated by a dark middle band). Densely scattered among the quills are many long, thin bristle-quills with the yellowish base and tip separated by a broad dark brown or blackish middle band. The pale tips produce a streaked effect over the whole of the upperparts other than the rump, which is covered only with short bicolored quills and a few woolly hairs. The underparts are covered by very coarse bicolored or tricolored hairs. The hands and feet are covered with coarse blackish hairs. The tail averages more than three-quarters of the length of the head and body on specimens, although the extended tail of a living individual appeared to be almost as long as the head and body. The upper side of the base of the tail is densely covered with short bicolored quills, but on some individuals tricolored bristles form an indistinct whitish or yellowish

HB:	26–29 cm
Tail:	21–25 cm
Wt:	No info. available

chevron near the middle of the tail. The upper side of the prehensile tail tip is naked and calloused, while the rest of the tail is densely covered by blackish bristles.

Similar species: None in range.

Habitat: Humid tropical forests, including a white-water river floodplain in the Amazonian lowlands, at 230–500 m.

Distribution: Lowland rainforests of E Ecuador and NE Peru, but may occur also throughout north-west Amazonia. Menezes *et al.* (2020) report on a new record from the Japurá River, extending the distribution of the species significantly to the east of its previously known range. Reported also to occur in Peru along both banks of the Tambopata River between the Refugio Amazonas and the Tambopata Research Center, and on the Manu River near Cocha Cashu. In Ecuador there are records from Sucusari River, a tributary on the north/east bank of the Napo River in Loreto. This species has been found at every site in the Peruvian Amazon where comprehensive arboreal camera trapping has been carried out, suggesting that it is probably widespread.

LC Bahian Hairy Dwarf Porcupine

Coendou insidiosus

Description: A variable porcupine. Medium-sized and smoky gray-brown with long, thick dorsal fur other than on the head, where some white may be visible at the base of the quills. The face is dark brown with uniformly black whiskers. The thick, sharp, defensive quills are normally bicolored (white and dark brown, occasionally with faint orange tips), but largely hidden by the long fur. The prehensile tail is roughly 50–75% of the length of the head and body, and concolorous with the rest of the upperparts for the first third, darker brown or black for the middle third and sparsely haired for the remaining third. The underparts and feet are mid-brown to dark brown, although they can be sparsely covered with black and yellow hairs.

Similar species: Broomstraw-spined Porcupine (*p. 115*) is considerably larger.

Habitat: Mainly in primary and mature secondary rainforest, but occurs also in restinga forests, palms, mangroves, degraded thickets without large trees, cacao plantations with an intact canopy of native trees, and dense young secondary growth.

HB:	29–35 cm
Tail:	18.0–22.2 cm
Wt:	No info. available

Distribution: Atlantic coastal region of E Brazil from southern Sergipe to northern Rio de Janeiro, including easternmost Minas Gerais.

LC Black-tailed Porcupine

Coendou melanurus

Description: A medium-sized porcupine, with long, blackish dorsal fur streaked with long, yellow-tipped guard hairs over the bicolored quills. The fur is pale and thickest on the shoulders. The head is finely grizzled with slender, fine whiskers. The underparts are pale gray-brown, frosted with white, and the legs and feet are grizzled dark gray-brown or blackish. The rump is covered by thick, yellow quills that may be hidden by the fur. The tail is relatively long and black, and 80–95% of the length of the head and body.

Similar species: None in range.

Habitat: Lowland primary and secondary forest, where found mainly in the canopy.

Distribution: Throughout the north-eastern Amazonian lowlands from E Venezuela into Guyana, Suriname, French Guiana and N Brazil.

HB:	33.0–43.5 cm
Tail:	28–38 cm
Wt:	1.5–2.4 kg

LC **Paraguay Hairy Dwarf Porcupine** *Coendou spinosus*

Description: A small, variably colored, hairy porcupine having dark brown- or black-based dorsal hairs with grayish, orange or yellow tips, and long tricolored quills over the head, shoulders, middle back, flanks, thighs, and sides of the tail. These quills have yellow and black bands with orange tips, some specimens having yellow-tipped quills. The quills on the rump are short and bicolored, each with a pale base and short black tip. There are no quills on the underparts. The whiskers and the ventral bristles on the tail are conspicuously bicolored, with dark bases and pale tips. Some specimens have light brown hair tips, and most have yellow or orange tips to the bristles on the underside of the tail. The form previously split as **Orange-spined Hairy Dwarf Porcupine** *C. villosus* has several long, thick hairs that completely cover the dorsal and lateral quills. The tail is rusty-orange at its base, with the last 10 cm of the upper surface naked, and ranges in length from as little as 40% to almost 100% of the length of head and body.

Similar species: None in range.

Habitat: Humid tropical and subtropical forests from near sea level to 1,150 m, but is

HB:	28.5–47.0 cm
Tail:	20.0–37.8 cm
Wt:	approx. 1 kg

generally found in primary forest at lower altitudes.

Distribution: SE Brazil (from Espírito Santo south to Rio Grande do Sul), E Paraguay, N Uruguay and NE Argentina.

LC Frosted Porcupine

Coendou pruinosus

Description: A medium-sized porcupine with dark brown to blackish fur with pale grayish or silver tips, creating a frosted effect the degree of which varies between individuals. The fur is long and dense enough to cover the quills other than on the head. The head is gray-brown with tricolored quills, the latter having a pale base and tip and a dark center. The head has fine white speckling and the whiskers are black and stiff. The bicolored body quills on the upperparts are short, each with a pale yellow or ivory-white base and a dark brown tip, while the bristle-quills which are interspersed throughout the upperparts are long, thick and tricolored with a long pale tip. The underparts are brown, thickly furred without quills, and heavily frosted with white or pale gray. The feet are brown and grizzled with white. The prehensile tail is 50–60% of the head-and-body length on specimens from montane regions, but those from lowlands appear to have a relatively longer tail. The dorsal surface of the proximal half of the tail is covered by a mixture of quills and soft fur like that on the rump. Pale-tipped bristles along each side of the tail converge posteriorly to form a conspicuous yellowish or whitish chevron. The prehensile tail tip is naked and calloused above, but the rest of the tail is covered by blackish bristles.

HB:	32–38 cm
Tail:	19 cm
Wt:	No info. available

Similar species: None in range.

Habitat: Lowland rainforest and cloud forest from 54 m to 2,600 m.

Distribution: Foothills, mountains and lowlands of N Colombia, and N Venezuela.

DD Roosmalen's Porcupine *Coendou roosmalenorum*

Description: One of the smallest porcupines with long, dull dark brown or grayish-brown dorsal fur. Specimens with brownish dorsal fur have pale grayish or silvery tips to the hairs, but these do not produce a distinctly frosted appearance. The fur is long enough largely to conceal the short bicolored quills with yellowish bases and dark brown tips which cover the dorsum. Long tricolored bristle-quills, pale yellowish or ivory-white at the bases and tips and with dark brown central bands, are interspersed across the upperparts other than the rump. Some of these quills may be bicolored, with a pale base and dark tip. The head also is covered with quills and the whiskers are fine. The underparts are densely covered with coarse bicolored and tricolored hairs. The hands and feet have coarse dark brown or black hairs on the dorsal surface. The tail is almost 90% of the combined head and body length, and is woolly at the base with bicolored quills on the upper surface. The middle section of the tail is covered with blackish bristles, and the prehensile tip is naked, with calluses on the upper tail.

Similar species: None in range.

Habitat: All three known specimens were captured in lowland rainforest.

HB:	29 cm
Tail:	24–26 cm
Wt:	approx. 600 g

Distribution: Known from three localities along the Madeira River in central W Brazil, but thought likely to be more widely distributed south of the Amazon River in western Amazonia.

DD Blackish Hairy Dwarf Porcupine

Coendou vestitus

Description: A small porcupine with soft, dull blackish-brown hair which largely covers the quills other than on the head and face. The short quills are bicolored, pale yellow or ivory-white for 80–85% of their length and dark brown or blackish for the remaining 15–20%. The pale quill bases are exposed when the fur is parted or ruffled. Scattered throughout most of the dorsal fur other than on the rump are longer bristle-quills that emerge from the fur like fine wires; these are yellowish at the base, with dark brown tips. The rump has only fur and short, sharp quills. None of the quills on the body has pale tips. The whiskers are long and blackish. The underparts are uniformly dark brown and densely haired. The feet are dark brown. The tail is relatively short, averaging about half the length of the head and body. The proximal half of the upper tail is covered with quills and woolly fur as on the rump, while the tip has a naked, calloused dorsal surface; the remainder of the tail is covered above and below with blackish bristles.

HB:	29–37 cm
Tail:	17.0–19.5 cm
Wt:	No info. available

are believed to be from lower montane moist forest.

Similar species: Quichua Porcupine (*p. 101*) has shorter fur which does not cover the quills and has a spiny rather than hairy appearance.

Habitat: Specimens were collected within an altitudinal range of 1,300–2,600 m and

Distribution: Known from only two localities in Cundinamarca, namely Quipile and San Juan de Rioseco on the lower eastern Andean slope in Colombia.

Pernambuco Dwarf Porcupine · *Coendou speratus*

Description: A small-bodied, long-tailed porcupine that appears to be completely spiny because it lacks long dorsal fur. The dorsal quills have conspicuously brownish-red tips that contrast with the blackish dorsal background color.

Similar species: Yellow Quill-tipped Porcupine (*p. 103*) is much larger.

Habitat: Remnant patches of sub-montane Atlantic Forest. All records of the species are of individuals or pairs observed at night 10–20 m above the ground in trees.

Distribution: Pernambuco and Alagoas, in NE Brazil.

HB:	33–44 cm
Tail:	28.6–32.0 cm
Wt:	1.4–1.6 kg

Broomstraw-spined Porcupine

Chaetomys subspinosus

Description: A medium-sized pale to dark brown porcupine. The head is rounded and covered with quills that largely conceal the ears. The muzzle is brown and almost naked, with a bulbous nose. The head and shoulders are densely covered with short, sharply tipped quills. The short quills extend on to the back, rump, legs and base of the tail, but become much longer, stiff and slightly wavy bristles, lacking the sharp tips of the more defensive quills on the head. The quills have a pale yellow base, dark brown center and pale brown tip. The thick-based tail curls downwards and is roughly 60–70% of the length of the head and body; it tapers, becoming slender towards the tip. The upper 20% of the tail is dark brown above and rusty below, the remainder of the tail being thinly haired and becoming almost naked towards the tip.

Similar species: Bahian Hairy Dwarf Porcupine (*p. 108*) is considerably smaller.

Habitat: Humid Atlantic coastal lowlands and mountains to 650 m. Primary and secondary rainforests, gallery forests, restinga forests including palms and mangroves, and also degraded thickets. Feeds on average 10 m above the ground and is most frequently found along forest edge.

HB:	36–54 cm
Tail:	26.0–27.5 cm
Wt:	1–2 kg

Distribution: The main populations occur in Bahia, E Brazil, along the coast in the north and south-east of the state and in Espírito Santo. Found also in Sergipe, but has disappeared from some areas following widespread deforestation.

Acouchis & Agoutis

FAMILY | **Dasyproctidae**

Acouchis and agoutis are medium-sized, short-legged rodents. Twelve species occur in South America. Acouchis have an obvious short tail (4–8 cm), while agoutis have a tiny inconspicuous tail. They are terrestrial and mainly diurnal, most active in the early morning and late afternoon. Agoutis are normally encountered singly or in pairs, but small groups of some species, most notably Central American Agouti, do occur where there is an abundance of food. Red Acouchi occurs in family groups and Green Acouchi is colonial. Acouchis and agoutis both feed on seeds and nuts and, where food is abundant, frequently store these for future consumption.

LC Red Acouchi

Myoprocta acouchy

Description: Larger than Green Acouchi and with dark chestnut-red or orange upperparts grizzled with black, with some yellow on the crown and neck. The center of the back and the rump are glossy black or red, with long rump hairs overhanging the rear of the body in a fringe. Underparts appear orange. The tail is short and thin, and whitish with a tuft at the tip.

HB:	33.5–39.0 cm
Tail:	5.1–7.8 cm
Wt:	1.1–1.4 kg

Similar species: Green Acouchi is much smaller and is less orange in appearance. **Common Red-rumped Agouti** (*p. 122*) is much larger.

Habitat: Undisturbed forested habitats, but less common in secondary forest.

Distribution: Guyana, Suriname, French Guiana, and NE Brazil east of the Branco River and north of the Amazon River.

Green Acouchi

Myoprocta pratti

Description: The smallest acouchi, with considerable color and size variation across its range; more than one species may in fact be involved. The upperparts and legs are mainly uniform grizzled olivaceous, the rump sometimes appearing darker, and the throat, chest and belly are white. Individuals from the Venezuelan Amazon are usually gray-brown on the sides and have the thighs grizzled with white hairs. The tail is short and thin, and whitish with a tuft at the tip.

Similar species: Red Acouchi is larger and has an overall orangey appearance, including reddish underparts. **Brown Agouti** (*p. 127*) is considerably larger and darker, without an obvious tail.

Habitat: Occurs at 50–1,200 m in mature terra firme forest with dense undergrowth, but can be found also in disturbed areas close to natural forest.

Distribution: S Venezuela, Brazil, Colombia, Ecuador and Peru. South of the Amazon River in Peru, Bolivia and Brazil to at least the Madeira River. North of the Amazon River from the eastern foothills of the Andes to west of the lower Negro River.

HB:	29.8–38.3 cm
Tail:	4.0–5.8 cm
Wt:	800 g–1.2 kg

DD # Azara's Agouti

Dasyprocta azarae

Description: The smallest agouti. Gray, with the upperparts washed dull tawny, to bright orange. It is bright to pale orange below. Animals from drier areas tend to be mainly gray or olivaceous.

HB:	43.0–57.5 cm
Tail:	1.0–3.5 cm
Wt:	2.4–3.2 kg

Similar species: Common Red-rumped Agouti (*p. 122*) has a reddish-orange rump.

Habitat: Forest patches, *e.g.* palm forest, within savannas and in lowland Atlantic Forest. Common in the Cerrado in eastern Bolivia and central Brazil at up to 700 m above sea level.

Distribution: South-central and SW Brazil, SE Bolivia, E Paraguay and extreme NE Argentina (Teta & Reyes-Amaya, 2021).

ⓒ Black Agouti *Dasyprocta fuliginosa*

Description: The largest agouti, with black upperparts finely grizzled with white. The rump hairs are black, often with indistinct white tips, but do not overhang the rump except in Venezuela, where the hairs have long white tips. The throat is white but the remainder of the underparts are blackish, and the tail, feet and legs are black. There can be some variation in coloration, some individuals appearing to be washed with orange.

Similar species: Overlaps in range with **Kalinowski's** (*p. 121*) and **Brown** (*p. 127*) **Agoutis** but unlikely to be confused owing to its dark coloration.

Habitat: Up to 1,000 m in mature and disturbed rainforest, deciduous forest and montane forest at lower levels on the eastern slope of the Andes.

Distribution: Widespread in the western Amazon Basin of Brazil west of the Negro River and the Madeira River, in S Venezuela, SE Colombia, E Ecuador, and E Peru north of Junín Department.

HB:	54–76 cm
Tail:	2–4 cm
Wt:	3.5–6.0 kg

NT Orinoco (Guamara) Agouti

Dasyprocta guamara

Description: A medium-sized agouti with uniform brown to black upperparts with ochraceous shading. The underparts are tawny-yellow.

Similar species: May overlap in range with **Common Red-rumped Agouti** (*p. 122*), which is larger and has a distinctive reddish-orange rump.

Habitat: Wet, marshy, dense forests in the Orinoco Delta, including rainforests and mangroves.

Distribution: Known only from two lowland localities in the state of Delta Amacuro, Venezuela.

HB:	46.7–55.7 cm
Tail:	2.0–2.3 cm
Wt:	3.0–4.4 kg

DD Kalinowski's Agouti

Dasyprocta kalinowskii

Description: A large agouti with upperparts covered by banded black and yellowish-rufous hairs. Black-tipped white hairs overhang the rump. The underparts are yellowish with brown grizzling, and the feet are blackish. The black-tipped white rump hairs are diagnostic.

Similar species: Black Agouti (*p. 119*) and **Brown Agouti** (*p. 127*) do not have black-tipped white rump hairs.

Habitat: Upper tropical forest on the steep slopes of the eastern Andes at 1,000–2,000 m.

Distribution: Restricted to S Peru, where known only from the eastern slopes of Andes in the Urubamba and Marcapata drainages.

HB:	63 cm
Tail:	2.4 cm
Wt:	No info. available

121

LC Common Red-rumped Agouti

Dasyprocta leporina

HB:	47–65 cm
Tail:	1–3 cm
Wt:	2.1–5.9 kg

Description: Considerably variable in size and coloration, but a medium-sized to large agouti, with populations north of the Amazon River normally larger. The head and forequarters are finely grizzled olivaceous, with a distinctive dark red to yellowish-orange rump with long, straight overlapping hairs. The top of the head, the neck and the middle of the back are blackish, with a crest of longer hairs on the top of the head and neck. Includes the form *cristata*, from Guyana and Suriname, which is sometimes treated as a separate species, the **Crested Agouti** *D. cristata*.

Similar species: **Red Acouchi** (*p. 116*) is much smaller. Overlaps in range also with **Black-rumped** (*p. 125*), **Orinoco** (*p. 120*), **Orange** and **Azara's** (*p. 118*) **Agoutis**. (See those species for details of distinguishing features.)

Habitat: A wide range of forested habitats, including mature, disturbed and secondary evergreen forests and deciduous and gallery forests. In French Guiana primarily in open forest.

Distribution: Venezuela, Trinidad, Guyana, Suriname, French Guiana and Brazil north of the Amazon and east of the Negro River, then south of the Amazon, east of the Madeira River, to central coastal Brazil. It is absent from NE Brazil.

DD Orange Agouti

Dasyprocta croconota

Description: A medium-sized agouti with distinct bright orange coloring on the hindquarters and rump and lacking any black or brown coloration on the hindquarters. The front half, including the head, is gray and it has a distinct dark dorsal streak.

Similar species: Common Red-rumped Agouti is generally larger and has less extensive orange-red hindquarters, with the red coloration restricted to the rump and overlain by long straight dark hairs.

Habitat: Tropical lowland moist rainforest, including second growth and forest edge, up to 200 m above sea level. In coastal scrub and forest on Ilha de Marajó.

Distribution: The eastern Amazon of Brazil from the lower Tapajós River to the left bank of the lower Tocantins River, and on Ilha de Marajó and Ilha Mexiana at the mouth of the Amazon River.

HB:	46.5–56.0 cm
Tail:	1.0–2.5 cm
Wt:	2.2–2.8 kg

DD Iack's Red-rumped Agouti

Dasyprocta iacki

Description: A medium-sized agouti. The top of the head is brown, grizzled yellow-orange, the cheeks are yellower than the crown, and the underside of the neck is yellow with white bands. The remainder of the upperparts is largely gray-brown, with a wide darker band speckled with orange down the center of the back; this band becomes paler on the sides. The rump is dark brown, speckled with orange; the sides are orange.

Similar species: Black-rumped Agouti has a black crown and a noticeable black rump.

Habitat: Atlantic Forest.

Distribution: Known only from two localities in Paraíba and Pernambuco States, in NE Brazil.

HB:	44.5–50.5 cm
Tail:	No info. available
Wt:	2.3–3.8 kg

LC Black-rumped Agouti

Dasyprocta prymnolopha

Description: A medium-sized agouti with a black crown, a crest of longer black hairs on the nape, and a distinctive wedge of long black hairs on the rump. The remainder of the upperparts is grizzled yellow-orange, with blackish foreparts that become reddish-orange from the middle of the back and over the thighs. The black crown, black stripe over the rump and reddish-orange upper thighs are distinctive.

Similar species: Overlaps in range with **Common Red-rumped** (*p. 122*) and **Iack's Red-rumped Agoutis**, both of which have a red rather than black rump.

Habitat: Up to 900 m in deciduous forest and scrub, including the Cerrado and Caatinga, and probably coastal rainforest habitats. The only agouti in the Caatinga biome.

Distribution: NE Brazil from eastern Pará State south of the Amazon River along the coast to Alagoas and Bahia States. Occurs inland into Minas Gerais and Tocantins States.

HB:	45.0–52.5 cm
Tail:	1.8–3.0 cm
Wt:	No info. available

LC Central American Agouti

Dasyprocta punctata

Description: A medium-sized agouti. Individuals from the Atlantic slope of N Colombia are dark reddish-brown with long, straw-colored, black-tipped hairs on the rump and a blackish crown and nape, while individuals from the Pacific slope are orangey-brown, red-brown or yellow-brown and finely grizzled with black.

Similar species: None in range.

Habitat: Found from sea level to 1,600 m. A lowland tropical forest species, found in mature deciduous and evergreen forest, secondary forest, gardens and plantations. Although diurnal, it is encountered at night more often than are other agoutis.

Distribution: N Ecuador, Colombia, and W Venezuela in the Sierra de Perijá and on the western slopes of the Sierra de Mérida, south along the Pacific coast of Colombia and Ecuador. Its range extends to the eastern slope of the eastern Andes in Colombia and into the headwaters of the Sarare River in Venezuela. Range extends through Central

HB:	48–60 cm
Tail:	2.0–5.5 cm
Wt:	3.2–4.2 kg

America and into E Mexico, including the Yucatán Peninsula.

DD Brown Agouti

Dasyprocta variegata

Description: A medium-sized, heavy-set agouti with finely grizzled upperparts ranging in color from black and tawny-yellow, through yellowish-brown, to black and orange. The head is blackish and the back may be darker along the midline. The chin, throat and occasionally the vent are white. There appears to be a cline in coloration, individuals becoming brighter from east to west and paler from north to south.

Similar species: Green Acouchi (*p. 117*) is considerably smaller and paler, with an obvious tail. **Black Agouti** (*p. 119*) is much larger, with all-blackish upperparts. **Kalinowski's Agouti** (*p. 121*) has black-tipped white rump hairs that are diagnostic.

Habitat: Mature, disturbed and secondary forest, and gardens and plantations, particularly in terra firme forest with many Brazil nut trees or *Attalea* palms.

Distribution: Lowland rainforests of S Peru, W Brazil and N & central Bolivia to NW Argentina (Teta & Reyes-Amaya, 2021).

HB:	44.5–54.0 cm
Tail:	1.1–3.8 cm
Wt:	3.0–5.2 kg

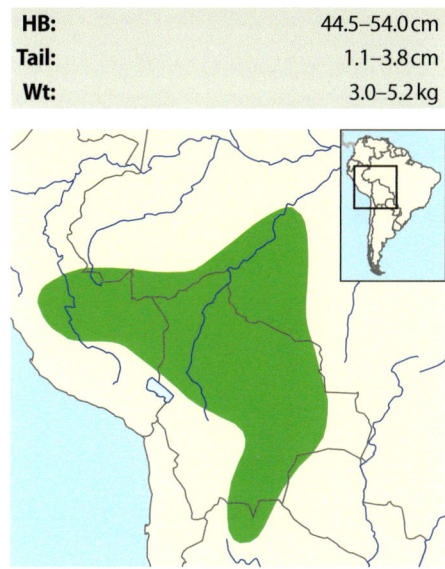

Pacas & Pacaranas

FAMILIES | **Cuniculidae & Dinomyidae**

Pacas and Pacarana are short-legged medium-sized agouti-like rodents with conspicuous white spots. Three species occur in South America. Pacas are short-tailed, while Pacarana has a long conspicuous tail. They are terrestrial and mainly nocturnal, being most active just before dawn and shortly after dusk. Normally solitary, but small groups do occur where there is an abundance of food. They feed on leaves, stems, fruits, rhizomes, seeds, nuts and crops.

LC Lowland Paca

Cuniculus paca

Description: Medium-sized rodent with white spots which merge to form 3–4 broken lines that run along the flanks, from the side of the neck and onto the rump. The upperparts range in color from chestnut-brown to dark brown or grayish; the underparts are white.

HB:	50.0–77.4 cm
Tail:	1.3–3.5 cm
Wt:	5–14 kg

Similar species: Mountain Paca is longer-haired with a less square-looking head. **Pacarana** (*p. 130*) is generally blackish (females are brown) with more extensive white spotting, shorter legs and a conspicuous tail.

Habitat: From sea level to 2,000 m in a wide range of forest types, including both humid and dry forests. Found also in mangroves near rivers, lagoons and tidal creeks. Will venture out into open areas such as cultivated fields.

Distribution: Colombia, Venezuela, Trinidad, the Guianas, Ecuador, Peru, Bolivia, Paraguay, NE Argentina and most of Brazil. Range extends through Central America to E Mexico.

NT Mountain Paca

Cuniculus taczanowskii

Description: Medium-sized rodent. Slightly darker and longer-haired than Lowland Paca and with a shorter, more rounded head. The upperparts are dark brown with whitish tips to the hairs, and with four lines of white spots along the top as well as the sides of the back and flanks. The underparts are white but washed with pale brown towards the rear.

Similar species: Lowland Paca is shorter-haired with a more elongated square-looking head. **Pacarana** (*p. 130*) is blackish (females are brown) with more extensive white spotting, shorter legs and a conspicuous tail.

Habitat: Poorly known, but found in mature, tropical highland forests, including cloud and montane forests, and has been found also in páramo and open fields.

Distribution: From 2,000 m to 4,000 m in mountainous areas in Colombia, NW Venezuela, Peru, Ecuador and Bolivia.

HB:	50–80 cm
Tail:	1.1–3.0 cm
Wt:	3.2–5.2 kg

129

LC Pacarana

Dinomys branickii

Description: A relatively large, heavy rodent with a large head, short rounded ears, short limbs, short stumpy tail, and thick blackish fur with white spots or blotches on the sides of the body, flanks and thighs which can form an almost continuous stripe on some individuals. The underparts are paler than the upperparts. Females are brown rather than blackish. Walks with an ungainly waddling gait.

Similar species: Lowland Paca (*p. 128*) has less extensive white spotting, longer legs and an inconspicuous tail. **Mountain Paca** (*p. 129*) is superficially similar but less stocky, longer-legged, paler and less extensively spotted, and does not have a conspicuous tail.

Habitat: From 250 m to 3,400 m, but at lower altitudes in Bolivia. Wet Andean and tropical montane forests, both primary and secondary, but will also enter crops and coffee plantations. Favors steep slopes with large rocks and tree trunks and can climb trees with considerable agility.

HB:	73–79 cm
Tail:	14–23 cm
Wt:	7.3–15.0 kg

Distribution: NW Venezuela, Colombia, both Andean slopes in Ecuador, the eastern Andes and the western Amazon Basin in Peru and Bolivia, and W Brazil.

Maras

Long-legged herbivores which feed largely on grasses, but Chacoan Mara will also eat other green plants. Maras are diurnal, with activity peaks early and late in the day. Usually encountered in pairs or in small groups, although a number of pairs may use the same communal warren during the breeding season.

NT Patagonian Mara   *Dolichotis patagonum*

Description: Long-legged, rather long-eared and short-tailed with a large, broad head, grayish upperparts, and a distinctive white patch on the rump separated from the grayish upperparts by a broad blackish line. The face is grayish, while the long ears, cheeks, chin and the rest of the underparts are orangey-brown, apart from the whitish vent.

HB:	60–80 cm
Tail:	2.5–4.0 cm
Wt:	7–9 kg

Similar species: Superficially resembles the introduced **European Hare** (*p. 452*), but that species has much longer ears, a more angular head, and is uniformly colored.

Habitat: Open lowland habitats, including grassland, scrublands and other semi-arid open areas, such as creosote bush flats in NW Argentina. Prefers habitats with some shrub cover, but will also use heavily grazed areas in the Monte Desert biome.

Distribution: Central Argentina from Catamarca Province in the north to northern Santa Cruz Province in the south.

LC Chacoan Mara

Pediolagus salinicola

Description: Much smaller than Patagonian Mara, and with long thin legs, a large head and ears, brownish-gray to dark gray upperparts with diffuse speckling, and whitish underparts. The head is grayish with orangey-brown cheeks and a white chin. Each eye has a white eye-ring which is broader in front of and behind the eye and narrower above and below. Digs large burrows, each made conspicuous by the extensive piles of dirt outside the entrance, or utilizes abandoned burrows of Common Plains Viscacha (*p. 139*).

Similar species: None in range.

Habitat: Found in low, flat arid Chaco thorn scrub. Often encountered along trails or under tall thorn trees in the early morning and early afternoon.

Distribution: Chaco ecoregions of extreme S Bolivia, Paraguay, and NW Argentina.

HB:	42.0–48.5 cm
Tail:	2 cm
Wt:	1.8–2.3 kg

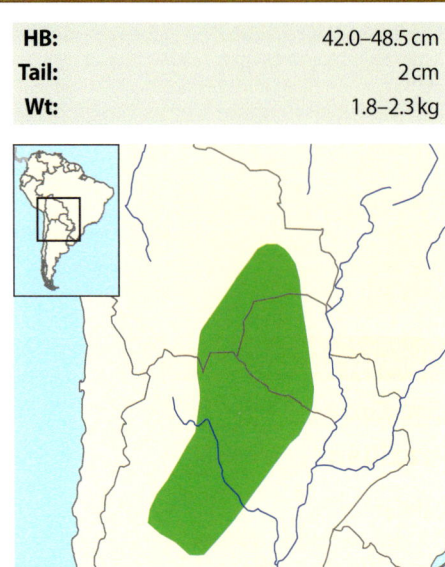

Guinea Pigs & Cavies
FAMILY | **Caviidae**
(SUBFAMILIES | Caviinae & Hydrochoerinae)

Guinea pigs and cavies are small to medium-sized rodents, with 16 species in four genera within two subfamilies recognized by IUCN. Three additional species have been described but have yet to be recognized by IUCN and are therefore not included here:

Jayat's Mountain Cavy *Microcavia jayat* and **Thomas's Mountain Cavy** *M. maenas* (Teta *et al.*, 2017).

Sorojchi's Mountain Cavy *M. sorojchi* (Teta *et al.*, 2022)

Although most guinea pigs and cavies are diurnal, many, particularly the mountain cavies in the genus *Microcavia*, are extremely difficult or impossible to identify in the field, hence the reason for simply providing a summary list of the species and photos of a few examples here.

SUBFAMILY | Caviinae

GENUS *Cavia*

LC **Brazilian Guinea Pig** — *Cavia aperea* — HB: 21.5–39.5 cm / Wt: 500–700 g

Habitat: Grasslands, savannas, Cerrado woodlands, gallery forests and agricultural areas at 400–3,000 m. **Distribution:** Colombia, Venezuela, Guyana, Suriname, Brazil, Bolivia, N Argentina, Uruguay and Paraguay.

LC **Montane Guinea Pig** — *Cavia tschudii* — HB: 22–27 cm / Wt: 290–390 g

Habitat: High-elevation grasslands, bushy habitats, deserts, riparian habitats, humid Pampas and cultivated areas to 4,500 m. **Distribution:** Peru, Bolivia, NW Argentina and NE Chile.

LC **Shiny Guinea Pig** — *Cavia fulgida* — HB: 22–27 cm / Wt: Not known

Habitat: Flooded marshland and semi-mountainous coastal habitats. **Distribution:** Coastal areas from Minas Gerais to Santa Catarina, in SE Brazil.

DD **Sacha Guinea Pig** — *Cavia patzelti* — HB: 28–29 cm / Wt: 700–725 g

Habitat: High level páramo including marshy areas and *Polylepis* forest at 3,000–3,800 m. **Distribution:** Known only from the type locality in the highlands of Chimborazo, Ecuador.

Continued on next page...

GENUS *Cavia*: **Brazilian Guinea Pig**   GENUS *Cavia*: **Montane Guinea Pig**

GENUS *Cavia* (continued)

LC **Greater Guinea Pig**   *Cavia magna*   HB: 22.0–34.5 cm
Wt: 440–840 g

Habitat: Wetlands and coastal marshes. **Distribution:** Extreme SE coastal Brazil and
E Uruguay.

CR **Santa Catarina's Guinea Pig**   *Cavia intermedia*   HB: 27.5–31.0 cm
Wt: 495–680 g

Habitat: Rocky island coastlines with grasses, shrubs and trees.
Distribution: Serra do Tabuleiro State Park, on Moleques Island do Sul, in the state of Santa
Catarina, Brazil.

GENUS *Microcavia*

LC **Southern Mountain Cavy**   *Microcavia australis*   HB: 17.0–24.5 cm
Wt: 240–400 g

Habitat: Dry grasslands, thorn scrub, riparian woodlands in dry gullies, rock walls and piles
and cultivated areas below 2,000 m. **Distribution:** Found in Argentina between highland areas
of Mendoza Province in the west and southern Buenos Aires Province in the east. The range
extends south to Santa Cruz Province and adjoining parts of S Chile. May occur also in high
mountain areas of San Juan Province.

LC **Northern Mountain Cavy**   *Microcavia niata*   HB: 19–20 cm
Wt: 380 g

Habitat: At 3,500–4,000 m in salt flats in Bolivia and boggy areas in Chile.
Distribution: Found in the altiplano of SW Bolivia and NE Chile.

LC **Shipton's Mountain Cavy**   *Microcavia shiptoni*   HB: 18.6–22.0 cm
Wt: 150–220 g

Habitat: Pre-Andean and Andean shrublands at 3,000–4,000 m.
Distribution: NW Argentina in Tucumán, Catamarca and Salta Provinces.

GENUS *Galea*

LC **Spix's Yellow-toothed Cavy**   *Galea spixii*   HB: 22.5–23.4 cm
Wt: 400–520 g

Habitat: Semi-arid Caatinga thorn scrub and cultivated areas.
Distribution: Widespread in E Brazil.

LC **Eastern Yellow-toothed Cavy**   *Galea flavidens*   HB: 20.5–23.1 cm
Wt: 150–330 g

Habitat: Cerrado savanna woodland at up to 1,700 m. **Distribution:** Known only from a small
region in the Cerrado savanna woodlands of Brazil.

DD **Common** (Highland) **Yellow-toothed Cavy**   *Galea musteloides*   HB: Not known
Wt: Not known

Habitat: High-elevation grasslands in the central Andes.
Distribution: S Peru, throughout much of Bolivia, and extreme NE Chile.

DD **Southern Highland Yellow-toothed Cavy**   *Galea comes*   HB: 24.3 cm
Wt: Not known

Habitat: High-elevation prickly scrub in the Andes.
Distribution: Andes of S Bolivia and N Argentina.

LC **Lowland Yellow-toothed Cavy**   *Galea leucoblephara*   HB: 19.8–23.5 cm
Wt: 180–280 g

Habitat: Arid to mesic grasslands. **Distribution:** Lowlands from central Bolivia through Paraguay south to Chubut Province in S Argentina.

SUBFAMILY | **Hydrochoerinae**

GENUS *Kerodon*

LC **Rock Cavy**   *Kerodon rupestris*   HB: 29.7 cm
Wt: 612–950 g

Habitat: Rocky outcrops, with green vegetation for foraging, in semi-arid Caatinga.
Distribution: NE Brazil from the state of Minas Gerais to Ceará.

DD **Acrobatic Cavy**   *Kerodon acrobata*   HB: (mean) 38.4 cm
Wt: approx. 1 kg

Habitat: Rocky limestone outcrops in seasonally dry tropical forests. **Distribution:** Cerrado regions of Brazil in a small area near the type locality in Goiás State, extending into Tocantins State, west of the Espigão Mestre, Serra Geral de Goiás.

GENUS *Cavia*: **Greater Guinea Pig**

GENUS *Microcavia*: **Southern Mountain Cavy**

GENUS *Galea*: **Common Yellow-toothed Cavy**

GENUS *Kerodon*: **Rock Cavy**

Capybaras

The world's largest rodents, some authorities disputing whether there are one or two species. They can be active at any time of day or night, but are most active after dawn and in the early evening. Largely nocturnal where heavily hunted. Lesser Capybara appears to be shy and mostly nocturnal in South America. Greater Capybara normally occurs in groups of 2–30 individuals, but groups of up to 37 animals have been recorded and males can be solitary at times, while Lesser Capybara normally occurs singly or in small groups. Capybaras feed mainly on grass but will also feed on other aquatic vegetation, such as water hyacinths, along with other plants, grain and fruits.

DD **Lesser Capybara**   *Hydrochoerus isthmius*

Description: A smaller version of Greater Capybara; some individuals show black on the rump and hind legs.

Similar species: None in range.

Habitat: Reported to be abundant along the edges of streams and in swamps and permanent lagoons in Colombia.

Distribution: W Colombia and NW Venezuela, where scarce. Range extends into E Panama.

HB:	approx. 103 cm
Tail:	1–2 cm
Wt:	No info. available

LC **Greater Capybara** *Hydrochoerus hydrochaeris*

Description: The world's largest rodent, with short legs and a large blunt head with small ears set well back; unlikely to be confused with any other species (it does not overlap in range with Lesser Capybara). It is brown to reddish-brown, although it may appear paler on the flanks. The fur is long. The partially webbed feet and the outside of the hind legs are dark brown to black and largely unfurred. A good swimmer, normally swimming with the head and back showing above water, and can stay underwater for up to five minutes at a time.

Similar species: None in range.

Habitat: Tropical and subtropical habitats, including grasslands and forested areas close to water. Found both in flowing water along rivers and streams and in seasonally flooded grasslands, ponds, lagoons, marshes and estuaries.

HB:	107–134 cm
Tail:	1–2 cm
Wt:	35–65 kg

Distribution: East of the Andes from Colombia and Venezuela south to NW and E Argentina. Introduced to Trinidad.

Coypus

FAMILY | **Myocastoridae**

A large aquatic rodent. Normally nocturnal but can be active by day, particularly at dawn and dusk, occurring singly and in small groups.

LC Coypu (Nutria)

Myocastor coypus

Description: Large and robust, with long, coarse, dark brown fur, a distinct white patch on the tip of the muzzle and chin, long whiskers, and prominent bright orange-yellow incisors. Short-legged, with webbed hind feet. The tail is long, scaly and rounded (rat-like). Usually swims with the full length of the body visible above the water's surface.

HB:	47.2–57.5 cm
Tail:	34.0–40.5 cm
Wt:	up to 6.7 kg

Similar species: Common Muskrat (*p. 454*) is smaller, has a laterally flattened tail, and normally swims with the top of the head, the back and sometimes the tail exposed, creating a double- or even triple-humped appearance. **North American Beaver** (*p. 453*) is larger, has a paddle-shaped tail, and normally swims with just the forehead, eyes and nose, or occasionally the full head and neck, exposed. **Otters** (*pp. 182–185*) can also be confused when swimming, but have a more curved form in the water and dive more frequently.

Habitat: Found at up to 1,200 m in the Andes, but mainly within 100 m of water in lowland wetlands, including lakes, swamps, freshwater marshes, irrigation channels and slow-moving streams and rivers with aquatic and semi-aquatic plants. In S Argentina and S Chile, found most frequently in the bays of coastal islands and estuaries of glacier-fed streams.

Distribution: Patchily distributed from Bolivia and Paraguay through SE Brazil into Uruguay, Argentina and coastal central Chile. Occurs throughout the Patagonian steppe (where possibly introduced), although absent from the seemingly suitable habitat of the Pantanal.

Viscachas & Chinchillas

Herbivorous rodents feeding on a wide variety of plants, including grasses, herbaceous shrubs, seeds, fruit and bark. Six species occur in South America. Mountain viscachas are diurnal, while Common Plains Viscacha is largely nocturnal. The smaller chinchillas are crepuscular and nocturnal. Viscachas and chinchillas are highly gregarious and normally live in colonies.

Mountain viscachas are extremely agile, leaping between boulders, and they use communal latrines, which makes it relatively easy to locate them. They look superficially like rabbits with a longer, bushy tail.

LC Common Plains Viscacha
Lagostomus maximus

Description: Large, short-eared and large-headed, and unlikely to be confused with any other species. Males are considerably larger than females. They range in color from grayish to grayish-brown, the flanks generally being paler than the back. The chin, throat, chest and belly are white. The head is broad, with a thick black band across the nose and a less distinct band across the eyes, with a white chin, and a whitish band below and above each eye. The lower whitish band extends on to the cheek. Males have thick, dark whiskers which are much more conspicuous than those of females and create a mustached appearance. The tail is short and furred.

Similar species: None in range.

Habitat: A variety of habitats in the lowlands, including subtropical, humid grasslands in north-east Argentina, dry thorn scrub in Bolivia, Paraguay and parts of Argentina, and desert scrub in the south-west of its range.

HB:	39.5–61.5 cm
Tail:	13.5–20.5 cm
Wt:	3.5–8.8 kg

Distribution: N, central & E Argentina, S & W Paraguay, and SE Bolivia.

DD Ecuadorian Mountain Viscacha · *Lagidium ahuacaense*

Description: A medium-sized viscacha with brown-gray upperparts and yellowish-gray underparts. It has a pale yellow stripe on the throat and a white stripe along the breast. The ears are blackish with cream-colored fringes. The tail is much longer than that of other viscachas. The upper tail has long, coarse maroon and cream hairs and the under-tail has short, blackish-brown hairs. The tip of the tail is covered with long hairs. The whiskers are largely black.

Similar species: None in range.

Habitat: The type locality is an isolated granite inselberg with dry montane secondary scrub and forest and extensive rocky areas with numerous large boulders. Field observations suggest that this species is timid and remains close to its den in deep rock crevices.

Distribution: Currently known only from the type locality in the Cerro el Ahuaca, Loja Province, Ecuador, at 1,950–2,480 m.

HB:	approx. 40 cm
Tail:	approx. 40 cm
Wt:	approx. 2 kg

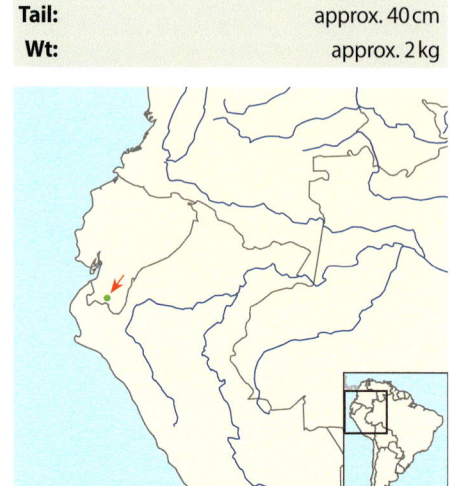

Common Mountain Viscacha *Lagidium viscacia*

HB:	29.5–46.4 cm
Tail:	21.5–37.6 cm
Wt:	750 g–2.10 kg

Description: Superficially resembles a rabbit, but has a long, bushy tail. The forelimbs are relatively short, while the hind limbs are long and muscular. Soft dense fur covers its whole body. Highly individually variable in coloration. The upperparts are generally buff, brown or dark gray, with a dark dorsal stripe, the rump and upper thighs often buff or orange. The underparts are pale yellow or tan. The tail has a crest of coarse pale hairs and the tip of the tail is black or orange.

Similar species: None in range.

Habitat: Rocky habitats in mountain areas and on isolated rocky outcrops in open steppe habitat, including high Andean plateau, high-elevation cold deserts and Patagonian steppes.

Distribution: The most widespread of the viscachas, occurring patchily from 700 m in W Argentina up to 4,800 m in NW Argentina, N Chile, S & W Bolivia and S Peru.

DD Wolffsohn's Mountain Viscacha

Lagidium wolffsohni

Description: A large viscacha, generally appearing more orange than Common Mountain Viscacha (*p. 141*). The overall coloration is brownish-gray with a slightly darker dorsal stripe, and the long fur being heavily suffused with orange. The underparts from the lower cheek to the belly are tawny to reddish in color. The ears are short, with short, thick hairs, and are black on the outer surfaces and yellowish to whitish on the inner surfaces. The forelegs are yellowish, with the hind legs slightly darker. The tail is bushy, with long mixed black and buff hairs creating a crested appearance. The underside of the tail is black with fine ochre-buff grizzling.

Similar species: None in range.

Habitat: Restricted to rocky outcrops in mountain areas up to 4,000 m.

Distribution: The Sierra de los Baguales in Argentina (Santa Cruz Province) and Chile (Magallanes Region). One locality in the Aysén Region in Chile, and from Los Glaciares and Perito Moreno National Parks in Argentina.

HB:	approx. 48 cm
Tail:	approx. 31 cm
Wt:	approx. 2 kg

EN Short-tailed Chinchilla

Chinchilla chinchilla

Description: Larger and stockier than Chilean Chinchilla (*p. 144*), with a thicker neck and shoulders, shorter tail, smaller ears, and longer silvery-gray dorsal hair suffused with black. The underparts and inner sides of the legs and feet are white. The bushy tail has two dark bands on the upper surface.

Similar species: None in range.

Habitat: In the Atacama found in areas with streams, boulders and caves, with sparse, scrubby vegetation.

Distribution: Formerly found throughout the Andes of Bolivia, Peru, NW Argentina and Chile, but now restricted to relict populations in the Atacama Region and the highlands of Antofagasta, in Chile. May still exist in border areas of Bolivia and NW Argentina.

HB:	30–38 cm
Tail:	up to 10 cm
Wt:	500–850 g (but rarely more than 600 g)

143

EN Chilean Chinchilla

Chinchilla lanigera

Description: A relatively small chinchilla, with dense, silky, bluish-gray or silvery-gray fur and yellowish-white underparts. The relatively long tail is covered with long, coarse, gray and black hairs. The face is pointed, with large eyes, large rounded ears and very long whiskers.

Similar species: None in range.

Habitat: Rock crevices and boulder piles in barren, arid and rugged montane areas connecting the coastal mountain ranges and the Andes. Typically found in rocky or sandy areas with a sparse cover of thorn shrubs, herbaceous plants, scattered cacti, and patches of succulent bromeliads.

Distribution: A declining species in the foothills of the Andes and coastal mountains from 400 m to 1,650 m in N Chile north of Talca. Half of the population is in Reserva Nacional Las Chinchillas.

HB:	22–24 cm
Tail:	14–17 cm
Wt:	369–493 g

Cats

The latest taxonomic review by the IUCN Cat Specialist Group (Kitchener *et al.,* 2017) recognized 11 species of cat as occurring within South America. Subsequently, however, do Nascimento & Feijó (2017) published a taxonomic review of the Oncilla (*Tigrina*) complex and concluded that it is in fact comprised of three rather than two species, Northern Oncilla now being split into Northern Oncilla and Eastern Oncilla. This treatment is followed here.

Garcia-Perea (1994) proposed that Pampas Cat be treated as three separate species based on morphological differences, but this was not accepted by the IUCN Cat Specialist Group. Subsequently, however, do Nascimento *et al.* (2020) have proposed that Pampas Cat should be split into five separate species, and this treatment, which has been widely accepted, is followed here, the descriptions of these species drawing heavily on this paper.

Ruiz-Garcia (2023) described the discovery of a possible new species of cat, **Nariño Cat** *Leopardus narinensis*, from the southern Colombian Andes. This treatment has not been followed here pending wider acceptance of the proposal.

Most of the species found in South America have dark markings on a paler background, although Puma and Jaguarundi are generally uniform in color. Many of the spotted forms have melanistic variants in which the spots or blotches can be seen only at close range and/ or in bright light.

Cats are carnivorous, Jaguars and Pumas taking prey as large as tapirs and Guanacos respectively. The smaller cats feed on a wide range of smaller mammals, birds and in some cases reptiles.

Most cats are solitary, except for females with young. They are generally crepuscular or nocturnal, although Jaguarundis are predominantly diurnal; larger species such as Jaguar and Puma can often be encountered by day, particularly in areas where they are not persecuted, and pampas cats have been found to be more diurnal in areas where larger cats are particularly active by night.

Most cats are primarily terrestrial, although many also climb well and several are good swimmers. Margays are largely arboreal.

Northern Oncilla

 **Puma** *Puma concolor*

HB:	♂ 107–168 cm; ♀ 95–141 cm
Tail:	♂ 63–96; ♀ 57–92 cm
Wt:	♂ 39–80 kg; ♀ 22.7–57.0 kg

Description: A large heavily built, long-tailed, light to dark tawny-brown cat with creamy-white underparts. Individuals living in colder temperate regions tend to be larger and heavier and can appear light grayish, particularly in winter, while those in tropical lowlands are often more richly colored, with some reddish tones. The tip of the tail is dark brown to black.

Similar species: Adults are unlikely to be confused with any other species, but young animals are spotted and blotched and could, when on their own, be mistaken for one of the smaller cats or even, given a poor or brief view, a **fox** (*pp. 163–171*).

Habitat: Found in a wide range of temperate, subtropical and tropical habitats, including rainforests, seasonally flooded savannas, montane areas and semi-arid scrub, although tending to avoid areas without rocks or vegetation for cover. From sea level up to 5,800 m in Peru.

Distribution: Widespread throughout South America, although now absent from large areas of NE and central-eastern Argentina, and from central and N Chile. Range extends through Central America and W North America, and also Florida, USA. Now re-establishing itself south of Buenos Aires in Argentina where no longer hunted; since the precise details of this part of the range are uncertain, it has been excluded from the map.

Additional photo *p. 4*

LC Jaguarundi

Herpailurus yaguarondi

Description: Short-legged and long-bodied, with a long narrow tail. The head is small, with a blunt face and small eyes, and the ears are small, rounded and widely spaced. Two color forms: a dark form which ranges in color from pale slate-gray to dark blackish-gray, and a reddish form which ranges from tawny-yellow to chestnut; the latter is more common in drier, more open habitats. Appears flat-backed when walking or running.

Similar species: Unlikely to be confused with any other cat. May be mistaken for an **otter** (*pp. 182–185*) or, more likely, **Tayra** (*p. 190*), but the latter is generally darker with a pale head or obvious white throat patch, has a shorter, bushy tail and moves with a bouncy gait.

Habitat: A predominantly lowland species occurring below 2,000 m, although it has been reported at up to 3,200 m in Colombia. Occurs in a wide range of both open and closed habitats, from Monte Desert, semi-arid thorn

HB:	52.5–94.0 cm
Tail:	27.5–59.0 cm
Wt:	3.0–7.6 kg

scrub, restinga, swamp and savanna woodland to primary rainforest.

Distribution: Throughout much of South America as far south as SE Brazil and central Argentina to approximately 39°S. Range extends through Central America and as far as N Mexico.

Reddish form (INSET); **dark form** (BOTTOM)

LC Geoffroy's Cat

Leopardus geoffroyi

HB:	♂ 44–88 cm; ♀ 43–74 cm
Tail:	23–40 cm
Wt:	♂ 3.2–7.8 kg; ♀ 2.6–4.9 kg

Description: The largest of the small spotted cats in South America, those from the south of its range being larger than those farther north. Appears longer legged in the field than the other small cats. Ranges in color from pale buff and silvery-gray to rich yellow-brown, northern animals tending to have richer tawny or reddish tones while southern ones tend to be paler. The belly is normally whitish or creamy. The body is covered in small dark brown or black spots which can merge to form bands on the nape, chest, sides of the body and the limbs. The long, black-tipped tail has up to 12 dark rings, with small spots on the pale rings. Melanism is common in Uruguay and parts of S Brazil and Argentina.

Similar species: Other spotted cats within its range tend to have rosettes rather than spots, but Geoffroy's Cat is known to hybridize with **Southern Oncilla** (*p. 157*) in S Brazil, creating potential identification difficulties. **Kodkod** is much smaller, more compact and shorter-legged and has a bushier tail.

Habitat: A wide range of habitats, including woodland, and dry forests, Pampas grasslands, marshes, mesquite brush, semi-arid xeric scrub, and high-altitude saline deserts.

Distribution: Up to 3,300 m in the Andes of S Bolivia and NW Argentina, in the Chaco in Paraguay, in S Brazil and south to the Strait of Magellan in S Argentina. In S Chile found only in the eastern parts.

VU Kodkod

Leopardus guigna

Description: Tiny and compact, with relatively short limbs and a thick bushy tail. The head is small and rounded, and the face has a distinct dark stripe across each cheek, a dark marking above each eye, and prominent dark stripes under the eyes and bordering the muzzle. Generally grayish-brown to russet-brown, with small dark spots forming broken lines on the nape and back. Melanism is common and may account for as much as two-thirds of some populations.

Similar species: Geoffroy's Cat is larger and obviously longer-legged, with a long cylindrical rather than bushy tail, and has smaller and more clearly defined dark brown or black spots.

Habitat: Up to 2,500 m in montane and coniferous forests on both slopes of the southern Andes. In the southern part of its range, the species is strongly associated with moist temperate mixed forests of the southern Andean and coastal ranges, particularly the Valdivian and monkey-puzzle (*Araucaria*) forests of Chile, which are characterized by the presence of southern beech (*Nothofagus*), and

HB:	37.4–51.0 cm
Tail:	19.5–25.0 cm
Wt:	♂ 1.7–3.0 kg; ♀ 1.3–2.1 kg

dense bamboo in the understory. In Argentina recorded in moist montane forest with bamboo, numerous lianas and epiphytes.

Distribution: Restricted to a relatively small area of central and S Chile from Santiago Province south to the islands of Chiloé and the Guaitecas (30°–48°S), and a small area of the eastern slopes of the Andes in SW Argentina in the provinces of Neuquén, Río Negro, Chubut and Santa Cruz (39°–46°S, west of 70°W).

Melanistic form (ABOVE LEFT); **spotted form** (BOTTOM)

NE **Central Chilean Colocolo** *Leopardus colocola*

Description: Ash-gray with rusty-cinnamon hairs on forehead, crown, nape, sides of body and tail. It has cinnamon cheek stripes and rusty-cinnamon gular stripes, and the ears are cinnamon with a blackish tip and margins. It has a dark gray dorsal line with cinnamon hairs. The sides of the body are rusty-cinnamon with dark grayish lines. The chest and abdomen have irregular rusty-cinnamon stripes, the legs have dark rusty rings, and the tail has dark gray rings.

Similar species: None in range.

Habitat: Mediterranean forests, woodlands and shrublands of the Chilean matorral.

Distribution: Restricted to the western slope of the Andes in central Chile, from sea level to 1,800 m. Occurs between the Atacama Desert in the north, the Valdivian temperate forests in the south, and the Andes in the east.

HB:	55.9–67.0 cm
Tail:	28.0–32.5 cm
Wt:	No info. available

Southern Colocolo   *Leopardus pajeros*

Description: The forehead, crown, nape and dorsal line are brownish-gray or dark brownish-gray. The sides of the body are brownish-gray, often unmarked but can have indistinct dark brown or dark yellowish-brown lines. The tail is unringed and brownish-gray. The gular stripes can be black, dark brown, yellowish-brown or dark yellowish-brown. Younger animals may show conspicuous dark markings running from the shoulders and along the flanks.

Similar species: None in range.

Habitat: Widely found in open habitats, normally in dry scrub and Pampas grassland, but also in dry woodland, Yungas montane forest, wetlands and rocky areas, including Andean steppes.

HB:	46.4–72.0 cm
Tail:	24.0–28.5 cm
Wt:	No info. available

Distribution: Occurs from NW Argentina (Catamarca Province) to the Strait of Magellan, Chile.

NE **Pantanal Cat** (Brazilian Pampas Cat) *Leopardus braccatus*

Description: The forehead, crown and nape are brown, the chin white or pale yellowish-brown, and the dorsal stripe is dark brown to black. The sides of the body are brown with slightly darker, often indistinct, brown lines. The feet are blackish, and the unringed tail is uniform brownish except for an obvious black tip. Young animals are brown and show conspicuous dark markings running from the shoulders along the flanks, these fading in adults.

HB:	54–65 cm
Tail:	23.0–27.9 cm
Wt:	No info. available

Similar species: None in range.

Habitat: Occurs in open habitats, including the Cerrado, savanna grasslands, scrublands, wetlands and even cultivated areas.

Distribution: Occurs in Brazil from the Pantanal to south-west Piauí in NE Brazil, Paraguay, and the department of Beni in the Bolivian lowlands, with a single record from Formosa in N Argentina.

Muñoa's (Uruguayan) **Pampas Cat**   *Leopardus fasciatus*

Description: Has a yellowish-gray forehead, crown and nape, two dark yellowish-brown cheek lines, narrow dark yellowish-brown gular stripes, and a dark yellowish-gray dorsal stripe. The chest and abdomen have dark brown to black longitudinal stripes, while the sides of the body are yellowish-gray with dark yellowish-gray lines. The feet are pale above and blackish below, and the tail has a few discontinuous rings towards the tip and a small black tip.

Similar species: None in range.

Habitat: Occurs in open habitats, including savanna grasslands, scrublands and wetlands.

HB:	38–62 cm
Tail:	24–34 cm
Wt:	3.7–4.0 kg

Distribution: S Brazil (southern Rio Grande do Sul State), Uruguay, and Corrientes in NE Argentina. Possibly separated geographically from Pantanal Cat and Southern Colocolo (*p. 151*) by the Paraná River in the west. Also separated from Pantanal Cat to the north by the forested areas of S & SW Brazil.

NE **Northern Colocolo** *Leopardus garleppi*

Description: The forehead, crown and nape are brownish-gray speckled with orange. It has black, dark brown, yellowish-brown or dark yellowish-brown gular stripes, at least one of which is markedly wider than the others. The dorsal line is dark brownish-gray with a few orange hairs. The sides of the body are normally pale brownish-gray but some individuals are pale yellowish-brown. It has well-marked rosettes with reddish-brown borders and orangey-brown centres. These form small bands on each side of the body. The tail has reddish-brown rings. A melanistic individual has been reported from the south-west of Parque Nacional Yanachaga-Chemillén, Pusapno, in Peru.

Similar species: Andean Mountain Cat has a distinctive long tail, is normally less heavily marked, and has a dark rather than a pink or reddish nose.

Habitat: Montane habitats on both slopes of the Andes, including Pampas grassland, dry woodland, Yungas montane forest, and rocky areas including Andean steppes at up to 5,000 m.

HB:	46.7–56.0 cm
Tail:	23–33 cm
Wt:	2.9 kg (one specimen)

Distribution: Occurs from Ecuador to NW Argentina (Catamarca and Córdoba) and N Chile (Tarapacá Region). May occur also in Colombia, but there are no proven records to date.

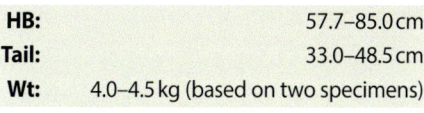 Andean Mountain Cat

Leopardus jacobita

Description: A small, stocky cat with thickset legs and an extremely long, thick and fluffy tail with 6–9 wide dark brown or black rings. The long tail draws comparisons with that of the Asian Snow Leopard *Panthera uncia*, and Andean Cat is often said to resemble a small Snow Leopard in shape. The thick fur is mainly ashy-gray with russet blotches which form vertical lines on both sides of the body, giving the appearance that it has continuous stripes. The legs also have dark and narrower blotches or stripes, but they do not form complete rings. Juveniles appear lighter-colored, with smaller but more extensive blotches. The face has broad dark stripes above the eyes and dark cheek stripes, and the nose is dark.

Similar species: Northern Colocolo lacks the distinctive long tail of Andean Cat, is normally more heavily marked, and has a diagnostic pink or reddish nose.

Habitat: Found mainly in rocky and steep terrain in arid and sparsely vegetated areas of the high Andes above the timberline. In scrub and steppe habitats within the Andean foothills of central Argentina and the Patagonian steppe. Appears to prefer areas with steep rocky slopes in the vicinity of *bofedales* (wet grasslands formed from glacial meltwater), often near viscacha colonies.

HB:	57.7–85.0 cm
Tail:	33.0–48.5 cm
Wt:	4.0–4.5 kg (based on two specimens)

Distribution: Patchily distributed between central Peru (10°13'S) and N and central Argentina (38°23'S) in the south. Mostly above 3,600 m in the central Andes in Argentina, Bolivia, Chile and Peru. As low as 1,800 m in the southern Andes in Argentina, and recent records have extended its range in Argentina into Patagonian steppe and scrub habitats at elevations as low as 650 m. Found at 2,200 m in the Atacama Region in Chile.

VU **Northern Oncilla**

Leopardus tigrinus

Description: A dark brown to light yellowish or orangey-brown oncilla with light buff body sides and white or light gray underparts. Medium-sized rosettes on each side of the body usually merge to form small and/or medium-sized oblique bands. The rosettes and the oblique bands have a black or very dark brown rim, the color inside the rosettes and bands being similar to that of the dorsum. Melanistic individuals do occur.

Similar species: Probably reliably separated from **Southern** and **Eastern** (*p. 158*) **Oncillas** only on geographical range. **Margay** (*p. 159*) closely resembles the oncillas. If seen well, however, that species has a much longer tail exceeding the length of the hind legs, has larger ears, and the hairs on the nape face forwards (backwards on oncillas). **Ocelot** (*p. 160*) is normally noticeably larger.

HB:	45.2–55.6 cm
Tail:	24.0–34.5 cm
Wt:	approx. 2.5 kg

Habitat: Mainly lowland, sub-montane and montane forests from sea level up to 3,000–3,200 m, although there are records to 4,800 m. May occur also in more open areas of the Llanos in Venezuela.

Distribution: Amapá in N Brazil, the Guianas, Venezuela, Colombia, Ecuador, Peru and Bolivia but the southern limits of the range are unclear. Occurs also from W Panama to W Costa Rica.

Melanistic form (TOP); **spotted form** (BOTTOM)

Additional photo *p. 145*

Southern Oncilla

Leopardus guttulus

Description: Small, slender and the size of a house cat, usually with dark yellowish-brown upperparts, slightly lighter sides of the body, and white or with very little gray on underparts. The upperparts range in color from yellowish-brown to ochraceous-buff. The upperparts and sides of the body are covered in small dark rosettes each with a thick and continuous black rim. The rosettes occasionally merge into small oblique bands. Melanistic individuals are common in some parts of its range, especially in humid forests. Hybridizes with Geoffroy's Cat (*p. 148*) in the state of Rio Grande do Sul, in Brazil. Hybrids have characteristics intermediate between these two species.

Similar species: Probably reliably separated from **Northern Oncilla** only on the basis of range. See **Eastern Oncilla** (*p. 158*) for details of the differences from that species. **Margay** (*p. 159*) closely resembles the oncillas; if seen well, however, Margay has a much longer tail exceeding the length of the hind legs, has larger ears, and the hairs on the nape face forwards (backwards on oncillas). **Ocelot** (*p. 160*) is normally noticeably larger.

HB:	36.5–53.9 cm
Tail:	22.8–35.0 cm
Wt:	1.03–2.50 kg

Habitat: Forested habitats, but has been found also in more open and degraded habitats, including plantations, semi-arid thorn scrub and forest fragments in savannas.

Distribution: NE Argentina, Paraguay and SE, S and central-western Brazil.

NE Eastern Oncilla

Leopardus emiliae

Description: A small and slender cat with pale yellow to light yellowish-brown upperparts and white, very light gray or slightly yellowish underparts, although some individuals are uniformly dark yellow. There is no evidence of melanism in Eastern Oncilla. It has both medium-sized and small dark spots. The sides of the body have small dark circular rosettes surrounded by small black spots or thin black lines that encircle the rosettes, but rarely surround them completely (*i.e.* the rosette rims are narrow and usually discontinuous). The rosettes do not merge to form bands. Some individuals show distinctive continuous black lines that extend from the rear half of the dorsum to the base of the tail, while on others these lines are not continuous or are barely perceptible.

Similar species: Southern Oncilla (*p. 157*) has thick and continuous rims to the rosettes and appears to be darker and less yellowish in overall coloration. **Margay** closely resembles the oncillas, but if seen well has a much longer tail that greatly exceeds the length of the hind legs, has larger ears, and the hairs on the nape face forwards (backwards on oncillas). **Ocelot** (*p. 160*) is normally noticeably larger.

HB:	41.5–51.0 cm
Tail:	26–32 cm
Wt:	1.27–3.50 kg

Habitat: Humid Amazonian and Atlantic rainforests, and in savannas and thorny scrub in the Cerrado and Caatinga regions.

Distribution: N, NE and central Brazil, in the states of Pará, Tocantins, Maranhão, Ceará, Rio Grande do Norte, Paraíba, Pernambuco, Alagoas, Bahia and Goiás.

Margay

Leopardus wiedii

Description: Small and slender, with a very long (just over two-thirds of the length of the body), thickly furred tail which extends well beyond the length of the hind legs. The tail has 10–12 dark rings and a black tip. The fur is yellowish-brown to clay-brown on the upperparts and the sides of the body, and white to buff on the underparts. The fur is covered with dark brown or black, pale-centered open spots (rosettes), and streaks forming longitudinal rows. The head is distinctly rounded and the eyes are very large (generating a bright eyeshine at night). Melanistic individuals have been recorded.

Similar species: Ocelot (*p. 160*) closely resembles Margay, but tends to be larger and more heavily built, with a less obviously rounded head and smaller eyes (eyeshine less bright than Margay). It has a shorter tail that does not exceed the length of the hind legs and consequently does not reach the ground. **Oncillas** (*pp. 157–158*) look very similar to Margay but are smaller and generally less richly marked, with a much shorter tail and relatively smaller ears. **Geoffroy's Cat** (*p. 148*) has spots rather than rosettes.

Habitat: Generally found at up to 1,500 m, but to 3,000 m in the Andes. It is more closely associated with primary forest than are other small cats, and found in both evergreen and

HB:	♂ 49.0–79.2 cm; ♀ 47.7–62.0 cm
Tail:	30–52 cm
Wt:	♂ 2.3–4.9 kg; ♀ 2.3–3.5 kg

deciduous forests from lowland rainforest to montane cloud forest. Occurs also in gallery forest and forest fragments in both dry and wet savanna habitats.

Distribution: Colombia and Venezuela through the Amazon Basin to S Brazil, Paraguay, NW Uruguay and NE Argentina, although the distribution within this area is highly fragmented. Range extends through Central America and into Mexico.

LC Ocelot

Leopardus pardalis

Description: A medium-sized, robustly built cat with a relatively short ringed tail that rarely exceeds the length of the hind legs. The fur ranges in color from creamy, through tawny-yellow, cinnamon and reddish-brown to gray, and is boldly marked with solid spots or blotches, or rosettes with russet-brown centers that may form lines across the body. The legs and feet are covered in smaller solid spots, and there are one or two bars on the inside of the legs. The spots on the shoulders and back may merge to form four or five stripes running from the neck to the base of the tail. The undersides of the neck and belly are whitish.

Similar species: **Margay** (*p. 159*) looks very similar, but tends to be much smaller and less heavily built; it has large eyes and a much longer tail that exceeds the length of the hind legs and consequently reaches the ground. (See the Margay account for information on the differences in eyeshine at night.) The three species of **oncilla** (*pp. 157–158*) are smaller still and with a tail length intermediate between the two. **Geoffroy's Cat** (*p. 148*) has spots rather than rosettes. Seen poorly, a small **Jaguar** can resemble a large male Ocelot

Habitat: A wide range of habitats, including mangrove forests, coastal marshes, savanna grasslands, thorn scrub, and most types of tropical and subtropical forests. Usually

HB:	♂ 67.5–101.5 cm; ♀ 69.0–90.9 cm
Tail:	25.5–44.5 cm
Wt:	♂ 7.0–18.6 kg; ♀ 6.6–11.3 kg

below 1,200 m, but has exceptionally been recorded at up to 3,000 m.

Distribution: Widespread across northern South America, including Trinidad, as far south as N Argentina and S Brazil. Absent from Chile, and it is unclear whether the species still occurs in Uruguay. Range extends through Central America and W & E Mexico as far north as extreme southern Arizona and southern Texas, USA.

Jaguar

Panthera onca

Description: South America's largest cat, with a muscular body, although size varies considerably across the range. The upperparts are tawny-orange, yellow, cinnamon or buff-gray, with large black rosettes normally with darker brown interiors and small black spots. The underparts are white or creamy-white. The head, legs, tail and underparts are covered with black spots. Melanistic individuals with rosettes visible in certain lights occur most frequently in lowland rainforest south of the Amazon River.

Similar species: A large male **Ocelot** seen poorly can resemble a small Jaguar.

Habitat: A range of forested and wooded habitats from cloud forest to tropical lowland rainforest, seasonally flooded savanna woodlands and dry Chaco forests. Closely associated with water, where it is an excellent swimmer.

Distribution: Up to 2,500 m across most of northern and central South America as far south as the Brazilian highlands. More patchily distributed farther south in Paraguay,

HB:	♂ 120–170 cm; ♀ 116–147 cm
Tail:	♂ 44–80 cm; ♀ 49–67 cm
Wt:	♂ 68–158 kg; ♀ 51–100 kg

S Bolivia, SE Brazil and N Argentina. Range extends discontinuously through Central America and into Mexico, with occasional records as far north as southern Arizona, USA.

Dogs & Foxes

Dogs and foxes, other than the Bush Dog, are generally medium- to long-legged, with an elongated muzzle, a flat back and a bushy tail. Eleven species are found in South America. Recent studies have suggested that Pampas Fox and South American Gray Fox are the same species, but at present they continue to be recognized as separate species by IUCN.

Most foxes are omnivorous, feeding on small mammals, birds, reptiles, insects, crustaceans, fruits and nuts. They will also take carrion. Culpeo will take lambs and occasionally adult sheep, while Darwin's Fox has been reported as killing Southern Pudu (*p. 442*). Hoary Fox is largely insectivorous.

Foxes are normally solitary, and are diurnal in areas where they are protected but largely nocturnal where they are hunted and/or where their prey is nocturnal. They can also be encountered in pairs or family groups, particularly near rich sources of food. Northern Gray Fox is more arboreal than other foxes and will forage up to 18 m above the ground.

Maned Wolf is normally solitary and feeds on a wide range of fruits, birds, reptiles, and mammals as large as small armadillos, although it has been recorded as occasionally taking prey as large as deer. It will also scavenge on roadkill and human refuse.

Short-eared Dog is mainly solitary and largely diurnal, although nocturnal activity has also been recorded.

Bush Dog hunts in family groups of 2–12 individuals and is mainly carnivorous, hunting large rodents such as pacas and agoutis, other small mammals such as armadillos, rats, rabbits and opossums, and reptiles and ground-nesting birds. It has been recorded as taking much larger prey, including deer and capybara. It readily follows prey into water and is extremely vocal, with a range of whines, screams, barks and growls.

Maned Wolf

LC Northern Gray Fox

Urocyon cinereoargenteus

Description: A medium-sized fox with relatively short legs. Grizzled-gray above, with a dark stripe along the top of the thick, bushy, black-tipped tail. Rusty-colored neck, flanks, legs and underside of tail. Dark gray face with white cheeks. White chin, throat, chest, belly and front of hind legs.

Similar species: Crab-eating Fox (*p. 171*) is larger and largely uniform dark gray, with distinctive black socks.

Habitat: Forested montane habitats within South America, but in a wide range of forested habitats elsewhere in its range.

Distribution: Colombia and N Venezuela. Range extends through Central America, Mexico and much of the USA and into Canada.

HB:	♂ 58–66 cm; ♀ 52.5–58.0 cm
Tail:	♂ 33.3–44.3 cm; ♀ 28.0–40.7 cm
Wt:	♂ 3.4–5.5 kg; ♀ 2.0–3.9 kg

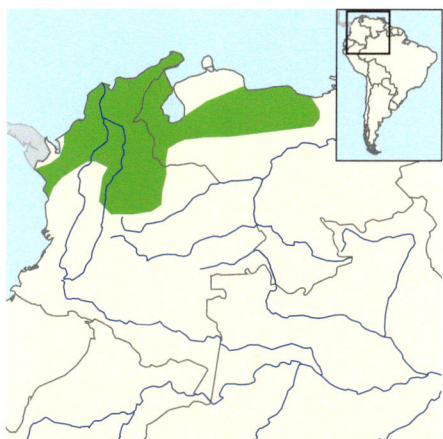

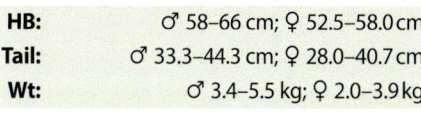

LC Culpeo

Lycalopex culpaeus

HB:	44.5–92.5 cm
Tail:	30.5–49.3 cm
Wt:	3.4–13.8 kg

Description: The largest fox in South America, with six subspecies. Northern subspecies tend to be smaller and lighter-colored. A powerful-looking fox with a broad head and wide muzzle. Males are larger and significantly heavier than females. The upperparts vary with subspecies, but are generally grayish with a darker rump. The tail is long and bushy, with a dark dorsal patch at the base, and a dark tip. The head, neck, flanks and legs are tawny or rufous, with no black on the feet or legs, and the underparts including the chin are whitish.

Similar species: Owing to Culpeo's variation in color, some individuals can be difficult to separate from **South American Gray Fox** (*p. 168*), but that species is generally smaller and has a narrower head and muzzle. Overlaps in range also with **Sechuran Fox** (*p. 167*), **Pampas Fox** (*p. 166*) and **Darwin's Fox** (see those species accounts for differences).

Habitat: Occupies a greater range of habitats than any other South American fox. Found in rugged mountainous terrain at up to 4,800 m, in deep valleys and open deserts, scrubby grasslands, scrubland and broadleaf temperate southern beech (*Nothofagus*) forest in the south.

Distribution: The Andes from S Colombia in the north, through Ecuador, Peru, W Bolivia and Chile to Hoste Island in the south. In Argentina to the east of the Andes from Jujuy Province in the north to the Atlantic coast at Río Negro and south to Tierra del Fuego.

Darwin's Fox

EN

Lycalopex fulvipes

HB:	♂ 48.2–56.1 cm; ♀ 48.0–59.1 cm
Tail:	♂ 19.5–25.5 cm; ♀ 17.5–25.0 cm
Wt:	♂ 1.90–3.95 kg; ♀ 1.80–3.70 kg

Description: A small, stocky fox with relatively short legs, dark gray-brown upperparts, white underparts including the chin, rufous lower legs, and a relatively short, dark gray bushy tail. The muzzle is short and thin, and it has a rounded forehead and rufous markings on the back of each ear.

Similar species: **South American Gray Fox** (*p. 168*) and **Culpeo** are both larger and have a longer, distinctly black-tipped tail.

Habitat: Dense monkey-puzzle (*Araucaria*)–southern beech (*Nothofagus*) forest, open southern beech forest and occasionally in open pastures. Has been found also in unmanaged eucalyptus plantations and highly fragmented forests. Most common in areas with low densities of humans (and dogs).

Distribution: Chile, where until recently thought to be restricted to Chiloé Island, and Nahuelbuta National Park 600 km to the north on the mainland. New records confirm the presence of the species in the Valdivian coastal range, near Maullín and

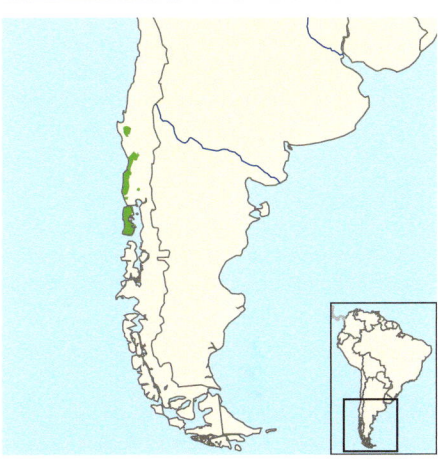

north of Llanquihue Lake in the foothills of the Andes.

 **Pampas** (Azara's) **Fox**

Lycalopex gymnocerca

Description: A medium-sized fox, having gray upperparts with a blackish band along the back and upper tail. The tail is relatively long, bushy and gray, with a black tip. The front of the head is reddish and the back pale gray to whitish. The muzzle is narrow, reddish to black above and largely pale below, with a black chin. The relatively large and broad ears are reddish above and white below. The legs appear reddish.

Similar species: Culpeo (*p. 164*) is significantly larger. **South American Gray Fox** (*p. 168*) is smaller and has a distinctive black patch on the chin and a black patch across the thigh.

Habitat: Found in Pampas grassland of southern South America. Common in tall-grass plains and occurs also on ridges, in dry scrub, coastal sand dunes and open woodlands, as well as in pastures and agricultural areas. In the southerly and easterly parts of its range it is replaced in drier habitats by South American Gray Fox. In areas of overlap with Crab-eating Fox (*p. 171*) it usually occupies more open areas, although the two species occur together in parts of Argentina.

HB:	♂ 59.7–74.0 cm; ♀ 50.5–72.0 cm
Tail:	♂ 28–38 cm; ♀ 25–41 cm
Wt:	♂ 4–8 kg; ♀ 3.0–5.7 kg

Distribution: E Bolivia south through Paraguay and S Brazil into NE Argentina and Uruguay. Chemisquy *et al.* (2019) suggest that the species' range is largely restricted to the east of the Paraná, Paraguay and de la Plata Rivers although a few specimens have been collected in Santa Fe, close to the Paraná River.

NT Sechuran Fox

Lycalopex sechurae

Description: A relatively small fox, with a small head, short muzzle, and rather long ears about two-thirds the length of the head. Gray-faced, with rufous-brown rings around each eye, a dark muzzle often with paler hairs around the lips, and the ears can be reddish on the back. The upperparts are grayish (sometimes with a dark dorsal stripe), with a black-tipped tail. The underparts are tawny or cream-colored. The front legs and the lower part of the hind legs are normally reddish in color.

Similar species: Culpeo (*p. 164*) is generally paler and much larger, with a broad head and muzzle giving it a more powerful-looking appearance.

Habitat: A range of habitats from sandy deserts to agricultural areas and dry forests.

Distribution: Coastal NW Peru and SW Ecuador, including the Andean foothills to 1,000 m.

HB:	50–78 cm
Tail:	27–34 cm
Wt:	2.6–4.2 kg

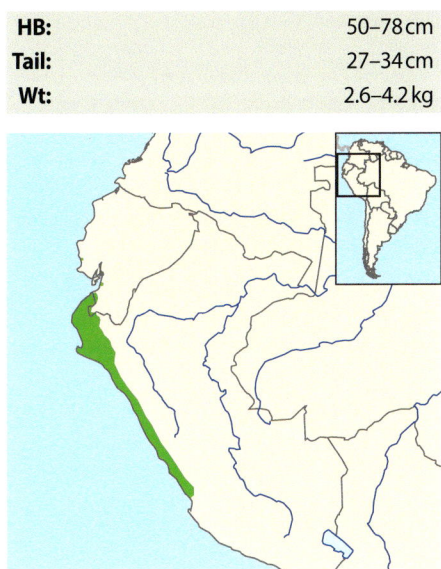

LC South American Gray Fox   *Lycalopex grisea*

Description: A small, large-eared fox, with a rufous-gray head and a conspicuous black chin spot. Largely grizzled gray above, with pale gray underparts. The legs and feet are pale tawny, with a black patch across the thighs. The tail is long and bushy, with a black dorsal stripe and tip, and the underside of the tail is a mixture of pale tawny and black fur contrasting with the gray upper tail.

Similar species: Culpeo (*p. 164*), **Pampas Fox** (*p. 166*) and **Darwin's Fox** (*p. 165*) (see those species accounts for differences).

Habitat: Generally found in steppes, Pampas grassland, and scrublands on plains and low mountains, but has been reported as high as 4,000 m. Appears to prefer shrubby open areas rather than areas with dense vegetation. Where the range overlaps with that of Darwin's Fox, the present species is generally found in more open areas rather than the dense patches preferred by Darwin's Fox. Where it occurs alongside Pampas Fox, it tends to favor drier habitats.

Distribution: Widespread on both sides of the Andes, from N Chile and NW Argentina down to Tierra del Fuego, where it was introduced. A newly discovered population, separated from the Chilean population by the

HB:	50.1–66.0 cm
Tail:	31.7–34.7 cm
Wt:	2.5–5.0 kg

Atacama Desert, has been found along the coast of S Peru from Lima to Tacna. A study by Chemisquy *et al.* (2019) suggests that the previously documented range of this species is incorrect and that it extends east into N and central Argentina towards the west and south of the Paraná, Paraguay and de la Plata Rivers, but they note that further genetic analysis is required to substantiate this.

NT Hoary Fox

Lycalopex vetula

Description: A small and slender fox, with a short, pointed muzzle and large ears. Generally pale grizzled gray above with buff or chestnut underparts, including the neck, chest and legs. The back of the neck is buff-white, giving it an almost collared appearance in some lights. The underside of the lower jaw is black, while on most other foxes it is white. The tail has a dark base and tip and often shows a dark dorsal spot.

Similar species: Crab-eating Fox (*p. 171*) is larger and darker, with dark legs.

Habitat: Open Cerrado habitats, but also in agricultural areas, including insect-rich pastures, crops and plantations. Rarely encountered in densely wooded Cerrado, on floodplains or in dry or gallery forests.

Distribution: Brazil, where widespread and common in Cerrado habitats.

HB:	♂ 49.0–71.5 cm; ♀ 51–66 cm
Tail:	♂ 27–38 cm; ♀ 25–31 cm
Wt:	♂ 2.5–4.0 kg; ♀ 3.0–3.6 kg

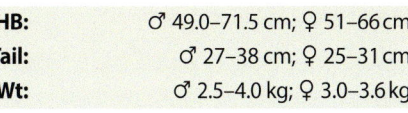

NT Maned Wolf

Chrysocyon brachyurus

Description: An unmistakable tall, long-legged canid with reddish-orange fur, a blackish mane running from the back of the head to the shoulders, distinctive black socks extending halfway up each leg, and a bushy white-tipped rufous tail. It is long-eared, with a blackish muzzle and a white throat patch. Melanism has been recorded in Brazil.

Similar species: Unlikely to be confused with any other canid, but could be taken for **Marsh Deer** (*p. 443*) if seen poorly.

Habitat: Mainly in tall grasslands, the Cerrado and Pampas savannas, scrub forest and seasonally flooded wetlands. Avoids thick forest, and its range in Brazil has expanded as forest has been converted to grasslands, ranch land and agricultural areas.

Distribution: Patchily distributed in Brazil south of the Parnaiba River in NE Brazil, east of the Paraguay River in E Paraguay, extending into northern Rio Grande do Sul State in Brazil, south to N Argentina (Santa Fe and Entre Ríos) and west through N Bolivia to the

HB:	95–115 cm
Tail:	38–50 cm
Sh:	70–74 cm
Wt:	20.5–30.0 kg

Pampas del Heath in Peru. Possibly extinct in Uruguay: last confirmed record being in 1990.

Additional photo *p. 162*

LC Crab-eating Fox

Cerdocyon thous

Description: A medium-sized fox, with coarse, generally dark grizzled gray fur giving it a dark bristly appearance, although coloration can range from almost black to silver-gray or pale yellowish-gray. A black dorsal line runs from the head to the tip of the tail. The tail is moderately bushy, with a dark base and black tip. The throat and belly are cream or whitish. The head and the front of the legs are dark gray-brown or reddish-brown, the back of the legs appearing black; often gives the impression of having black socks.

HB:	57.0–77.5 cm
Tail:	22–41 cm
Wt:	4.5–8.5 kg

Similar species: Pampas Fox (*p. 166*) and **South American Gray Fox** (*p. 168*) are generally paler and less uniform in appearance. **Northern Gray Fox** (*p. 163*) is smaller, has a rusty-colored neck, flanks, legs and underside of tail, and no black 'socks'. **Hoary Fox** (*p. 169*) is smaller and lighter, with paler legs.

Habitat: Up to 3,690 m in a wide range of habitats, including marshes, savanna, Cerrado, Caatinga, Chaco, and dry and semi-deciduous gallery, Atlantic, monkey-puzzle (*Araucaria*), lowland and montane forests.

Distribution: Colombia and N Venezuela, with a few records from Ecuador, Suriname and Guyana. Largely absent from the Amazon Basin. Found also from the eastern Andean foothills of Bolivia and Argentina to the Atlantic Forest of E Brazil and south to Argentina (Entre Ríos and northern Buenos Aires). Range extends into Panama.

NT **Short-eared Dog** *Atelocynus microtis*

Description: A medium-sized fox-like canid with a long muzzle but relatively short rounded ears compared with other foxes. Ranges in color from black or brown to rufous-gray, with a dark dorsal line from the head to the tail. The coat is generally short, contrasting with the bushier tail, which is dark above and slightly lighter below.

Similar species: Crab-eating Fox (*p. 171*), which may occur in the south of its range, is smaller, paler and has a longer coat.

Habitat: Appears to prefer undisturbed Amazonian lowland rainforest away from humans and closely associated with rivers. Reported from terra firme forest, swamp forest, stands of bamboo and primary forest along rivers.

Distribution: The range is fragmented and poorly known: within lowland Amazonia from Colombia, Ecuador, Peru, Bolivia and Brazil.

HB:	72–100 cm
Tail:	25–35 cm
Wt:	9–10 kg

 Bush Dog

Speothos venaticus

HB:	57.5–75.0 cm
Tail:	12.5–15.0 cm
Wt:	5–8 kg

Description: A distinctive small, stocky, short-legged dog with a long body and short bushy tail. Broad-headed with a short-muzzle, small dark eyes and short rounded ears. Coloration varies from tawny to dark brown or even black, but the head and neck are usually reddish-tan or tawny, becoming dark brown or black on the hindquarters and legs. The underparts also are dark, but it may exhibit a pale/white throat or chest patch.

Similar species: More likely to be confused with **agoutis** (*pp. 118–127*) than with any other canid.

Habitat: Often near water and has been recorded in lowland forested habitats, including primary and gallery forest, semi-deciduous forest and seasonally flooded forest, at up to 1,500 m. Occurs also in Cerrado and Pampas areas and ranch land, and there are reports from Caatinga, Chaco and coastal mangroves.

Distribution: Northern South America east of the Andes from Colombia south to Paraguay and NE Argentina, but rare and poorly known and reported to be common only in Peru and Guyana. Isolated populations may occur also west of the Andes in Ecuador and Colombia. Range extends through Panama and just into Costa Rica.

Bears

Only one species of bear, the Andean (or Spectacled) Bear, occurs in South America. It is an omnivore that feeds mainly on fruits, bromeliads and other succulent plants. Known to raid crops, particularly maize and corn in some areas, and will also kill birds and small mammals, and has been recorded as chasing prey as large as a Mountain Tapir. Rests both in trees and in ground nests, and frequently feeds on fruits high in the tree canopy. At some sites in Ecuador its presence is totally dependent on the fruiting of Aguacatillo *Persea caerulea* (a type of wild avocado tree), and bears are absent from areas for months at a time. Generally active by day, but more nocturnal in areas where it is heavily persecuted.

VU **Andean** (Spectacled) **Bear**

Tremarctos ornatus

Description: The only bear in South America. Usually black, but some individuals are reddish-brown, with pale facial spectacles often extending on to the throat and chest. Some have an almost creamy-white face other than for black patches around the eyes.

Similar species: None in range.

HB:	130–190 cm
Tail:	<10 cm
Wt:	♂ 100–200 kg; ♀ 60–80 kg

Habitat: A variety of habitats, including desert-scrub forest in north-west Peru, but most commonly in high-elevation forests and humid grasslands at altitudes from 250 m to 4,750 m. As low as 250 m in Peru but no lower than 1,200 m in Colombia.

Distribution: Andes from Sierra de Perijá, Macizo de El Tamá and Cordillera de Mérida, in Venezuela, south through Colombia, Ecuador and Peru to the eastern slope of the Andes in Bolivia. Found in all three Andean ranges in Colombia and Peru and also in coastal desert-scrub forest in N Peru. Formerly in N Argentina.

Eared Seals (fur seals & sea lions)   FAMILY | **Otariidae**

Three species of eared seal breed or occur regularly around the coasts of South America, and a further three species, all fur seals, occur as vagrants.

Sea lions feed during the day and at night on sardines, lantern-fish, deep-sea smelts and squid, and South American Sea Lions have also been recorded as preying on fur seals and even juvenile elephant seals. Fur seals feed on a wide variety of fish and are active by day and by night. Sea lions and fur seals can both be found rafting in large groups close to the surface, often in association with seabirds and cetaceans.

LC South American Fur Seal   *Arctocephalus australis*

Description: Large and stocky, with long prominent ears, although size varies considerably with age and sex. Males are dark brownish-gray to dark olive-brown (showing silvery frosting as they mature) with a prominent mane of grayish hairs from the crown to the shoulders, including the neck and upper chest. Underparts are slightly paler. It has a pointed, flat-topped muzzle, a short forehead and a rounded crown. Females are uniform gray-brown or dusky above and paler tan or brown on the belly, and the chest and neck are generally paler. Most females have a dusky eye mask and a paler muzzle and ears. When swimming, fur seals will porpoise like sea lions and also leap clear of the surface.

HB:	♂ 190 cm; ♀ 140–150 cm
Wt:	♂ 90–160 (200 kg); ♀ 30–60 kg

Similar species: Most likely to be confused with **sea lions** (*pp. 178–179*), but with good views, particularly on land, identification should be straightforward: fur seals are smaller, with a less rounded head and a more pointed muzzle, and are longer-eared, with a less extensive mane in males and a darker coat in females. Sea lions have very large, broad fore flippers which curve distinctly backwards. Three species of fur seal occur as vagrants along the coasts of South America (see *p. 177*). Identification of females and juveniles is extremely difficult. See Shirihai & Jarrett (2006) for further information.

Continued on next page...

FEMALE

South American Fur Seal (*continued*)

Habitat: Rocky coasts, in boulder-strewn areas and/or on ledges above the shoreline with some source of shade and easy access to the sea. Typically found as far out to sea as the edge of the continental shelf, although there are records at up to 600 km from shore.

Distribution: One population occurs along the Atlantic coast of southern Brazil, Uruguay and Argentina, and the Pacific coast of southern Chile as far north as Isla Guafo. On the Atlantic coast, although haul-outs exist farther north, the northern limit for breeding colonies is at Islas del Castillo, Uruguay. Breeding and non-breeding colonies exist throughout the coast of Argentina. A second population occurs in Peru and northern Chile, with the majority of the breeding population in Peru at 15°–17°S, with an isolated colony as far north as Isla Foca (5°20'S). South American Fur Seal breeds from mid-October to mid-January, the exact timing varying from colony to colony.

MALE (LEFT) AND FEMALE (RIGHT)

Vagrant fur seals

In addition to the three species of eared seal that breed or occur regularly around the coasts of South America, a further three species, all fur seals, occur as vagrants. These are included here for completeness.

Juan Fernández Fur Seal
Arctocephalus philippii

Recorded as a vagrant to the Pacific coast from Peru to Chile. Males up to 200 cm, females 120–140 cm in length. Males are dark brownish-black, the chest and neck being black, with a strongly contrasting buff-gray or golden-yellow crown and nape. Has a long pointed muzzle with a bulbous nose and convex crown.

Juan Fernández Fur Seal (SUB-ADULT MALE)

Antarctic Fur Seal
Arctocephalus gazella

Recorded as a vagrant to Brazil and the extreme south of the continent. Males up to 200 cm, females 120–140 cm in length. Extremely similar to **South American Fur Seal** (*p. 175*), although slightly smaller and paler. Males have a more grizzled coat and longer whitish whiskers, but field identification is difficult.

Antarctic Fur Seal (MALE)

Galápagos Fur Seal
Arctocephalus galapagoensis

Recorded as a vagrant to Ecuador, with unconfirmed reports of pups being born on the mainland coast. Males 150–160 cm, females 110–130 cm in length. Very similar to **South American Fur Seal** (*p. 175*), but not known to overlap in range.

Galápagos Fur Seal (MALE)

EN Galápagos Sea Lion

Zalophus wollebaeki

Description: A relatively small sea lion. Males have a small crest, creating a steep forehead. Females lack the crest and have a flatter head profile, with an indistinct forehead and a narrower neck. The muzzles of both sexes are relatively narrow and tapering. The coloration of males darkens with age and varies from dark brown to grayish or golden-brown, with paler areas around the muzzle and eyes. Females are buff to pale brown, with little contrast between the upperparts and the underparts.

Similar species: May overlap in range with vagrant **South American Sea Lion**, but that species is much larger and thicker-necked, and has a short muzzle, males having a distinct mane. Female South American Sea Lions also show more contrast between the upperparts and underparts.

Habitat: Breeds on sandy or rocky beaches from May to January.

Distribution: A colony was established in 1986 at Isla de la Plata, just offshore from

HB:	♂ up to 150 cm; ♀ up to 120 cm
Wt:	♂ up to 250 kg; ♀ up to 100 kg

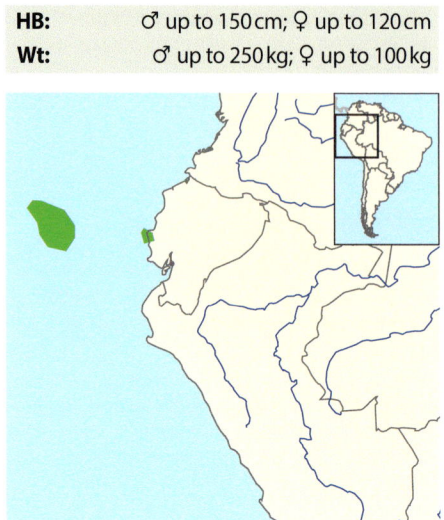

mainland Ecuador, but this site is not regularly used. Vagrants from the species' main range in Galápagos occur along the Ecuadorian and Colombian coasts.

SUB-ADULT MALE

South American Sea Lion *Otaria byronia*

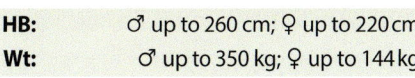

Description: A large sea lion. Males are much larger than females and are dark brown or duller tan-brown, older males having a bull-necked appearance with a rusty-brown mane extending down to the flippers, giving them a large-headed and small-bodied appearance. Females are slimmer, with grayish-brown upperparts and dull yellowish-buff underparts; they can be pale-headed or have patchy coloration. Juvenile males initially resemble females but darken and develop a denser coat and mane. Both sexes have a short, broad and blunt-tipped muzzle and inconspicuous ears.

Similar species: Males are easily separated from the much smaller **South American Fur Seal** (*p. 175*) on land, but identification of females and juveniles is less straightforward, as is identification at sea. Vagrants may overlap in range with **Galápagos Sea Lion** off the coasts of Ecuador and Colombia, but that species is much smaller and has a relatively small and narrow head with a longer muzzle, the males having a crest but lacking a distinct mane. Females and juveniles may be indistinguishable at sea, where their larger size and more contrasting upperparts and underparts are difficult to see. Sea lions should not be confusable with elephant seals (*p. 180*), but their tendency frequently to porpoise can result in their being misidentified as dolphins (*p. 416*) when seen briefly or poorly.

Habitat: Largely coastal, occurring in marine habitats over the continental slope and shelf

HB:	♂ up to 260 cm; ♀ up to 220 cm
Wt:	♂ up to 350 kg; ♀ up to 144 kg

and occasionally farther offshore, with records up to 320 km from the coast. Can be found in freshwater habitats, including rivers and near glacial meltwater. Breeding colonies may be on sandy beaches, stony beaches or rocky islands, and the breeding season is December–March.

Distribution: Widely distributed along the Pacific coast from N Peru south to Cape Horn, vagrants occasionally being seen farther north in Ecuador and Colombia. Found also along the Atlantic coast north to S Brazil, although the northernmost breeding colonies are in Uruguay. There are long stretches of coast without any breeding colonies.

MALE

FEMALE

Earless Seals (elephant seals & seals)   FAMILY | Phocidae

There are two species of elephant seal, only one of which occurs in South America. They submerge tail first, with the head held upwards, and often float with their head and hind flippers clear of the water. When feeding, they normally dive to depths of 300–500 m and exceptionally to 2,000 m. Dives normally last for 20–30 minutes but can last for up to 120 minutes.

LC Southern Elephant Seal   *Mirounga leonina*

Description: Males are huge, with a large square-shaped head and a conspicuous proboscis overhanging the mouth, and are unlikely to be confused with any other species in South America. Coloration varies from gray, through brown to rusty or even tan-colored, often heavily scarred as a result of fights with other bulls. Females are considerably smaller, with a rounder face, and lack the males' proboscis. Immature males have a partially developed proboscis after 5–6 years.

Similar species: Although females and immatures may overlap in length with vagrant Leopard, Weddell and Crabeater Seals, the latter are usually smaller and/or have a more mottled appearance. Fur seals and sea lions are readily identifiable by their distinctive structure and eared appearance.

Habitat: Open sandy or stony beaches. Males arrive on breeding beaches in August, pregnant females arriving in September or October. Young are generally weaned within a month.

Distribution: Although most haul-outs are on Subantarctic and Antarctic islands, individuals can be encountered around the Pacific and

| HB: | ♂ 450–650 cm; ♀ 250–400 cm |
| Wt: | ♂ 1,500–3,700 kg; ♀ 359–800 kg |

Atlantic coasts of the south of the continent, with records from as far north as Ecuador on the Pacific coast. A number of animals appear regularly at sites on the coasts of S Argentina and Chile, and breeding populations exist on Peninsular Valdez in S Argentina and at three sites in S Chile.

MALE (LEFT); FEMALE (RIGHT)

Vagrant seals

Apart from Southern Elephant Seal, three other species of earless seal have been recorded in South America, all of which are vagrants. These are included here for completeness.

LC Leopard Seal
Hydrurga leptonyx

An uncommon visitor to southern coasts. Long, up to 3.8m, sleek and long-necked with a distinctive large, flat, reptilian-like head and small eyes set well apart. The muzzle is long and broad with powerful jaws and a distinctive broad gape. Generally silver to dark blue-gray but can appear blackish-gray. Usually darker on the upperparts and sides, with darker gray and black spots and some paler markings. Paler below. The head is often darker above the eye, with a broad silver-gray band across the lower cheek and gape. The fore-flippers are long and broad, and centrally positioned.

Leopard Seal

C Weddell Seal
Leptonychotes weddellii

Recorded as a vagrant north to Uruguay and Juan Fernández Island. A plump, rather rotund seal up to 3.3 m long, with small flippers and a small head with a blunt, cat-like face with a barely protruding muzzle and large, close-set eyes. Bluish-black to dark silvery-gray short, dense fur, irregularly streaked and blotched gray-white. The patches increase in frequency between the sides and pale abdomen. The fore-flippers are short and angular, and the hind-flippers are proportionately small.

Weddell Seal

C Crabeater Seal
Lobodon carcinophaga

Recorded as a vagrant as far north as SE Brazil. A slim, medium-sized seal up to 2.6 m long, with a long neck, square head and a dog-like protruding muzzle with a long mouth. It has a slight forehead and slightly upturned snout and small dark eyes set well apart. Adults are grayish-brown above, paler below, with chocolate-brown patches on the silvery-gray sides to the body. The patches are largest near the head and sides of the neck. The coat becomes uniformly paler prior to molting.

Crabeater Seal

Otters

Otters are medium-sized to large mustelids with long bodies, short legs and normally long tails. Four species are found in South America and they are primarily aquatic.

Most otters are largely solitary other than females with young, but Giant Otters are highly gregarious and can be encountered in family groups of up to 15 individuals. Marine, Neotropical and Giant Otters are primarily diurnal, being most active in the early morning and from middle to late afternoon. Southern River Otters are largely nocturnal and solitary, but are active during the day in undisturbed areas. Otters eat mainly fish, although there is great variation in their diet in different areas and also among species, and they will also eat crustaceans, molluscs, reptiles, insects and even small mammals and birds. Giant Otters are known to prey also on anacondas, caimans and turtles.

EN Marine Otter

Lontra felina

Description: A small otter with mid- to dark brown upperparts when dry, but appears shiny silvery-gray when wet. The underparts, cheeks and throat are paler fawn. It is broad-headed and long-whiskered, with a shorter muzzle than other South American otters, and has a relatively short tail.

Similar species: Southern River Otter tends to occupy more sheltered coastal areas and can be identified by its larger size, longer slender tail, more uniform finer fur, and more contrasting grayish cheeks, throat and upper chest.

HB:	53–79 cm
Tail:	30.0–36.2 cm
Wt:	3.2–5.8 kg

Habitat: Prefers exposed rugged rocky coastlines, where it uses crevices and caves above the high-water level as holts, although it will occasionally travel up freshwater inlets. Has been found also around harbors in urban areas. Spends much of the day lying up out of the water.

Distribution: Patchily distributed in rocky areas along the Pacific coast from around Chimbote, in N Peru, south to the southern tip of Chile and east to Argentina (Isla de los Estados). Extremely rare in Argentina.

EN Southern River Otter

Lontra provocax

Description: A relatively small, short-furred otter, which is dark brown with paler silvery underparts and a grayish neck and throat. It has a long, muscular tapering tail.

Similar species: Occasionally overlaps in range with the smaller, darker **Marine Otter** and occurs also alongside the introduced **American Mink** (*p. 454*) in some areas, but the latter is much smaller and has a short bushy tail.

Habitat: Freshwater and marine habitats at up to 300 m. Primarily in inland waters in the northern parts of its range, and marine habitats in the south. Freshwater populations use a range of well-vegetated wetlands, including rivers, lakes (including Andean lakes), marshes and estuaries, while marine populations occupy rocky coasts with abundant vegetation but prefer less exposed, less windy coasts than Marine Otters.

Distribution: Chile and Argentina. In Chile freshwater populations occur south of the Imperial River (38°S). In Argentina largely restricted to the Limay watershed, mainly within the Nahuel Huapi National Park.

HB:	57–61 cm
Tail:	35–40 cm
Wt:	5–10 kg

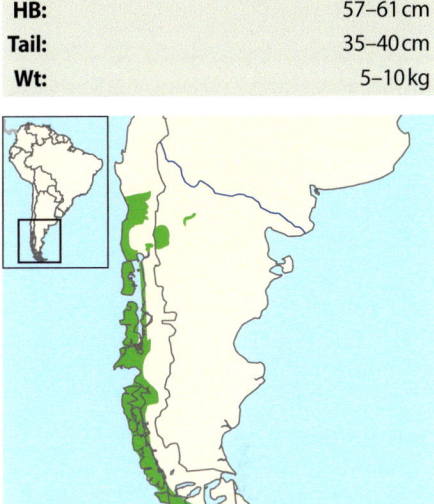

Coastal populations along the Pacific coast from 46°S to Tierra del Fuego in Chile, and in the Archipelago Fueguino on De los Estados Island and in the Beagle Channel in Argentina.

NT Neotropical Otter

Lontra longicaudis

Description: A relatively small but heavily built otter with a long stocky body, short legs, a long thick-based tapered tail, and dark brown pelage with a gray neck and throat. The head is small with a short, broad muzzle and long whiskers.

HB:	36–66 cm
Tail:	37–84 cm
Wt:	5–15 kg

Similar species: **Giant Otter** is much larger and has a creamy-white chin and throat, and a paddle-like tail.

Habitat: Up to 3,000 m in a variety of habitats with well-vegetated banks, including lakes, clear fast-flowing rivers and streams, rainforests, marshes, coastal swamps and rocky coastlines.

Distribution: Northern South America, including Trinidad, south as far as Buenos Aires Province in Argentina, although absent from W Peru, W Bolivia and a small area of arid NE Brazil within the Caatinga biome. Range extends through Central America and into Mexico.

EN Giant Otter

Pteronura brasiliensis

Description: The world's largest otter, with a broad, flattened head and large eyes. Normally dark chocolate-brown but can appear reddish, with large and distinctive white or creamy markings on the upper chest, neck, throat and lips. These can form a solid patch or be streaked with chocolate-brown. The tail is large and flattened like that of a beaver, creating a paddle-like appearance. Unlikely to be confused with any other species, particularly as it is highly gregarious and rarely seen other than in family groups.

Similar species: **Neotropical Otter** is much smaller, lacks a prominent creamy-white chin and throat, and has a tapered tail.

Habitat: Large, slow-moving rivers, streams, lakes and swamps, especially areas with gently sloping riverbanks and dense, overhanging vegetation. Has been recorded also in agricultural canals, reservoirs, dams and drainage channels along roads.

Distribution: Discontinuously east of the Andes from N Venezuela in the north to Misiones in Argentina in the south. Significant

HB:	♂ 100–130 cm; ♀ 100–120 cm
Tail:	45–65 cm
Wt:	♂ 26–32 kg; ♀ 22–26 kg

populations still exist in the Pantanal and parts of the Amazon. Probably extinct in Uruguay and Argentina, and restricted to a single area in Paraguay.

Weasels

Weasels are short-legged, long-bodied mustelids. Eight species occur in South America, including the introduced American Mink (*p. 454*). They can be active by day and night and are largely solitary, although in some areas Lesser Grisons are commonly seen in groups of up to 10–15 individuals. Tayras, grisons and Long-tailed Weasel swim and climb well. Weasels feed mainly on rodents and other small mammals, but will also take birds, reptiles, insects, eggs and fruit. Tayras are omnivorous, taking mammalian prey up to the size of agoutis and monkeys, although they have been seen even to chase small deer. They will also feed on carrion, birds, eggs, fruits, honey and invertebrates.

Tayra (JUVENILE)

LC Patagonian Weasel

Lyncodon patagonicus

Description: Short-legged and slender-bodied, with a grizzled grayish-brown back. The top of the head is white, this extending as a broad stripe on each side of the head to the shoulder. The nape, throat, chest and limbs are dark brown and the rest of the underparts light brown to gray.

Similar species: Likely to be confused only with the larger **Lesser Grison** (*p. 188*), which has only a narrow white or creamy band running across the forehead and down the neck and shoulders.

Habitat: Herbaceous and shrub steppes and xerophytic woodlands.

Distribution: NW, central and S Argentina and S Chile, within arid and semi-arid areas at up to 2,000 m. There are only two recent records from Chile, where its range is poorly known.

HB:	30–35 cm
Tail:	6–9 cm
Wt:	80–250 g

LC **Lesser Grison**

Galictis cuja

Description: A rather slight mustelid. Long-bodied and short-legged, with grizzled yellow-gray to brownish-gray upperparts. The face and underparts are black, with a narrow white or creamy band across the forehead and running down the sides of the neck and shoulders. Generally appears quite long-haired.

Similar species: Patagonian Weasel (*p. 187*) is smaller and has more white on the sides of the head and neck. **Greater Grison** is considerably larger, with a proportionately longer tail, and has a shorter-haired appearance.

Habitat: A wide range of habitats at up to 4,200 m, including forests, woodland, grassland, savanna, marshes, steppe and desert.

Distribution: SE Peru, S and W Bolivia, central Chile, Paraguay, Uruguay, Argentina and E and SE Brazil.

HB:	27.3–52.0 cm
Tail:	12–19 cm
Wt:	1.0–2.5 kg

LC **Greater Grison**
Galictis vittata

Description: A muscular mustelid with a short grayish tail. The upperparts are grizzled gray, while the underparts, including the legs and feet, are black. The face is largely black and, as with the much smaller Lesser Grison, it has a white line across the forehead which extends to the ears and down each side of the neck. Generally appears quite short-haired.

Similar species: Lesser Grison is much smaller, with a proportionately shorter tail, and has a longer-haired appearance.

Habitat: A wide range of habitats from tropical forests to savanna, grasslands, wetlands and open country, including plantations and cultivated areas, at up to 1,500 m.

Distribution: From NW Colombia southward as far as Bolivia and central & NE Brazil. Range extends through Central America and into SE Mexico.

HB:	45–60 cm
Tail:	13.5–19.5 cm
Wt:	1.4–4.0 kg

LC Tayra

Eira barbara

Description: Large and generally blackish-brown to black, normally with a white or creamy throat patch, and often with a pale head and neck which can range from pale gray-brown to pale yellow. Appears long-legged in the field, and has a bounding movement with an arched back and long bushy tail.

Similar species: Pale-headed individuals are distinctive, but dark-headed individuals can be confused with the dark form of **Jaguarundi** (*p. 147*): that species is, however, flat-backed, lacks a pale throat patch and has a long narrow tail. **Otters** (*pp. 182–185*) can resemble Tayra but are usually lighter-colored, with more extensive pale areas on the throat, and have a tapered tail. **Small Indian Mongoose** (*p. 455*) (introduced to Trinidad) is much smaller and paler.

Habitat: A wide range of forested habitats, including tropical and subtropical forests, secondary rainforests, gallery forests, gardens, plantations, cloud forests, and dry scrub forests, at up to 2,000 m but has been recorded to 2,400 m. Found also in agricultural areas close to human settlements.

HB:	55.9–71.2 cm
Tail:	36.5–47.0 cm
Wt:	2.7–7.0 kg

Distribution: Throughout much of South America, including Trinidad, as far south as N Argentina, although apparently absent from the Caatinga of NE Brazil and the high Andes. Range extends through Central America and into Mexico.

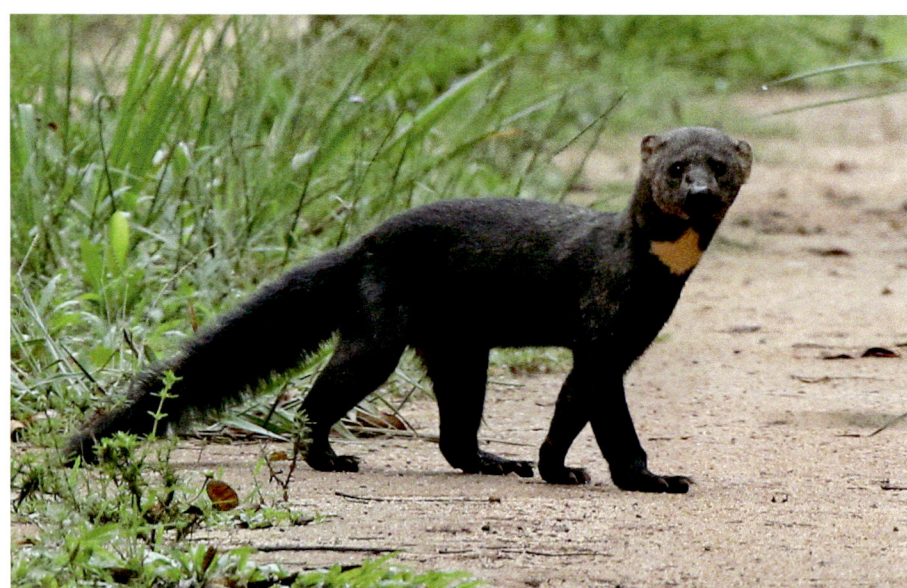

Additional photo *p. 186*

Long-tailed Weasel

Neogale frenata

Description: Slim-bodied, with a relatively long tail measuring up to two-thirds of the length of the body. Upperparts are rich rusty-brown to chocolate-brown, the underparts creamy white to yellow. The black-tipped tail is otherwise uniform with the upperparts. (South American individuals lack the distinctive white or yellow facial markings of animals in S USA and Central America.)

Similar species: Amazon and **Colombian Weasels** (*pp. 192–193*) have an all-dark tail and dark markings on the underparts.

Habitat: A wide range of habitats, including open woodland, brushland, grassland and marshes.

Distribution: The Andes from N Bolivia through Peru, Ecuador and into Colombia and Venezuela, and the Llanos. Range extends through Central America, Mexico, much of the USA and into Canada.

HB:	♂ 22.8–26.0 cm; ♀ 20.3–22.8 cm
Tail:	7.6–15.2 cm
Wt:	♂ 160–450 g; ♀ 80–250 g

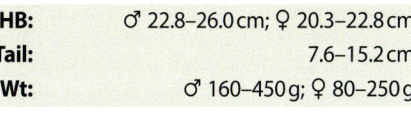

LC **Amazon Weasel** *Neogale africana*

Description: Larger than other *Neogale* weasels with uniform reddish-brown to dark brown upperparts and tail, and pale buff to cream underparts with a dark brown stripe extending from the lower belly to the chest or throat. The stripe may be discontinuous on some individuals. Young animals tend to be darker than adults.

Similar species: Long-tailed Weasel (*p. 191*) has a black tip to the tail and lacks the dark stripe on the underparts. There appears to be no overlap in range with Colombian Weasel.

Habitat: Poorly known. There are records from tropical lowland forest, and specimens have also been taken in rubber plantations and houses.

Distribution: Recorded in the Amazon Basin of Ecuador, Peru and Brazil, and possibly N Bolivia, but thought likely to occur elsewhere in the Amazon Basin, including S Colombia.

HB:	24–38 cm
Tail:	16–21 cm
Wt:	100–300 g

VU Colombian Weasel

Neogale felipei

Description: Small and uniformly dark brown on upperparts, with light buff-orange underparts fading to whitish on the chin; dark patch on the throat or upper chest. The feet have extensive webbing between the toes.

Similar species: Long-tailed Weasel (*p. 191*) has a black tail tip and lacks dark markings on the throat or chest. (See Amazon Weasel.)

Habitat: Mid- to high-elevation Andean forests.

Distribution: Six specimens are known from five localities in W Colombia (Chocó–Valle del Cauca, Huila and Cauca) and N Ecuador (Napo). There is also a sight record from La Cuelata National Park in Venezuela. Records are from altitudes between 1,525 and 2,700 m. Highly likely to be under-recorded, given that weasels in this part of the Andes are extremely challenging to identify in the field.

HB:	21.7–22.5 cm
Tail:	11.1–12.2 cm
Wt:	120–150 g

Skunks

Black-and-white carnivores that feed largely on invertebrates, but will also prey on birds, eggs, rodents and small reptiles, and will scavenge on carrion. Solitary, and mainly crepuscular and nocturnal, being active throughout the night. Humboldt's Hog-nosed Skunk is slightly more diurnal in late autumn and winter, but Striped Hog-nosed Skunk is rarely seen during the day.

Three species of skunk occur in South America, although Schiaffini *et al.* (2013) suggested that, based on molecular and morphological studies, Molina's and Humboldt's Hog-nosed Skunks are conspecific. Kasper *et al.* (2009), however, had previously concluded that they are genetically and morphologically distinct, occur in different habitats, and do not overlap in range. Burgin *et al.* (2020) concur with Schiaffini *et al.* (2013) in lumping the two species, but they note that further investigation is required. IUCN still recognizes the two species and this treatment is followed here.

Humboldt's Hog-nosed Skunk

LC Striped Hog-nosed Skunk

Conepatus semistriatus

Description: The largest of the South American skunks but with a gradation in size, the species becoming progressively smaller from north to south. The upperparts are typically black from the head to the rump, with a single white stripe along each side of the body. The underparts are black. There is, however, considerable variation in patterning. The tail is white apart from black fur at its base.

Similar species: None in range.

Habitat: Grasslands with shrubby cover, woodland savanna and deciduous forest, but appears largely absent from lowland rainforest. Occurs also in disturbed habitats such as plantations and pastures. Closely associated with Cerrado and Caatinga biomes in NE Brazil.

HB:	♂ 35–50 cm; ♀ 33–45 cm
Tail:	13.5–30.9 cm
Wt:	1.4–3.5 kg

Distribution: Occurs south from W Colombia through W Ecuador into NW Peru west of the Andes and east from N Colombia into N Venezuela and south to the east of the Andes in Colombia. There are isolated populations in central and NE Brazil. Range extends through Central America from W Panama to E Mexico.

LC Humboldt's Hog-nosed Skunk *Conepatus humboldtii*

Description: The smallest skunk. Has two parallel white stripes of variable width from the side of the crown to the base of the tail, and black, dark brown, reddish-brown or, on some individuals, creamy-brown fur on the back and the base of the tail. The remainder of the bushy tail is white.

Similar species: Molina's Hog-nosed Skunk is larger than Humboldt's Hog-nosed Skunk, but Schiaffini *et al.* (2013) were unable to find a color pattern that could be used to separate the two species reliably, and it is unlikely that they can be safely identified in the field other than on geographical range.

Habitat: A wide range of habitats from rocky deserts to forest at 200–700 m above sea level, but does seem to prefer open grasslands. Frequently found around human habitation.

Distribution: Restricted to central and southern Argentina and southern Chile. In Santa Cruz, Chubut, and southern and central parts of Río Negro Provinces in Argentina, and in adjacent areas of Chile.

HB:	♂ 20–32 cm; ♀ 20–30 cm
Tail:	16.5–20.2 cm
Wt:	500 g–2.5 kg

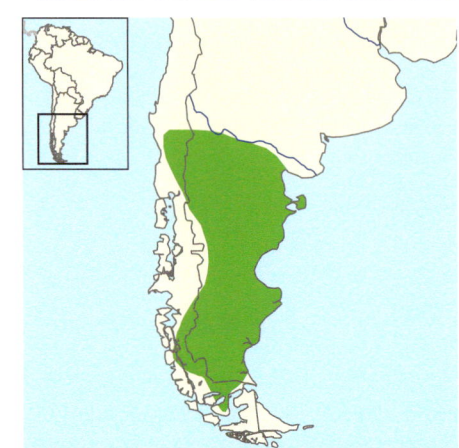

Additional photo p. 194

Molina's Hog-nosed Skunk
LC

Conepatus chinga

Description: A highly variable species, being black, brown or dark reddish with white stripes running down each side of the body. These stripes may join on the head. Schiaffini *et al.* (2013) identified four types of dorsal-stripe pattern in Molina's and Humboldt's Hog-nosed Skunks: no stripes present; stripes extending from the crown to the mid-dorsal region; stripes extending from the crown to the hips; and stripes extending the full length of the body and on to the base of the tail. They also identified three head patterns: plain, striped and spotted.

Similar species: Schiaffini *et al.* (2013) were unable to find a color pattern that could separate Molina's and **Humboldt's Hog-nosed Skunks**, and it is unlikely that they can be safely identified in the field other than on geographical range.

Habitat: A variety of habitats, including grasslands, savannas, steppe, and Chaco and other shrub forests, from sea level to above 4,000 m in the Bolivian altiplano. Kasper *et al.* (2009) questioned whether reports of its occurrence in heavily forested areas are accurate, and suggested that it is restricted

HB:	♂ 35–49 cm; ♀ 30–45 cm
Tail:	13.3–29.0 cm
Wt:	1–3 kg

to forest borders and monkey-puzzle (*Araucaria*) forests.

Distribution: S Peru through Bolivia south to Uruguay, W Paraguay, and S Chile and Argentina. Recorded occasionally in S Brazil.

Raccoons & allies (kinkajous, olingos, raccoons & coatis)

FAMILY | **Procyonidae**

Small to medium-sized omnivores with a long body, short legs and a relatively long banded tail, although Kinkajou has a uniform tail. Nine species are currently recognized in South America. See Helgen *et al.* (2013) for a comprehensive review of Olingo taxonomy and identification.

Olingos and Kinkajou are largely arboreal and mainly nocturnal, although diurnal activity has also been recorded and they will descend to the ground to cross open areas. They are mainly solitary, although olingos have been recorded as sharing fruiting trees with Kinkajous. They feed mainly on fruits and nectar, but will also take small rodents and lizards, birds, insects and eggs. They will also visit birdtables and hummingbird feeders.

Kinkajous can move rapidly, but they frequently hang from branches to reach fruit, often using the hind legs and tail to hold on to the branch.

Coatis are mainly diurnal and both terrestrial and arboreal. They frequently forage on the ground for invertebrates, but will also eat vertebrates and climb to feed in fruiting trees. They rest during the day, and at night sleep on branches of trees. Adult males are generally solitary outside the breeding season, but females, juveniles and young males of White-nosed and South American Coatis often form noisy groups of 10–30 and exceptionally up to 60 individuals. Mountain Coatis are largely terrestrial and feed mainly on insects, but they will also eat vertebrates and fruit.

Crab-eating Raccoons are largely nocturnal, spending the day in hollow trees and at night foraging on the ground around waterways. Although omnivorous, they feed primarily on crabs, crayfish, insects and fish over much of their range. They are largely solitary, but are occasionally seen in pairs or small family groups.

Western Mountain Coati

LC Kinkajou

Potos flavus

Description: Normally golden-brown above but can also be pale gray-brown or dark brown, and can have a dark brown line down the spine. Underparts range from creamy yellow to orange. The fur is dense and woolly. The head is broad, with a short blunt muzzle. The ears are short and rounded, and it has large eyes which give out bright orange eyeshine. The long, tapered prehensile tail has a dark brown tip. Males have a bare patch on the throat and are larger than females.

Similar species: Eastern Lowland Olingo (*p. 201*) is much smaller, less bulky, and has a narrower face and a shorter tail.

Habitat: Widely in deciduous and evergreen forests, including secondary forests and dry scrub, at up to 2,500 m.

Distribution: Throughout the northern half of South America as far south as central Bolivia and SE Brazil. Range extends through Central America and into Mexico.

HB:	40.5–76.0 cm
Tail:	37–57 cm
Wt:	1.4–4.6 kg

LC **Western Lowland Olingo**

Bassaricyon medius

Description: A medium-sized olingo having rusty-brown or orange upperparts with an indistinct blackish dorsal stripe. The underparts are pale orange or yellow and the head is gray-brown. The tail is long, faintly banded and slightly bushy, with a dark tip. Individuals from higher elevations have a shorter tail and longer fur and appear more brownish with less orange tones than lowland animals. The northern subspecies, *orinomus*, is larger than the southern subspecies, *medius*, and often has a reddish tail that contrasts with its less rufous head and body.

Similar species: **Eastern Lowland Olingo** is paler, with a more distinct dorsal stripe.

Habitat: Forests from sea level up to about 1,800 m. Spends the day in holes in trees.

Distribution: ssp. *medius* occurs west of the Andes in W Colombia and W Ecuador, its range overlapping with Olinguito (*p. 202*) in some places. ssp. *orinomus* occurs in NW Colombia, where it has been recorded in

HB:	31.0–41.5 cm
Tail:	35–52 cm
Wt:	900 g–1.2 kg

lowland forests, its range extending into Panama.

ssp. *orinomus*

Eastern Lowland (Allen's) Olingo

Bassaricyon alleni

Description: Upperparts are rusty-brown or orange with a well-defined blackish dorsal stripe. Underparts are pale orange or yellow. The head is gray-brown, with a short, pointed muzzle. The tail is long and faintly banded and is slightly bushy, with a dark tip. Individuals from higher elevations have a shorter tail and longer fur and appear more brownish, with less orange tones, than lowland animals.

Similar species: Western Lowland Olingo has a less defined dorsal stripe. **Kinkajou** (*p. 199*) is much larger and bulkier, has a broader face and has a long, tapering prehensile tail. (See also **Olinguito** (*p. 202*).)

Habitat: Most records are from lowland forest below 1,000 m, but recorded also at up to 2,000 m in Ecuador and Peru.

Distribution: The only olingo on the eastern slope of the Andes and in the lowlands east of the Andes, including Venezuela, E Colombia, E Ecuador, E Peru, NW Bolivia, W Brazil (Amazonas and Acre) and Guyana (two

HB:	30.4–45.5 cm
Tail:	40.1–53.0 cm
Wt:	1.1–1.6 kg

records). It may occur more widely in Brazil and possibly in Suriname and French Guiana.

NT **Olinguito** (Andean Olingo) — *Bassaricyon neblina*

Description: First described in 2013 and the smallest of the olingos. Identified by its richly colored black-tipped, tan to strikingly orange to reddish-brown dorsal pelage. Has an indistinctly banded, bushy and relatively short tail. It has a rounded face, which varies from gray to golden-brown depending on the subspecies, with a blunter, less tapering muzzle, and smaller and more heavily furred ears than other olingos.

Similar species: Olinguito can be confused with higher-elevation populations of **Eastern Lowland Olingo** (*p. 201*), but at present there is no evidence that the two species' ranges overlap. Eastern Lowland Olingo is less obviously rufous and has larger ears and a longer tail than Olinguito.

Habitat: Cloud forest at 1,500–2,750 m, where it occupies the canopy and is an adept jumper.

Distribution: Both slopes of the western Andes of Colombia and Ecuador, and along both slopes of the central Andes in Colombia.

HB:	32.5–40.0 cm
Tail:	33.5–42.4 cm
Wt:	750 g–1.07 kg

Crab-eating Raccoon

Procyon cancrivorus

Description: The only raccoon occurring in South America. Slender, with short coarse grayish-brown or reddish-brown fur and with orange or whitish underparts. A distinctive black mask extends beyond the eyes and partway across each cheek. It has blackish legs and feet, and the clearly ringed tail is roughly half the length of the head and body.

Similar species: None in range.

Habitat: A wide range of habitats, including Amazon rainforest, wooded grasslands, wetlands and along coasts, usually associating with water.

Distribution: East of the Andes in the lowlands from N Colombia and Venezuela, and Trinidad, south to N Argentina and Uruguay. Range extends through Panama to SE Costa Rica.

HB:	54–76 cm
Tail:	25–38 cm
Wt:	3.1–7.7 kg

LC **White-nosed Coati** *Nasua narica*

Description: The largest coati, with a long, low profile. When encountered, generally has its head down and the long tail held in an upright position. Usually dark brown, reddish-orange or yellow-brown, with creamy grizzling on the shoulders. The long muzzle has a diagnostic white lower section, and it has a white throat and white spots above and below each eye. The snout is blackish. The very long tail is narrow and tapered and of the same color as the body, with or without indistinct dark bands. The feet and legs are dark brown or blackish.

Similar species: The much smaller **Western Mountain Coati** (*p. 207*) lacks the white facial markings and has a much shorter tail.

Habitat: Deciduous and evergreen forest and secondary forest in the Chocó biogeographic region.

Distribution: The Pacific region of NW Colombia at altitudes of up to 3,000 m. Past reports from W Ecuador and NW Peru appear to be erroneous. Range extends through Central America and Mexico into SW USA.

HB:	43–68 cm
Tail:	42–68 cm
Wt:	3.0–5.6 kg

South American Coati

LC

Nasua nasua

Description: Individually variable in color, although the base color is brown with the tone ranging from orange or reddish to very dark brown and almost black. Color can vary considerably within the same group and even within the same litter, although most have a white lower jaw and buff throat. The muzzle is dark. The strength of the bars in the long-tapered tail is also highly individually variable.

Similar species: None in range.

Habitat: Occurs widely in forested habitats, including rainforest, gallery and cloud forest, Cerrado, dry Chaco scrub and wooded areas within savanna, at up to 2,500 m.

Distribution: From W Ecuador through E Colombia to Venezuela in the north, to Uruguay and N Argentina in the south, although absent from eastern-central and NE Brazil and the Llanos of Venezuela.

HB:	43–59 cm
Tail:	42–55 cm
Wt:	2.0–7.2 kg

Mountain coatis

Ruiz-García *et al.* (2020) proposed, on the basis of phylogenetic studies, that Eastern Mountain Coati *Nasua meridensis* should be treated as a synonym of (Western) Mountain Coati *N. olivacea*. However, IUCN still recognize two species of mountain coati and this treatment is followed here. Further studies by Ruiz-García *et al.* (2021) also questioned the validity of the genus *Nasuella* and suggested that mountain coatis should be treated as members of the genus *Nasua*. This treatment is followed here.

EN Eastern Mountain Coati

Nasua meridensis

Description: Small and dark olive-brown, with a blackish dorsal stripe. It has a narrow, pointed face with a naked nose. The tail is relatively short and has 6–8 bands.

Similar species: None in range.

Habitat: Cloud forests and páramo in the Venezuelan Andes at 2,000–4,000 m.

Distribution: Restricted to the Andes of Venezuela and known from only five localities.

HB:	43–54 cm
Tail:	19.2–30.0 cm
Wt:	1.0–1.5 kg

NT Western Mountain Coati

Nasua olivacea

Description: A small, dark reddish-brown or blackish-brown coati with paler rufous-tinged throat and chest. It has a narrow, pointed face with a naked nose. The tail is relatively short (50–60% of body length), with 6–8 bands.

Similar species: South American (*p. 205*) and **White-nosed** (*p. 204*) **Coatis** are much larger and have a proportionately longer tail (in excess of 90% of body length).

Habitat: Inhabits cloud forest and páramo habitats, at elevations of 1,300–4,250 m.

Distribution: Occurs throughout the Andes of Colombia and Ecuador. In 2018, camera-trap photographs from the province of Mariscal

HB:	40.9–48.7 cm
Tail:	22–27 cm
Wt:	1.0–1.5 kg

Cáceres in the San Martín Department of north-central Peru extended the range approximately 318 km south of the previously known distribution (Hyde *et al*, 2021).

Additional photo *p. 198*

PRIMATES

The remarkable number of 184 species of primates occur in South America. These are broken down in this book into 13 main groups, one of which, the saki monkeys, comprises two distinct sub-groups. Since the accounts for primates form the bulk of the book – 44% of all the species of larger mammals illustrated – an image of a typical example from each of the groups is shown here to provide an indication of their relative shape and form for comparative purposes. The number of species within the group (or sub-group) is given, and a cross-reference is included to the relevant page where there is an introduction that covers any taxonomic uncertainties, and provides an overview of biology and ecology.

Marmosets (*p. 210*)
(25 species)

Tamarins (*p. 239*)
(25 species)

Lion Tamarins (*p. 266*)
(4 species)

Squirrel Monkeys (*p. 270*)
(8 species)

Capuchins (*p. 278*)
(23 species)

Night Monkeys (*p. 304*)
(11 species)

Titi Monkeys (*p. 316*)
(35 species)

Saki Monkeys (*p. 356*)
Genus *Pithecia* (16 species)

(Bearded) Saki Monkeys (*p. 356*)
Genus *Chiropotes* (5 species)

Uakaris (*p. 380*)
(8 species)

Howlers (*p. 389*)
(13 species)

Spider Monkeys (*p. 403*)
(7 species)

Woolly Monkeys (*p. 410*)
(2 species)

Muriquis (*p. 414*)
(2 species)

Marmosets

The IUCN Primate Specialist Group recognizes 25 species of marmoset, although Mittermeier *et al.* (2013) had previously recognized an additional species, **Rio Manacoré Marmoset** *Mico manicorensis*. The latter, however, is now believed to be conspecific with Marca's Marmoset *M. marcai* (Garbino, 2014). Twenty-one of the species are endemic to Brazil.

Marmosets are diurnal and predominantly arboreal. They (and tamarins, *p. 239*) are distinguished from other New World primates by their small size and their modified claws (rather than nails) on all digits except the big toe. They eat fruits, flowers, nectar, gum, sap and latex, and also animal prey including frogs, snails, lizards, birds' eggs, spiders and insects. They have morphological and behavioral adaptations for gouging the bark of trees and vines to stimulate the flow of gum, and for some species this is an important component of the diet. They live in extended family groups of normally 4–15 individuals, with home ranges of 10–40 hectares depending on the availability and distribution of habitat and food, but some species have been recorded in groups of up to 30 individuals.

Geoffroy's Tufted-ear Marmoset

Cebuella, Callimico & Callithrix **marmosets**

Northern Pygmy Marmoset
Cebuella pygmaea
(p. 214)

Southern Pygmy Marmoset
Cebuella niveiventris
(p. 215)

Goeldi's Monkey
Callimico goeldii
(p. 216)

Black Tufted-ear Marmoset
Callithrix penicillata
(p. 235)

Buffy-headed Marmoset
Callithrix flaviceps
(p. 233)

Buffy Tufted-ear Marmoset
Callithrix aurita
(p. 234)

Geoffroy's Tufted-ear Marmoset
Callithrix geoffroyi
(p. 236)

Wied's Tufted-ear Marmoset
Callithrix kuhlii
(p. 237)

Common Marmoset
Callithrix jacchus
(p. 238)

Mico marmosets

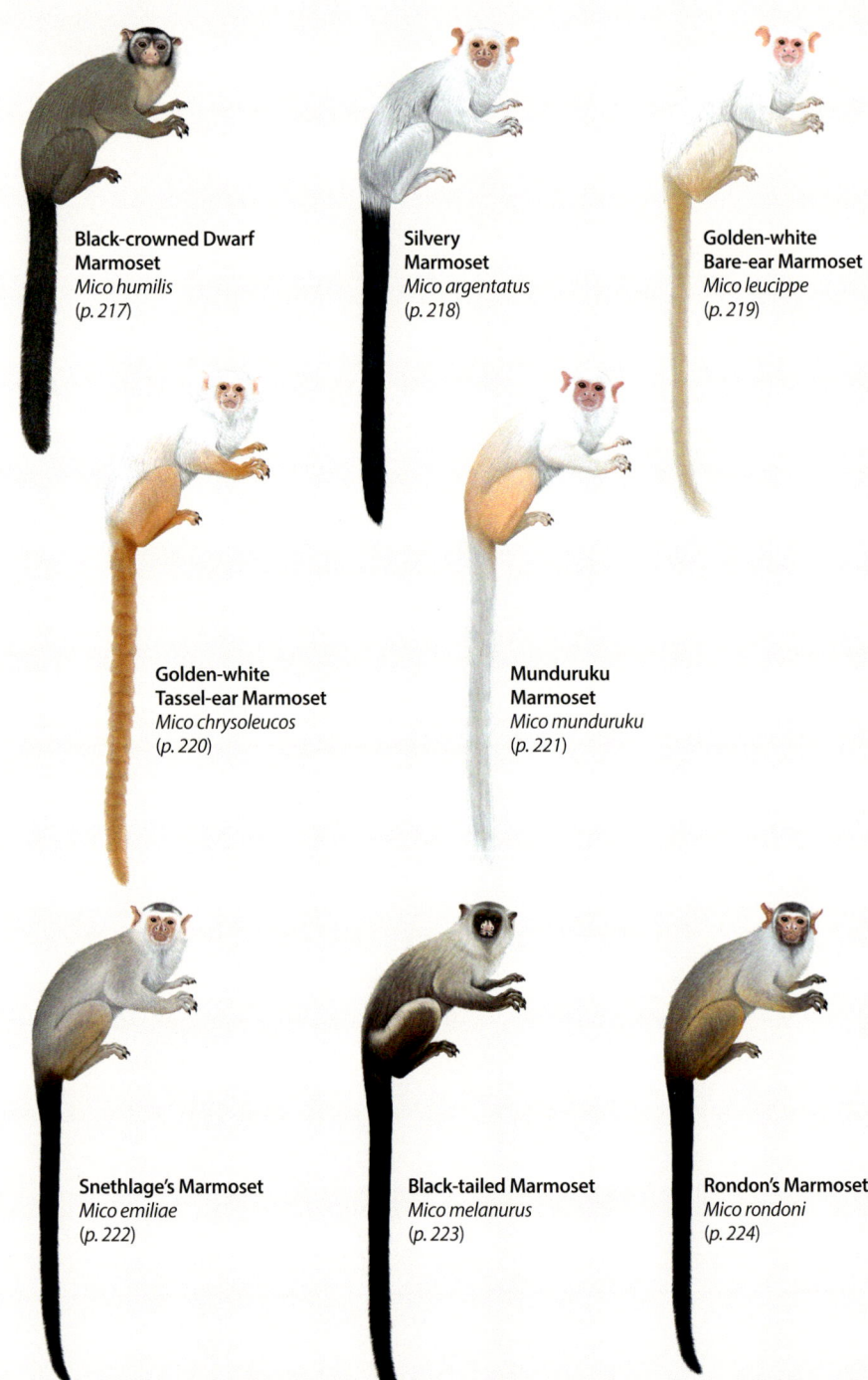

Black-crowned Dwarf Marmoset
Mico humilis
(p. 217)

Silvery Marmoset
Mico argentatus
(p. 218)

Golden-white Bare-ear Marmoset
Mico leucippe
(p. 219)

Golden-white Tassel-ear Marmoset
Mico chrysoleucos
(p. 220)

Munduruku Marmoset
Mico munduruku
(p. 221)

Snethlage's Marmoset
Mico emiliae
(p. 222)

Black-tailed Marmoset
Mico melanurus
(p. 223)

Rondon's Marmoset
Mico rondoni
(p. 224)

Mico **marmosets** (continued)

Black-headed Marmoset
Mico nigriceps
(p. 225)

Schneider's Marmoset
Mico schneideri
(p. 226)

Rio Aripuanã Marmoset
Mico intermedius
(p. 227)

Marca's Marmoset
Mico marcai
(p. 228)

Rio Acarí Marmoset
Mico acariensis
(p. 229)

Maués Marmoset
Mico mauesi
(p. 230)

Santarém Marmoset
Mico humeralifer
(p. 231)

Sateré Marmoset
Mico saterei
(p. 232)

VU **Northern Pygmy Marmoset**

Cebuella pygmaea

Description: Along with Southern Pygmy Marmoset, the smallest monkey in the world. Has grayish, black mixed with buff, or brownish-gold fur on the back, sometimes with a greenish tinge, and white, tawny or orange fur on the underparts. A longer mane of hair surrounds the face, covering the ears, and there are white marks at the edges of the mouth and a white vertical line on the nose. The hands and feet are yellowish or orange. The tail has indistinct dark rings on a paler background. The animal can produce a variety of vocalizations for communicating, including a high, sharp warning whistle and a clicking sound to indicate threats. Not known to overlap in range with Southern Pygmy Marmoset.

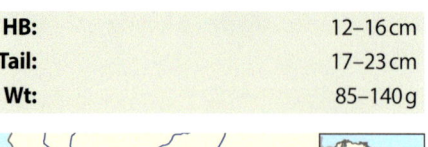

HB:	12–16 cm
Tail:	17–23 cm
Wt:	85–140 g

Similar species: None in range.

Habitat: Inundated forests, liana forest, and the edges of rivers and lakes, but occurs also in secondary forest and in isolated forest patches near human settlements, including areas affected by agricultural activities and hunting. Generally found in the lower layers of the forest, keeping to dense vegetation in the understory. In Ecuador, usually below 400 m but has been recorded from 200 m to 940 m.

Distribution: Occurs in the upper Amazon Basin, north of the Solimões River in Brazil to the west of Paraná do Aranapu and Paraná do Jarauá on the lower Japurá River. Found south of the Caquetá River in Colombia, but reports of its occurrence farther north in the upper Guaviare River region in Colombia remain unconfirmed. Found throughout Amazonian Ecuador, and in Peru north of the Solimões–Amazon–Marañón Rivers, west to the region of the Cerro Campanquiz and the basin of the Santiago River, and south to the Mayo River in the department of San Martín and to both sides of Marañón River. It is, however, currently unclear which pygmy marmoset occurs south of the Marañón

River and west of the Huallaga River. A small isolated population has been reported west of the Huallaga River in northern Peru at Gran Simacache, about 100 km west of the known distribution.

Southern Pygmy Marmoset *Cebuella niveiventris*

Description: Very similar to Northern Pygmy Marmoset, but paler on the hind legs and grayer on the hind parts. Does not overlap in range with Northern Pygmy Marmoset.

Similar species: None in range.

Habitat: Assumed to be similar to that of Northern Pygmy Marmoset.

Distribution: E Peru (east of the Mayo and Huallaga Rivers) and in the state of Amazonas in Brazil south of the Solimões River and

HB:	12–16 cm
Tail:	17–23 cm
Wt:	85–140 g

north of the Purús River. Found between the Madeira River in the east and the Andes in the west. In Bolivia occurs east to the region of Cobija (in N Bolivia) and extends south as far as to the south of the Tahuamanu River along the Muyumanu River, and east to at least Santa Rosa on the Abunã River. It is likely that this species is present throughout the interfluvium of the Purús and Madeira Rivers south to the Abunã River.

VU Goeldi's Monkey

Callimico goeldii

Description: Distinctive, with long, thick, soft black or blackish-brown fur, with a few pale marks on the nape and back, and occasionally two or three pale rings at the base of the otherwise bushy black tail. A distinctive cape (mane) covers the neck and the shoulders. The face, ears, hands and feet are black. When threatened, it arches its back and raises its bristles in defense to look larger.

Similar species: Black-mantled Tamarin (*p. 240*) is slightly larger, is much paler, and has a proportionately longer and less bushy tail.

Habitat: In forests along the edges of streams, bamboo forest and patches of secondary growth. Usually forages in the understory of the forest, but will occasionally feed on the forest floor or higher in the trees.

HB:	19–25 cm
Tail:	26–35 cm
Wt:	355–366 g

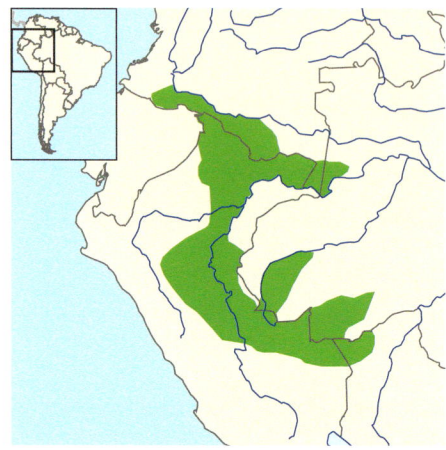

Distribution: Very patchily distributed in the upper Amazon from the Caquetá River, in Colombia, south through the Peruvian Amazon and the extreme western Amazon of Brazil into the Pando Department of N Bolivia. Not known to occur in Ecuador.

LC Black-crowned Dwarf Marmoset *Mico humilis*

Description: Very small, with a black crown which contrasts with the uniformly colored (not mottled) light to dark brown upperparts. The tail is dark brown with no obvious rings. The face is bare, with an unpigmented blackish area around the nose and each eye, and white eyebrows which extend around the sides of the face. The ears are bare, except for short white tufts in the center. The underparts, including the inner surface of the limbs, are creamy to pale honey-orange. The upper surface of the hands, feet and lower arms is orange mixed with black.

Similar species: None in range.

Habitat: Restricted to terra firme rainforest in the central Amazon, including forest edges near villages. Also found in seasonally

HB:	16–17 cm
Tail:	22–24 cm
Wt:	150–185 g

inundated riparian forests. May be heavily dependent on managed forests with a variety of species, fruit orchards, and gardens growing on anthropogenic soils.

Distribution: The Brazilian state of Amazonas, where restricted to the west bank of the Aripuanã River from its mouth, just south-west of the town of Novo Aripuanã, south at least to the village of Tucunaré, and west, along the right bank of the Madeira River to the mouth of the Mataurá River, and the right bank of the Uruá River. An isolated population occurs along the middle section of the Atininga River about 10 km east of the Manicoré River.

LC Silvery Marmoset

Mico argentatus

Description: Silvery-gray above, often with a brownish or blackish tinge to the lower back, and white to pale yellowish below. The tail is black. Silvery-gray limbs are slightly darker than the upperparts. The head is creamy white, the face and ears being pinkish-red and naked.

Similar species: May overlap in range with **Golden-white Bare-ear Marmoset**, which has a creamy-white, not black, tail.

Habitat: Common in terra firme primary forests and in extensive areas of secondary-growth forest near the mouth of the Tapajós River, but largely restricted to dense lowland floodplain forests between the Xingu and Tocantins Rivers. Has also been found in mixed open forest and in forest patches in Amazonian white-sand savanna.

HB:	20–22 cm
Tail:	26–33 cm
Wt:	350–410 g

Distribution: South of the Amazon River in Brazil between the mouth of the Tocantins River in the east and the Tapajós and Cuparí Rivers in the west, south to the Iriří River as far as the lower Curuá River. Does not occur south of Belo Monte on the Xingu River, and is found only north of the Tucuruí Dam reservoir on the Tocantins River.

LC Golden-white Bare-ear Marmoset *Mico leucippe*

Description: Similar to Silvery Marmoset, but creamy-white with no contrast between the upperparts and underparts. The tail is creamy or light beige. The limbs, including the hands and feet, are light beige but can be orange or gold. The face and ears are naked and are pinkish in color.

Similar species: May overlap in range with **Silvery Marmoset**, which has a black tail.

Habitat: Humid lowland tropical rainforest up to 200 m above sea level. Prefers degraded and secondary growth and edge habitat.

Distribution: A small area in the Brazilian state of Pará, between the Cuparí and Tapajós Rivers south to the Jamanxim River.

HB:	approx. 21.5 cm
Tail:	approx. 31 cm
Wt:	approx. 330 g

Marmosets compared *pp. 211–213*

LC Golden-white Tassel-ear Marmoset *Mico chrysoleucos*

Description: A yellowish-white marmoset with a pale cream, orange or gold belly, hands and feet. The tail is pale orange or gold at the base, the remainder of the tail being slightly darker and loosely ringed. The face and ears are pink, with prominent fan-shaped white ear-tufts. Unlikely to be confused with any other marmoset in range.

Similar species: None in range.

Habitat: Amazonian secondary lowland rainforest.

Distribution: Found south of the Amazon River in Brazil, between the Madeira and lower Aripuanã Rivers in the west and the Canumã River in the east. Occurs on the north bank of the Paraná Urariá River and has been observed at Santa Bárbara on the north bank of the Canumã River. Ranges

HB:	20–24 cm
Tail:	32–40 cm
Wt:	280–310 g

south to Prainha, on the east bank of the Aripuanã River north of the mouth of the Roosevelt River.

Munduruku Marmoset

Mico munduruku

Description: Largely white, with a beige-yellowish spot on the elbow. The undersides of the forelimbs are cream-colored and the saddle is beige-yellowish, becoming cream on the sides and on the belly. The hind limbs are cream, becoming dark yellow towards the posterior region. The rump also is dark yellow. This species is distinct from all other *Mico* marmoset species in having a white tail, feet and hands, white forearms with a beige-yellowish spot on the elbow, and a beige-yellowish saddle.

Similar species: None in range. (Golden-white Bare-ear Marmoset (*p. 219*) is similarly cream-colored and also occurs in the Tapajós–Jamanxim interfluvium, but only south of the Cururú River and south and east of the Novo River, and there is no evidence that the two species overlap in range.)

Habitat: Lowland primary and secondary terra firme forests.

Distribution: Restricted to the south-west part of Pará State in Brazil, where it is known from seven localities in the northern Tapajós–

HB:	18.0–27.6 cm
Tail:	27.7–33.0 cm
Wt:	250–435 g

Jamanxim interfluvium. Separated from other *Mico* species by the Tapajós River in the west and north, by the Jamanxim River in the east, and by the Novo and Cururú Rivers in the south.

LC Snethlage's Marmoset

Mico emiliae

Description: The head is white, apart from the pinkish-white muzzle and a blackish crown, and the ears are large with sparse brown hairs. The eyebrows, cheeks and chin are whitish. The upperparts are pale gray-brown, becoming a darker orangey-brown on the rump and outer thighs. The tail is black with an orangey-brown base. The underparts are silvery-gray and paler than the upperparts. The upper limbs are gray-brown to orangey-brown, and whiter on the inner surfaces. The hands and feet are dark brown to blackish. Photos of a presumed

HB:	21 cm
Tail:	32.5 cm
Wt:	approx. 340 g

young animal appear to show an orangey-brown belly and vent.

Similar species: Split from **Silvery Marmoset** (*p. 218*) but more closely resembles **Black-tailed Marmoset** (the ranges of the these species do not, however, overlap).

Habitat: Amazonian lowland rainforest and mixed open forest and forest patches in savanna/bush savanna (Cerrado).

Distribution: Brazil, where found in the south of the state of Pará and possibly adjacent areas in the north of the state of Mato Grosso. It occurs south from the Iriri River to as far south as the southern margin of the Peixoto de Azevedo River. The Teles Pires River is believed to be the western limit of its range, and it is replaced by Schneider's Marmoset (*p. 226*) to the west of that river. In the headwaters of the Teles Pires River it is replaced by Black-tailed Marmoset.

Black-tailed Marmoset

Mico melanurus

Description: The head, neck and breast are beige, except for the crown, which is dark brown. The nose has a rosy appearance, and the rest of the face and the ears are naked and black. The mantle is grayish-brown, becoming darker brown on the lower back and rump. There is a distinctive orange spot on the hip and a white stripe on the thigh. The tail is black with a chestnut base. The chest and neck are buff to creamy, and the rest of the underparts, including the inner sides of the limbs, are cream to ochre or orange. The limbs are dark brown.

HB:	22–24 cm
Tail:	30–34 cm
Wt:	approx. 330 g

Similar species: Rondon's Marmoset (*p. 224*) lacks a distinct pale thigh stripe and is paler and grayer.

Habitat: Amazonian lowland rainforest, in flooded savanna forests in the Pantanal, in tall forest along the ephemeral waterways in the north-eastern Chaco, and in primary and secondary-growth forest, dry deciduous forest, gallery forest and forest patches in Bolivia.

Distribution: In Brazil, its range is unclear owing to the identification of new species within what was previously thought to be its range. It does occur to the east of the Madeira River, from the mouth of the Aripuanã River along the east bank north to at least 10°S, south to beyond the Guaporé River and west to the Juruena River, or the Teles Pires River. Its range extends south through the Pantanal into Bolivia and Paraguay as far south as the north-eastern Paraguayan Chaco at approximately 20°S. In Bolivia it is found east of the Mamoré River, in the departments of Beni and Santa Cruz.

VU Rondon's Marmoset

Mico rondoni

Description: Pale silvery-gray, with a blackish crown, the black extending on to the back of the head, to the front of the ears and down the sides of the face, contrasting with a white forehead. The silvery-gray upperparts and underparts become darker grayish-brown on the lower back and reddish-brown on the rump, thighs and base of the tail. The remainder of the tail is darker, and almost black. The thighs become black towards the ankles, while the hands and feet are reddish-brown.

Similar species: May occur within the geographical range of **Black-tailed Marmoset** (*p. 223*), but the limits of the two species' ranges are unclear. The latter has a distinctive pale thigh stripe and is generally darker above than Rondon's Marmoset.

Habitat: Amazonian lowland rainforest, often occurring alongside Weddell's Saddleback Tamarin (*p. 249*). Tends to forage at least 10 m above the ground, while the tamarin travels at lower levels.

HB:	20–24 cm
Tail:	29.5–33.0 cm
Wt:	250–390 g

Distribution: Occurs in the north of Rondônia State in Brazil, where it is believed to occur from the Mamoré–Madeira Rivers and Jiparaná, in the north and west, to the Serra dos Pacáas Novos, in the south.

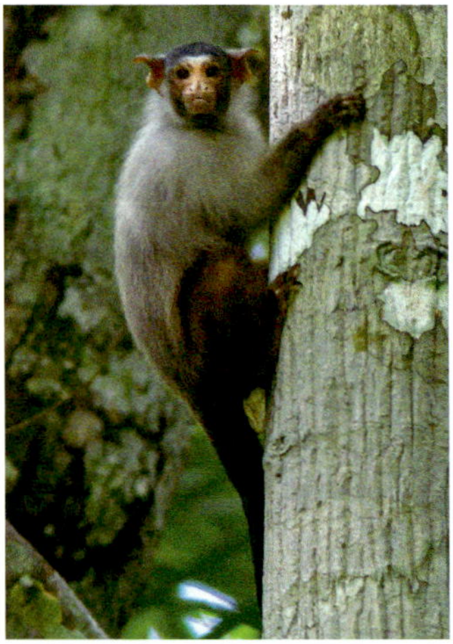

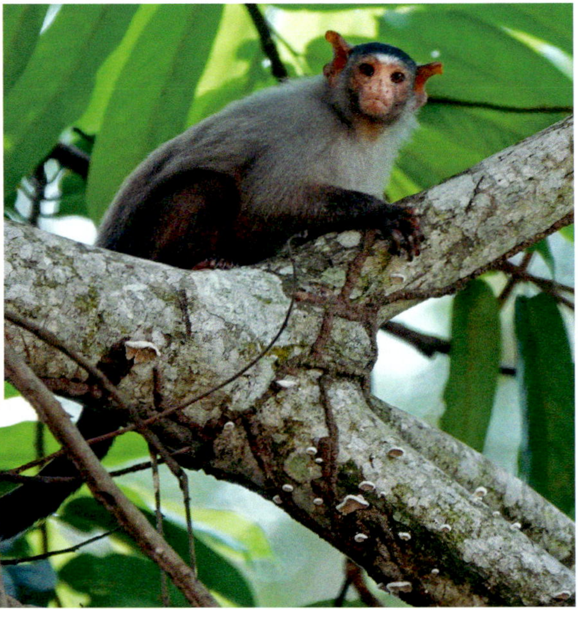

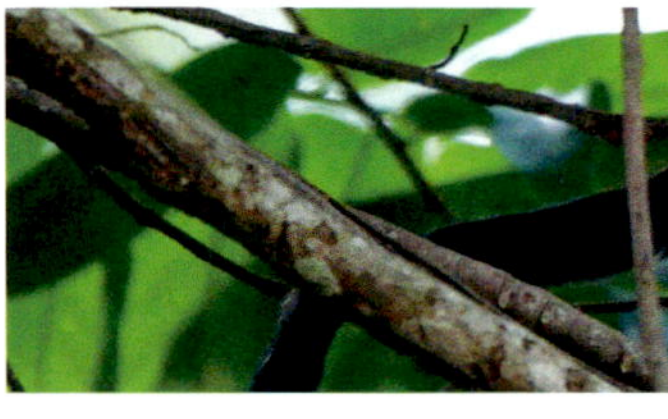

NT Black-headed Marmoset

Mico nigriceps

Description: Mainly light brown to brownish-gray above, becoming darker on the lower back and dark brownish-gray on the rump. The face is naked and blackish with some dark brown spotting, the crown and forehead also are black, and the ears are black and lack tufts. The underparts are yellow to orange, and the thighs are reddish-orange with an indistinct pale stripe. The hands and feet are blackish or rusty and the tail is black. Males have a white hairless scrotum.

HB:	19–22 cm
Tail:	31–33 cm
Wt:	330–400 g

Similar species: None in range.

Habitat: Humid tropical rainforest, preferring secondary growth and forest-edge habitat.

Distribution: Found in the south of the state of Amazonas and the extreme north of the state of Rôndonia in Brazil, where it is believed to occur in the area bounded by the Marmelos River in the north and east, the Madeira River in the west and the Jiparaná River in the south.

Schneider's Marmoset

Mico schneideri

Description: The head and face are white with a black crown, the mantle gray, the saddle and rump uniform lead-colored and the tail black. The forearms are grayish-cream each with a blackish-golden hand; the feet are golden-orange. The underparts are light grayish-cream and orange.

Similar species: None in range, but see **Snethlage's Marmoset** (*p. 222*).

Habitat: Primary and secondary terra firme forests along with forests in the Amazonia–Cerrado transitional zone.

No information on measurements and weight available

Distribution: Restricted to the Juruena–Teles Pires interfluvium in southern Amazonia, Mato Grosso State, in Brazil. Occurs west to the Juruena River, east to the Teles Pires River and north to their confluence. Replaced by Snethlage's Marmoset to the east of the Teles Pires River. In the headwaters of the Teles Pires it is replaced by Black-tailed Marmoset (*p. 223*). The southern limits of Schneider's Marmoset's distribution are poorly defined, but it extends to the headwaters of the Juruena and Teles Pires Rivers as far south as the city of Lucas do Rio Verde.

Rio Aripuanã Marmoset

Mico intermedius

Description: Distinctive, with a pink face and ears with some gray pigmentation. The tufted ears have white hairs on the outer surface. The head (apart from the grayish crown), neck and upper back are silvery-gray or whitish, the mantle becoming spotted with chestnut and brown. The lower back and rump are dark brown. The tail is dark reddish-brown at the base, becoming progressively paler, and is uniformly light beige. The throat and upper chest are creamy white and the remainder of the underparts are pale orange. The limbs are orange, being palest on the forelimbs, and it has a

HB:	20–24 cm
Tail:	32–40 cm
Wt:	280–310 g

yellow-gold thigh stripe. Will feed above ant swarms and even descend to the ground, but normally forages by searching clumps of dead leaves while moving slowly through dense vegetation.

Similar species: None in range.

Habitat: Humid tropical rainforest with a preference for secondary growth and forest edge, particularly areas with abundant secondary-growth patches and dense understory. Lower densities occur in areas of tall old-growth forest with a sparse understory and in riparian flooded forests.

Distribution: South-east of Amazonas State and the north-west of Minas Gerais State, in Brazil. Occurs between the Roosevelt and Aripuanã Rivers, including the entire basin of the Guariba River. The southern limits of the range are unknown.

VU Marca's Marmoset

Mico marcai

Description: Has a dark chestnut-brown crown which contrasts with the dark gray nape and mantle. The upperparts appear light gray mottled with dark gray. A white triangular patch in the center of the forehead extends as a circular white rim on to each side of the forehead and on to the face, up behind the eyes. The face is pigmented other than around the nostrils and between the eyes. The ears are unpigmented. The throat, chest and forelimbs are light brown and the remainder of the underparts, including the hind limbs, are reddish-brown to ochre-colored. The tail is completely black, some individuals having reddish-brown ochraceous hairs towards the base of the tail.

Similar species: None in range.

HB:	approx. 22 cm
Tail:	approx. 36 cm
Wt:	280–310 g

Habitat: The type locality is in dense tropical rainforest, but nothing more is known about the ecology of the species.

Distribution: Occurs between the Aripuanã, Marmelos and Madeira Rivers in Brazil. The southern limit of the range is unclear due to uncertainty regarding the taxonomy of marmosets located south of the savannah zone in the Campos Amazônicos National Park.

LC Rio Acarí Marmoset

Mico acariensis

Description: The face is bare and flesh-colored, with black spots or patches between the eyes and on the lower lip and chin. There are black spots on the side of the nose, on which there is also a distinctive triangular black patch. The outer surfaces of the ears are white. The upperparts are whitish on the crown and nape, becoming gray on the back and dark gray mixed with orange on the lower back. The tail is black with an orange patch at its base. The neck and chest are pale gray and the belly is orange. The upper sides of the limbs are gray mixed with orange, and the undersides are whitish-orange. The thighs have a distinctive pale stripe.

Similar species: None in range.

HB:	approx. 22 cm
Tail:	approx. 38 cm
Wt:	approx. 350 g

Habitat: Amazonian lowland terra firme primary and secondary rainforest, including forest around plantations.

Distribution: Occurs in Brazil along the east bank of the lower Acarí River, and is consequently assumed to occur between the Acarí River (in the west) and the Sucunduri River (to the east), south to a contact zone with Black-tailed Marmoset (*p. 223*) between the Aripuanã and Juruena Rivers. Recorded at three localities along the west bank of the Sucunduri River between 06°48'S and 07°17'S.

LC **Maués Marmoset**

Mico mauesi

Description: A distinctive dark marmoset having dark chocolate-brown upperparts with pale grayish-white marbling on the back and a silvery hip patch. The upper back and shoulders have a slight reddish tint. The limbs are silvery-gray, and the feet and hands are darker with an orange tint. The underparts are buff, with an orange tint on the belly. The tail is dark chocolate-brown or black with weakly defined pale rings. The head is chocolate-brown with silver-brown cheeks, and the face is pinkish with black spots on each side of the mouth. The ears are surrounded by large tufts of erect silvery-brown hairs.

Similar species: None in range.

Habitat: A poorly known species found in dense tropical rainforest, but has been recorded also in secondary forest.

HB:	20–23 cm
Tail:	34–38 cm
Wt:	315–405 g

Distribution: Found only south of the Amazon River in Brazil. It occurs along the west bank of the Maués River opposite the city of Maues and to the west as far as the Paraná Urariá and Abacaxis Rivers. Has been reported also from the east bank of the lower Abacaxis River in the state of Amazonas (3°54'S, 58°46'W) and in the vicinity of the town of Abacaxis, also on the east bank of the Abacaxis River (3°55'S, 58°45'W). Replaced by Sateré Marmoset (*p. 232*) on the west bank of the Abacaxis River.

Santarém Marmoset

Mico humeralifer

Description: The upperparts appear marbled, being blackish with irregular gray-white spots. The shoulders are light grayish-brown to white, and the thighs are brownish with a thin pale hip stripe. The limbs are blackish, apart from gray fur which extends down from the mantle over the outer surfaces of the upper arms. The tail is gray with black rings. The underparts are yellowish to orange. The face varies in color from yellow to light brown, with patches of pink skin around each eye and the mouth. The forehead is gray, while the front of the otherwise gray-brown crown is black, the black extending down the side of the face to the cheek and on to the nose. The ear-tufts are buff to pale gray in color and have a tassel-like shape that extends sideways from the head.

Similar species: None in range.

HB:	18–21 cm
Tail:	31–33 cm
Wt:	(means): ♂ 475 g; ♀ 472 g

Habitat: Dense vines and river edges in secondary Amazonian lowland rainforest, but apparently absent from flooded forests.

Distribution: South of the Amazon River in Brazil between the Maués River (and possibly its tributary the Parauari River) in the west and the Tapajós River in the east, and south though Amazônia National Park. The southern limits of the range are unclear.

LC Sateré Marmoset

Mico saterei

Description: Has a highly contrasting appearance, with a dark reddish-brown to almost black back, tail and limbs while the head, neck and belly are yellowish-white or golden-orange. The crown is dark gray and the nape is silvery. The thighs are reddish-brown with a fairly obvious golden-orange stripe. The face and ears are naked and pinkish, the back of each ear having an orange-red spot. Lacks ear-tufts, unlike most of the other *Mico* marmosets that occur in the area. The genitalia are bright orange, which is diagnostic among marmosets. Unlikely to be confused with any other marmoset in its range.

Similar species: None in range.

HB:	approx. 20 cm
Tail:	approx. 37 cm
Wt:	approx. 430 g

Habitat: Primary, secondary, and disturbed terra firme forest and black-water inundated forest (*igapó*).

Distribution: The state of Amazonas in Brazil, between the Abacaxis River in the east and the Canumã–Sucunduri Rivers in the west. Found also south of Paraná–Urariá River on the east bank of the Madeira River. The southern limit of the range is unclear, but may extend as far south as the Juruena River.

CR Buffy-headed Marmoset

Callithrix flaviceps

Description: Has a distinctive buffy-orange head and face, apart from a white muzzle and a blackish central mask surrounding the eyes and nose. The ear-tufts are long and buff-orange. The upperparts are orangey-brown with dark brown or black bands. The underparts are normally paler orange or yellowish with a black midline, but can appear darker and even blackish-brown. The tail is grayish-brown with lighter bands, and the feet and hands are tan. A solitary individual could, given a poor view, be mistaken for a squirrel.

Similar species: None in range. (But there appears to be a zone of hybridization with **Buffy Tufted-ear Marmoset** (*p. 234*) at the Serra do Brigadeiro, in south-east Minas Gerais.)

HB:	22–25 cm
Tail:	30–35 cm
Wt:	approx. 400 g

Habitat: Montane and sub-montane evergreen and semi-deciduous forests at altitudes from 270 m to 1,800 m. Prefers forests with dense vegetation and dense understory, and is rare in tall old-growth forest with sparse understory.

Distribution: Brazil, from Serra da Mantiqueira in southern Espírito Santo, south of the Doce River at least to the state boundary with Rio de Janeiro. Extends west into eastern Minas Gerais at scattered localities in the fragmented forests of the Manhuaçu River Basin.

Marmosets compared *pp. 211–213*

EN **Buffy Tufted-ear Marmoset**

Callithrix aurita

Description: Dark blackish-brown, with small reddish spots on the lower back and legs. The tail is yellowish-gray with distinct black bands, and the underparts are blackish-brown. The hands and feet are yellowish-brown. The face and chin are buff. The hairs around the ears and cheeks are black, forming a ring around the face, while the ear-tufts are short and whitish to pale buff. There is a characteristic white patch on the forehead and a reddish-buff patch on the crown. Some geographical variation is evident, individuals from Rio de Janeiro State being darker than those in São Paulo State.

Similar species: None in range, although there appears to be a zone of hybridization with **Buffy-headed Marmoset** (*p. 233*) at the Serra do Brigadeiro, in south-east Minas Gerais. There are reports that it appears also to be hybridizing with the introduced **Common Marmoset** (*p. 238*) in Rio de Janeiro and possibly São Paulo States.

Habitat: Montane rainforests at 500–1,300 m.

HB:	19–25 cm
Tail:	27–35 cm
Wt:	400–450 g

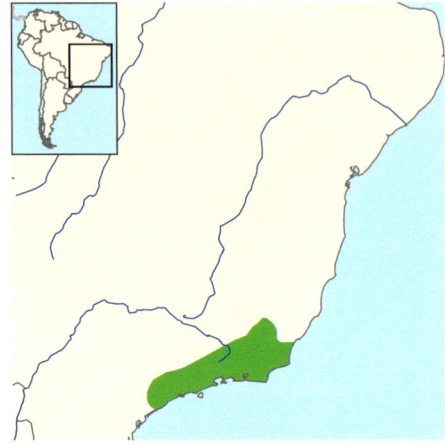

Distribution: Brazil, where found in the southern part of the state of Minas Gerais, the state of Rio de Janeiro, and the east and north-east of the state of São Paulo.

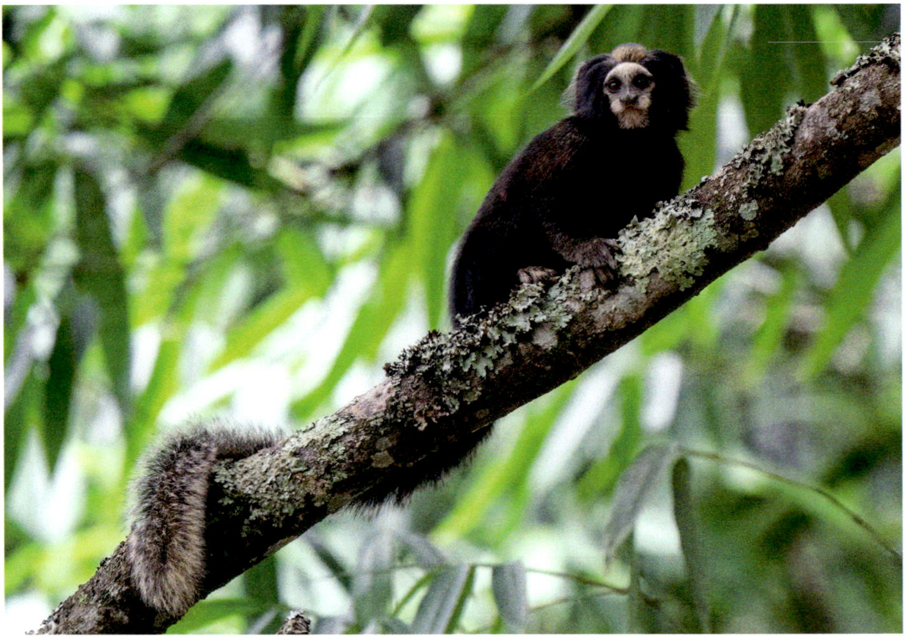

LC Black Tufted-ear Marmoset

Callithrix penicillata

Description: Grizzled gray and black above, with a banded appearance to the hindquarters. The underparts and limbs are buff-gray, but the feet can also be gray to black. The tail is gray-brown with pale gray or white rings. The head is dark yellowish-brown, with a black face, a triangular white patch on the forehead and long, drooping black ear-tufts. Has a conspicuous and distinctive black cape (mane) and collar around the neck and chest.

Similar species: None in range. (But hybridizes with both **Geoffroy's** (*p. 236*) and **Wied's** (*p. 237*) **Tufted-ear Marmosets**: see those accounts for details of hybrid zones.)

Habitat: Occupies a variety of forested habitats, including gallery forest, xeric scrub and forest, palm woodlands and some areas of mesophytic Atlantic Forest. Prefers disturbed and secondary-growth forest.

Distribution: Widespread across the Cerrado region of east-central Brazil in the states of Bahia, Minas Gerais, Goiás, extreme south-west Piauí, Maranhão, and northern São Paulo north of the Tieté and Piracicaba Rivers.

HB:	20–23 cm
Tail:	29–33 cm
Wt:	180–250 g

The full extent of the range is confused by the presence of introduced animals outside the natural range, and by the fragmentation of forests. May be replacing other species in localities east and south of its original range.

LC **Geoffroy's Tufted-ear Marmoset** *Callithrix geoffroyi*

Description: Distinctive and easily identified by the conspicuous white crown, forehead, cheeks and throat, which contrast strongly with the remainder of the black head, including the prominent black ear-tufts. The upperparts are banded gray and black with orangey-brown mottling, which can produce larger patches. The underparts are dark brown or black, except for the white throat. The tail is black with narrow pale gray or white rings. The hands and feet are black.

Similar species: None in range. but hybridizes with **Black Tufted-ear Marmoset** (*p. 235*) in parts of the Serra da Piedade

HB:	18–23 cm
Tail:	approx. 29 cm
Wt:	190–350 g

along the Piracicaba River, where the Atlantic Forest gives way to the Cerrado.

Habitat: Occurs in lowland and sub-montane forest, gallery forests and, in the north of its range, dry forest patches in desert scrub in the Jequitinonha Valley. Generally restricted to lowland areas below 700 m, but has been recorded at up to 1,274 m on the eastern slopes of Serra do Cipó. Overlaps in range with Buffy-headed Marmoset (*p. 233*) in southern Espírito Santo and south-east Minas Gerais, but generally occurs below 700 m, while Buffy-headed Marmoset occurs above 400 m.

Distribution: Brazil in the state of Espírito Santo and the forested eastern and north-eastern part of Minas Gerais, north as far as the Jequitinhonha and Araçuaí rivers and south to near the state border of Espírito Santo and Rio de Janeiro. Introduced at Belmonte, and now occurs east of there in gallery forests throughout the dry thorn scrub (Caatinga) to the south of the middle reaches of the Jequitinhonha River.

Additional photo *p. 210*

Wied's Tufted-ear Marmoset

Callithrix kuhlii

Description: The upperparts are banded gray and dark brown, often with reddish-brown tones, and it has reddish-brown upper thighs. The tail is blackish with pale gray or white rings, and the limbs are dark gray to black. The face, cheeks and throat are pale buff-gray to light brown, with a white patch on the forehead. The adult's crown is buff-gray, that of juveniles being black. The ear-tufts are long and black, and the cape is grizzled grey.

Similar species: None in range, but hybridizes with **Black Tufted-ear Marmoset** (*p. 235*) in the region of Almenara, in Minas Gerais. Individuals intermediate in

HB:	20–23 cm
Tail:	29–33 cm
Wt:	350–400 g

appearance between Wied's and Black Tufted-ear Marmosets have been recorded north from the Contas River along the coast to just south of the city of Salvador, Bahia. Individuals observed near Nazaré (just south of Salvador) lacked the white frontal blaze and, although retaining the pale cheek patches typical of Wied's Tufted-ear Marmoset, they were paler gray.

Habitat: Occurs in sub-montane humid rainforest, lowland coastal forest, and restinga forest with piaçava palms. Also degraded and secondary-growth forests, cacao plantations with enough tall forest trees remaining to provide shade, and abandoned rubber plantations.

Distribution: Brazil between the Contas and Jequitinhonha rivers in southern Bahia, just entering the north-east of Minas Gerais. The western boundary is probably defined by the inland limits of the Atlantic Forest. It is broadly sympatric with Golden-headed Lion Tamarin (*p. 267*).

LC **Common** (White-tufted) **Marmoset**   *Callithrix jacchus*

Description: Has grizzled pale gray, pale brown or gray and buff-orange upperparts, with the hindquarters and the tail banded gray and pale gray or white. The underparts and limbs are gray. The head and neck are dark brown, with a black crown and a patch of white on the forehead. The prominent ear-tufts are white and fan-shaped and contrast sharply with the rest of the dark head. Some regional variation has been recorded, with paler individuals found in drier areas and more chestnut ones in moister areas.

Similar species: None in range, but see **Buffy Tufted-ear Marmoset** (*p. 234*).

Habitat: Gallery forest, semi-deciduous and deciduous forest patches in dry Caatinga thorn scrub, and in humid Atlantic Forest. Highly adaptable, being able to live in urban parks and gardens and in rural villages where not persecuted and with sufficient food available.

HB:		16–21 cm
Tail:		24–30 cm
Wt:		318–322 g

Distribution: Scrub forest and Atlantic Forest of NE Brazil, in the states of Alagoas, Pernambuco, Paraíba, Rio Grande do Norte, Ceará and Piauí, Maranhão, Bahia and possibly north-eastern Tocantins. The distribution map shows only the native range and excludes introduced populations, including areas within the states of Bahia, Espírito Santo, Paraná, Rio de Janeiro, Santa Catarina, São Paulo and Sergipe.

Tamarins

The IUCN Primate Specialist Group recognizes 25 species of tamarin. They also propose splitting Spix's Saddle-back Tamarin into four species, but since the paper describing them has yet to be published, it would inappropriate to set a precedent by including them here and such treatment has not been followed.

Tamarins are diurnal and predominantly arboreal. They (and Marmosets, *p. 210*) are distinguished from other New World primates by their small size and their modified claws (rather than nails) on all digits except the big toe. They feed on fruits, flowers, nectar, gum and sap, but many also eat some animal prey, including frogs, snails, lizards, spiders and insects. They occur in groups of 2–15 individuals.

Cotton-top Tamarin

LC Black-mantled Tamarin *Leontocebus nigricollis*

HB:	21–25 cm
Tail:	31–35 cm
Wt:	(means): ♂ 484 g; ♀ 468 g

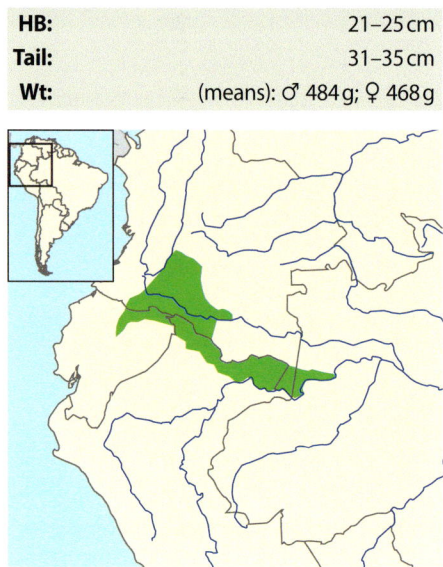

Description: Appears dark gray to black at long range. Three subspecies are recognized. ssp. *nigricollis* has a uniformly black mantle, predominantly reddish or mahogany lower back, rump and thighs and reddish tail more or less mixed with black. ssp. *graellsi* has a blackish to blackish-brown head, neck, mantle and forelimbs finely marked with gray or buff. The temporal patches are pale brown; the back, the rump and the thighs are grayish-brown, coarsely mixed with buff and black. The underparts are dark brown to black. The face is sparsely furred, with a whitish short-haired nose. ssp. *hernandezi* has an almost uniformly blackish nape and mantle, the mantle tapered towards the mid-back and with the black continuing usually as a mid-dorsal band or stripe across the lower back and upper part of the tail. The sides of the lower back are mixed blackish and orange. The sides and ventral surface of the upper 5–10 cm of the tail are orange/gray-brown, the remainder of the tail blackish. The neck and chest are predominantly orange/gray-brown and the belly is mixed orange and blackish.

Similar species: Goeldi's Monkey (*p. 216*) is superficially similar but is slightly smaller and much darker, with a proportionately shorter and bushier tail.

Habitat: Amazonian lowland seasonally flooded forest, dry forests in Colombia, remnant forests or forest patches and secondary forest. Has been observed in seasonal dry forest and spiny-leaved scrub. In Ecuador found in tropical and subtropical humid forests at 200–1,300 m, although most commonly below 400 m.

Distribution: Colombia south of the Caquetá, Caguan and Orteguaza Rivers, south into N Ecuador and N Peru to the east of the Andes. ssp. *nigricollis* occurs in Colombia but the range is unclear; it also occurs east from N Peru and E Ecuador to the Jiparaná River in Rondônia, Brazil. ssp. *graellsi* occurs in the upper Amazon of NE Peru, E Ecuador and SE Colombia. ssp. *hernandezi* occurs in E Colombia between the Caquetá, Caguan and Orteguaza Rivers, and the base of the Cordillera Oriental, Intendencia de Caquetá.

ssp. *graellsi*

LC Lesson's Saddle-back Tamarin — *Leontocebus fuscus*

Description: Dark-faced, with short gray hairs around the sides of the nose and mouth. The crown, forehead and cheeks are black, while the sides of the head beneath the ears are reddish-brown to blackish-brown. The mantle, arms, rump and thighs are orangey-brown and streaked with black, and the saddle on the lower back has black and buff mottling. The underparts, including the inner parts of the limbs, are rufous with dark brown or blackish hairs. The hands and feet are black mixed with orange, and the tail is black with the upper third of the under-tail rufous.

Similar species: None in range.

Habitat: Dense vegetation in both primary and secondary lowland forest up to 500 m above sea level.

Distribution: Amazonas State in NW Brazil and extending to SE Colombia. Occurs north of the Solimões River between the Japurá–Caquetá and Ica–Putumayo Rivers. The eastern limit of its range in Brazil is the Tocantin River. In Colombia it is found north of the middle Caquetá River, west of the Yarí River through the basin of the Caguán River, and lower parts of the Orteguaza River, west to the Andean foothills. Occurs north to the south bank of the Guayabero River and possibly east to the area around San José de Guaviare, on the southern bank of the Guaviare River.

HB:	22.3–22.9 cm
Tail:	34.6–35.0 cm
Wt:	350–400 g

NT Golden-mantled Saddle-back Tamarin *Leontocebus tripartitus*

Description: Small, with a uniformly black head, throat and neck. The nose and sometimes the whole face are white, and there is a broad bright yellow to gold collar extending from the neck to the upper back, chest and forelimbs. The remainder of the upperparts is mottled gray and orange, and the underparts are orange. The tail is largely black, although the underside of the upper third is yellowish-orange. The feet and hands are orange with blackish tints.

Similar species: Red-mantled Saddle-back Tamarin does not have a distinctive yellow-gold collar.

HB:	22–26 cm
Tail:	32–34 cm
Wt:	approx. 400 g

Habitat: Lowland Amazon tropical humid forest from 187 m to 330 m. Found in the middle to lower strata in terra firme, seasonally flooded and riparian forest.

Distribution: E Ecuador and N Peru between the south bank of the Napo River in Ecuador and the Putumayo River in Peru.

Red-mantled Saddle-back Tamarin   *Leontocebus lagonotus*

Description: Small, with a black to blackish-brown neck, shoulders, forelimbs and tail. The back from the back of the shoulders to the hips is blackish, mottled with yellow to orange tints, and the hind limbs are dark brown to reddish-brown or even bright reddish, although they can also be of the same color as the forelimbs. The underparts are dark reddish. The head is black, with a white, short-haired nose.

Similar species: Golden-mantled Saddle-back Tamarin has a distinctive yellow-gold collar.

HB:	17.5–27.0 cm
Tail:	25–32 cm
Wt:	330–400 g

Habitat: Eastern foothills of the Andes and adjacent Amazonian lowlands in the mid-lower strata of tropical and upper tropical forest at 200–1,800 m. Primary and secondary forest, but appears to prefer the transition between terra firme and seasonally flooded forests (Tirira 2017).

Distribution: E Ecuador and N Peru as far south as the Amazon/Marañón Rivers.

LC Andean Saddle-back Tamarin

Leontocebus leucogenys

Description: The head, mantle, forelimbs and throat are predominantly black or blackish-brown, but can occasionally be buffy gray-brown. The saddle is mottled black and buff, the rump is reddish-brown, and the tail is black with a reddish-brown base. The underparts other than the throat are reddish-brown, washed with black. Normally has a dark patch on the front of the orangey-brown thighs. The outer sides of the arms are similar in color to the forelimbs, but are striated with orange. The undersides of the hands and feet are black. The face is black, with short whitish-gray hairs around the mouth and nose.

HB:	20.5–23.0 cm
Tail:	30.5–33.0 cm
Wt:	350–400 g

Similar species: None in range.

Habitat: Primary, secondary and riparian forests, forest edge including around plantations, and human habitations along the eastern base of the Andes at up to 900–1,000 m.

Distribution: North-central Peru south of the Marañón River, from San Martín, through Huánuco and Pasco to the Perené River, northern Juno, west of the Huallaga River and upper Ucayali River in Loreto to as far north as the Pisqui River.

Illiger's Saddle-back Tamarin

Leontocebus illigeri

Description: The facial skin is black, with short pale hairs around the mouth and nose, and the forehead and crown are black. The relatively short mantle and outer-surfaces of the arms are chestnut. The saddle is black, mottled with gray or buff. The rump and thighs are reddish-orange and the base of the tail is likewise reddish-orange, the remainder of the tail being black. The upper sides of the hands and feet are black, mixed in with orange. The throat is blackish, and the rest of the underparts, including the inner surfaces of the limbs, are reddish.

HB:	approx. 20 cm
Tail:	approx. 31 cm
Wt:	(mean) 294 g

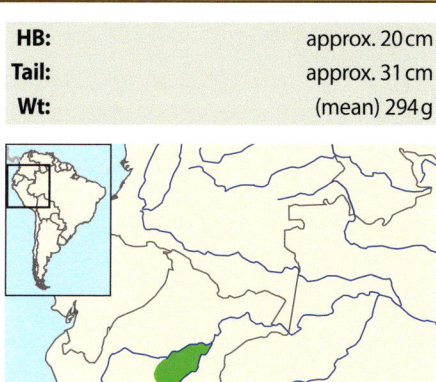

Similar species: None in range.

Habitat: Primary and secondary lowland rainforest, including seasonally flooded forests.

Distribution: Loreto, in E Peru between the lower Huallaga and Ucayali rivers, from the south bank of the Marañón River south to the Caxiabatay River or possibly to the Pisqui River.

LC Geoffroy's Saddle-back Tamarin *Leontocebus nigrifrons*

Description: Black-faced, with short pale hairs around the mouth and nose. The forehead, sides of the head and neck are black. The crown, nape and mantle are orange or buffy gray-brown with faint black striations. The outer surfaces of the upper arms are darker than the mantle, and the forearms and inner surfaces of the arms are blackish. The saddle is mottled or streaked black with gray or buff. The rump and thighs are of a similar color to the mantle, and the tail is black with a rufous base. The chest appears dark brown to black, but is intermixed with a few orange or reddish hairs. The hands are orange, with blackish hairs on the upper surfaces.

Similar species: None in range.

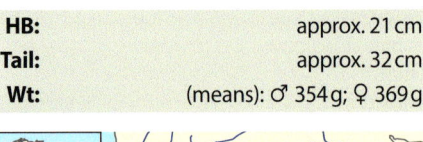

HB:	approx. 21 cm
Tail:	approx. 32 cm
Wt:	(means): ♂ 354 g; ♀ 369 g

Habitat: Found in tall secondary forest, including forest-edge habitats.

Distribution: Restricted to NE Peru south of the Marañón River between the Amazon and Yavarí Rivers, and along the east bank of the Ucayali River as far as the Blanco River, probably extending east as far as the Yavarí River on the Brazilian border, and south as far as the east bank of the Ucayali River opposite Pucallpa.

LC Cruz-Lima's Saddle-back Tamarin

Leontocebus cruzlimai

Description: Distinctive, with the typical tri-banded appearance of saddle-back tamarins (black, white/beige, and bright rusty reddish-orange) reasonably well defined, and extending from the crown to the saddle. The crown, mantle, underparts, arms and legs are bright orange-rusty red, the upper hands and feet tending to be more blackish than red. The saddle is mottled gray and buff and the tail is blackish. Has grayish-white hairs around the mouth, and the forehead has a distinctive whitish transverse band extending around the eyes. It has been recorded as traveling in mixed troops with Red-cap Mustached Tamarin (*p. 252*).

Similar species: None in range, but see **Weddell's Saddle-back Tamarin** (*p. 249*), which is much darker.

HB:	(mean) 25.1 cm
Tail:	(mean) 29.6 cm
Wt:	(mean) 390 g

Habitat: Encountered mainly in terra firme forest, although it has also been recorded in white-water inundated forest. Both forest types contain patches of vegetation dominated by bamboo *Guadua* sp.

Distribution: Restricted to western Amazonian Brazil, where surveys in the Purús National Forest have found it along the north bank of the Inauini River, a tributary of the middle Purús River, and also on the north bank of the Purús River north and south of the Inauini River. It may extend as far north as the Pauini River, with ssp. *primitivus* of Spix's Saddle-back Tamarin (*p. 248*) occurring to the north of the Pauini.

Spix's Saddle-back Tamarin

Leontocebus fuscicollis

Description: Highly variable, with black, brown or red head, arms, legs and upper and lower back (separated by its mottled saddle). The tail is reddish or brown at the base, the remainder being blackish or buff gray-brown. The forehead and crown are blackish or buff gray-brown, the nape being buff gray-brown. The face is black and has short gray hairs around the mouth and nose. Four subspecies are recognized.

ssp. *fuscicollis* generally has an orange-buff crown and forehead. The cheeks and face are black, giving it a masked appearance. The mantle and arms are dark buffy gray-brown to blackish-brown. The broad black saddle extends across the middle and lower back and is mottled with buff or orange. The tail is black apart from the short rufous base. The throat, neck and chest are dark brown or reddish-brown.

ssp. *avilapiresi* is largely dark brown, with the forehead, crown, mantle, rump, thighs and

HB:	21 cm
Tail:	32 cm
Wt:	(means): ♂ 341 g; ♀ 355 g

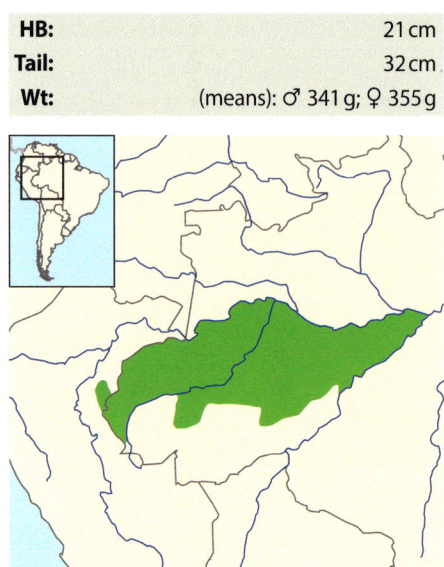

upper arms blackish-brown with orange hairs, except on the rump and thighs. The extensive saddle is mottled. The lower arms and legs are darker than the upper arms and thighs. The tail is black apart from the base, which is brown like the rump. The underparts are dark brown with some orange striations.

ssp. *mura* has a dark brown forehead with a few grayish hairs and does not have a white band. The broad saddle is dark brown or black with ochraceous mottling. The arms are black.

ssp. *primitivus* has a buffy gray-brown crown, mantle, limbs, rump and chest. There is a distinct grayish band across the forehead which is separated from the crown by a narrow black line. The mottled saddle is poorly defined. The chin is grayish and the throat and chest are largely black. The tail is buffy gray-brown at the base and blackish towards the tip.

Similar species: None in range.

Habitat: Amazonian lowland and lower montane rainforests, seasonally flooded forest, remnant forest patches and secondary forest, from 100 m to 1,200 m.

ssp. *fuscicollis*

Distribution: Three of the subspecies are found only in western Amazonian Brazil, while SSP. *fuscicollis* occurs also in E Peru.

SSP. *fuscicollis*: South of the Solimões River in Brazil, from the Javarí River in the west eastwards through the Jutaí River Basin to the west bank of the Juruá River. Occurs also in Peru, west of the Yavarí River as far as the Tapiche River, extending north from there as far as the west bank of the Blanco River.

SSP. *avilapiresi*: South of the Solimões River between the Juruá and Purús Rivers, including the basins of the Urucu and Coarí Rivers, and probably the Tefé River. Found also at Jaraqui,

on the east bank of the Juruá. The southern limits are not known.

SSP. *mura*: Restricted to the interfluvium of the Madeira River (in the east) and the Purús River (in the west). Occurs south to the Igapó-Açú River. The northern boundary may be delimited by the inundated forests (várzea) of the Amazon River.

L. f. primitivus: Known from Pauiní, below the mouth of the Pauiní River on the Purús River, and from an unspecified locality on the Juruá River. The range extends west between the Pauiní and Tapauá Rivers to the east bank of the Juruá and Tarauacá Rivers.

Weddell's Saddle-back Tamarin *Leontocebus weddelli*

SSP. *weddelli*

HB:	18–27 cm
Tail:	25–38 cm
Wt:	340–440 g

Description: Two distinct subspecies are recognized.

SSP. *weddelli* has a black or blackish-brown crown, sides of the head, mantle, throat, neck, chest, arms and upper surfaces of hands and feet. On many individuals the mantle is sparsely striated with orange. The saddle is mottled black and buff or gray. The rump, thighs and belly are reddish-orange. The tail is

Continued on next page…

Weddell's Saddle-back Tamarin (*continued*)

black, apart from a short rufous patch at its base. The inner arms are black, intermixed with orange or reddish. The forehead has a distinctive crescent-shaped transverse white band. The face is black, with short pale hairs around the mouth and nose.

SSP. *melanoleucus* is almost totally creamy white, often with a yellowish-buff wash or streaking except for the black facial skin, back of the ears and genitalia. The underparts, including the inner arms, are whitish or yellowish. It has short gray hairs around the mouth and nose.

Similar species: None in range.

Habitat: Tall forest with both sparse and dense understory and secondary forests with or without closed canopies.

Distribution: The south-western Amazon Basin in Brazil, between the Madeira–Mamore and Purús Rivers in the states of Amazonas, Rondônia and Acre, into Bolivia in the departments of Pando, northern La Paz, and Beni to the Mamore River. Extends into Peru, in the departments of Madre de Dios and Puno into the upper Ucayali Basin in Loreto and Cusco.

SSP. *weddelli*

Spix's Mustached Tamarin

Tamarinus mystax

Description: Head blackish; the skin around the cheeks and nostrils is largely unpigmented and covered with comparatively long white hairs, and the whiskers are well developed. The remainder of face is covered with black hairs. The mantle is blackish-brown, with an orange subterminal band and a black, terminal band to the hairs. The hairs have a white base which can sometimes be seen. The arms are black, and the hairs on the saddle, rump, and outer side of the thighs are blackish with an orange or ochraceous-orange subterminal band. The lateral fringe is like the saddle, although the base of the hairs is drab or whitish. The upper surface of the hands and feet are black. The underparts, including the inner sides of the limbs, are black to blackish-brown. The tail is black, except for the base which is the same color as the rump. There can be individual variation in the tone of the saddle, rump and hindlimbs.

Similar species: None in range.

Habitat: Amazonian lowland, seasonally flooded forest, remnant forest patches and secondary forest, and reported to prefer dense vegetation in secondary forest. Mainly in the upper canopy and often found alongside

HB:	(means): ♂ 24.2 cm; ♀ 24.6 cm
Tail:	(means): ♂ 38.2 cm; ♀ 38.7 cm
Wt:	♂ 380–681 g; ♀ 360–650 g

Geoffroy's Saddle-back Tamarin (*p. 246*) and/or Spix's Saddle-back Tamarin (*p. 248*).

Distribution: Peru and Brazil south of the Solimões–Amazon River, from the west bank of the Juruá River in Brazil as far west as the Ucayali River in Peru. The southern limit of the range is the Blanco River in Peru.

NE Red-cap Mustached Tamarin

Tamarinus pileatus

Description: Two subspecies are recognized. ssp. *pileatus*. The head is largely black with a broad reddish-orange cap covering the forehead and crown. There is a thin reddish-orange line extending down between the eyes. The area around the cheeks and nostrils is covered with comparatively long white hairs that form whiskers. The remainder of the facial skin is mostly dark-pigmented and covered with black hair. The mantle is blackish-brown with fine light brownish-yellow or orange hairs. The saddle and rump are light brownish with a yellow subterminal band to the hairs. The outer sides of the hindlimbs are brown and the upper surface of the hands and feet are blackish. The underparts, including the inner sides of the limbs, are blackish-brown. The tail is black except for the base which is the same color as the rump.

ssp. *pluto* lacks the reddish-orange cap and more closely resembles Spix's Mustached

HB:	(means): ♂ 24.2 cm; ♀ 24.6 cm
Tail:	(means): ♂ 38.2 cm; ♀ 38.7 cm
Wt:	♂ 380–681 g; ♀ 360–650 g

Tamarin (*p. 251*) but has a largely black head apart from the white whiskers, a black mantle with light brownish-yellow flecking, pale back and light brownish-yellow saddle, rump and limbs. The tail is black except for the base which is the same color as the rump.

Similar species: None in range.

Habitat: Amazonian lowland, seasonally flooded forest, remnant forest patches and secondary forest, and reported to prefer dense vegetation in secondary forest. Mainly in the upper canopy and often found alongside Spix's Saddle-back Tamarin (*p. 248*).

Distribution: Brazil. ssp. *pileatus* occurs south of the Solimões–Amazon River. In the north of its range found west to the Tefé River, and east to the Coari River. In the south of its range occurs west to the right bank of the Juruá River and east to the left bank of the Purús River. Also found on the left bank of the Tefé River, around its headwaters. The southern limit of its range is the Pauini River. ssp. *pluto* is found south of the Solimões–Amazon River, from the east bank of the Coari River to the west bank of the Purús River, south to the Tapauá River.

NE Kulina's Mustached Tamarin

Tamarinus kulina

Description: A newly described species, distinguished from all other species of *Tamarinus* tamarins by its light black-brown mantle and forelimbs with a yellow subterminal and black terminal band to the hairs. The base of the hairs is white which can sometimes be seen. The light black-with-brown saddle, rump and outer sides of the thighs with brown subterminal bands to the hairs are also diagnostic. The forehead is blackish and the crown blackish-brown. As with other mustached tamarins, the area around the nostrils and cheeks is covered with comparatively long white hairs and well-developed whiskers. The remainder of the face is covered with black hairs. The upper surface of the hands and feet are black. The underparts, including the inner sides of the limbs, are black to light black-brown. The tail is black except for the base which is the same color as the rump.

HB:	(means): ♂ 24.2 cm; ♀ 24.6 cm
Tail:	(means): ♂ 38.2 cm; ♀ 38.7 cm
Wt:	♂ 380–681 g; ♀ 360–650 g

Similar species: None in range.

Habitat: Amazonian lowland, seasonally flooded forest, remnant forest patches and secondary forest, and reported to prefer dense vegetation in secondary forest. Mainly in the upper canopy and often found alongside Spix's Saddle-back Tamarin (*p. 248*).

Distribution: Restricted to western Amazonian Brazil between the east bank of the Juruá River and the west bank of the Tefé River.

LC **Red-bellied Tamarin** *Tamarinus labiatus*

Description: Usually blackish, with reddish or orange underparts and a paler wash on the hind parts. The nape and mantle are buffy gray-brown. Has a reddish or orange patch on the underside of the base of the tail. Three subspecies are recognized. ssp. *labiatus* has a black crown with a well-marked reddish midline or 'Y'-shaped patch at the front of the crown, a well-defined triangular or diffused silvery patch behind the crown, a largely black throat, and the rest of the underparts reddish except occasionally for the upper part of the chest. ssp. *thomasi* has only a small pale spot behind the crown and has a black throat and upper chest. The rest of the underparts tend to be orange rather than reddish. ssp. *rufiventer* has a reddish crown patch which is 'Y'-shaped at the front, and a diffused silvery or silvery-buff behind the crown.

HB:	21–28 cm
Tail:	30–38 cm
Wt:	(means): ♂ 477 g; ♀ 515 g

Similar species: None in range.

Habitat: Generally in the middle strata of primary and secondary rainforest and semi-deciduous forests; it rarely descends to the understory. Normally avoids seasonally flooded forest, although will enter it when there is no flooding. The species has been encountered in mixed troops with Weddell's Saddle-back Tamarin (*p. 249*) and Goeldi's Monkey (*p. 216*) in NW Bolivia.

Distribution: Central and south-central Amazon of Brazil, Peru and Bolivia. South of the Solimões River in Brazil between the Madeira and Purús Rivers. The southernmost part of the range extends along the Abunã River into the Pando Department of northern Bolivia south as far the Tahuamanu River. In Peru, in the basin of the Acre River, south to the Tahuamanu River. The range of ssp. *thomasi* is well separated from the other two subspecies between the Japurá and Solimões Rivers.

Mottle-face Tamarin

Tamarinus inustus

Description: Black or very dark brown, with a largely unpigmented face except for dark blotches (which give the species its name) and a patch of white skin between the upper lip and the nose. The ears are naked and black, although some individuals have white patches. A mantle of lighter brown hair covers the dorsum, while the limbs and tail have darker pigmentation.

HB:	20.8–25.9 cm
Tail:	33–41 cm
Wt:	430–500 g

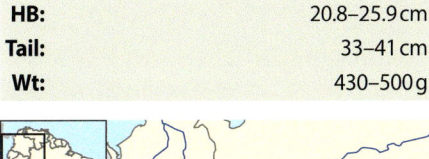

Similar species: None in range.

Habitat: Amazonian lowland, seasonally flooded forest, remnant forest patches and secondary forest. Reported to prefer primary white-sand forest (*campina* and *campinarana*), and successional forest. Often in the vicinity of villages. Forages in the middle and lower strata as low as 0.5 m above the ground.

Distribution: SE Colombia and Amazonian NW Brazil. In Colombia north of the Caquetá River and east of the Yarí River north to the Gayabero and Guaviare Rivers and enters Brazil between the Japurá and Negro Rivers. The eastern limit of its range is in the vicinity of the Padauarí River north of the Negro River.

LC Black-chinned Emperor Tamarin   *Tamarinus imperator*

Description: Readily identified by the extremely long mustache (considerably longer than that of any other mustached tamarin other than the recently split Bearded Emperor Tamarin). Dark buff gray-brown above, with a silvery-brown crown and a largely pale orange tail, although the lower third of the tail is similar in color to the rest of the upperparts. The underparts and inner surfaces of the arms are reddish-brown or orange. The face (apart from the conspicuous white mustache), hands and feet are black.

HB:	23–26 cm
Tail:	35–42 cm
Wt:	400–550 g

Similar species: None in range.

Habitat: Middle and lower canopy of Amazonian lowland and lower montane rainforests, seasonally flooded forest, remnant forest patches and secondary forest. Frequently in riparian forests in parts of its range.

Distribution: Recorded in the Brazilian state of Amazonas between the Purús and Acre rivers. In the state of Acre, specimens are known from Manoel Urbano, the Branco River, and the São Pedro River Basin. A number of surveys have failed to find the species along the right bank of the Acre River despite unsubstantiated reports of the species from locals at two locations along the southern Acre River in Bolivia: Buena Vista and Los Campos. The south-western limit of the species' range remains unclear.

Bearded Emperor Tamarin

Tamarinus subgrisescens

Description: Very similar to Black-chinned Emperor Tamarin but the underparts and inner surfaces of the arms are grizzled brown rather than reddish-brown or orange and the mustache extends down over the lower lip to create a bearded appearance.

Similar species: None in range.

HB:	23–26 cm
Tail:	35–42 cm
Wt:	400–550 g

Habitat: Middle and lower canopy of Amazonian lowland and lower montane rainforests, seasonally flooded forest, remnant forest patches and secondary forest. Frequently in riparian forests in parts of its range.

Distribution: Occurs in the Brazilian states of Acre and Amazonas, the Peruvian departments of Madre de Dios and Ucayali, and the Bolivian department of Pando. In Bolivia found in the Muyumanu River Basin, on the border of Peru, at the south Tahuamanu River, and at sites along the right bank of the Acre River. In Peru recorded from localities along the Madre de Dios and Manu Rivers, at the Curanja River (department of Ucayali), Atalaya, the mouth of Urubamba River, and at Inaperi. In Brazil occurs along the right bank of the upper Juruá River in Acre, and the Envira River in the municipalities of Feijo and Cruzeiro do Sul. The west bank of the Purús River at Pauini in the state of Amazonas appears to be the northern limit of its range.

VU Western Black-handed Tamarin

Saguinus niger

Description: All-black except for the lower back, which has buff mottling. Extremely similar to Midas Tamarin (*p. 260*), with which it was previously lumped, but that species is larger, has smaller ears and has contrasting orange or yellowish-orange, rather than black, upper surfaces of the hands and feet.

Similar species: None in range.

Habitat: Amazonian primary lowland, seasonally flooded forest, sub-montane forest, remnant patches of forest and secondary forest.

Distribution: South of the Amazon River in the state of Pará, in Brazil. From the east bank of the Xingu River to the west bank of Tocantins River, and in the south-western portion of Marajó Island in the Amazon delta. The range extends as far south as the Gradaús and Floresta Nacional de Carajás.

HB:	21–26 cm
Tail:	32–40 cm
Wt:	(means): ♂ 431 g; ♀ 428 g

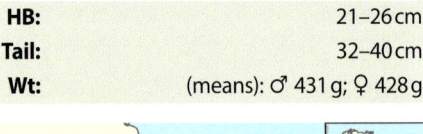

Eastern Black-handed Tamarin

Saguinus ursula

Description: Similar to Western Black-handed Tamarin, but distinct in having mid-dorsal hair with a wide intermediary band of bright golden-buff color, long dorsal hair in the interscapular region, and naked rather than obviously haired face, hands and fingers. The two are not known to overlap in range.

Similar species: None in range.

Habitat: Assumed to be similar to that of Western Black-handed Tamarin.

Distribution: South of the Amazon River in Brazil, from the east bank of the Tocantins River in the state of Pará eastwards to the limits of the Amazon Forest with the Cerrado and Caatinga biomes, in the state of Maranhão.

HB:	21–26 cm
Tail:	32–40 cm
Wt:	(means): ♂ 431 g; ♀ 428 g

LC Midas Tamarin

Saguinus midas

Description: Uniformly black, with buff mottling on the back from the shoulders to the base of the tail. Distinguished from the very similar Western Black-handed Tamarin (*p. 258*), with which it was previously lumped, by its orange or yellowish-orange hands and feet which contrast strongly with the rest of its body. Juveniles have a triangular patch of white on the forehead and whitish cheek patches.

Similar species: None in range.

Habitat: Primary and secondary forest, and also in disturbed forests and at forest edges. Also in savanna forest, swamp forests, white sand forests and in secondary habitats in close proximity to urban areas.

Distribution: Forested areas east of the Essequibo River in Guyana, Suriname and French Guiana, and in Brazil on the north bank of the Amazon River east of the Negro River as far as the coast. It appears to be absent from a large part of central Guyana. Occurs also west of the lower reaches of the

HB:	24.6–25.2 cm
Tail:	38.3–38.6 cm
Wt:	380–500 g

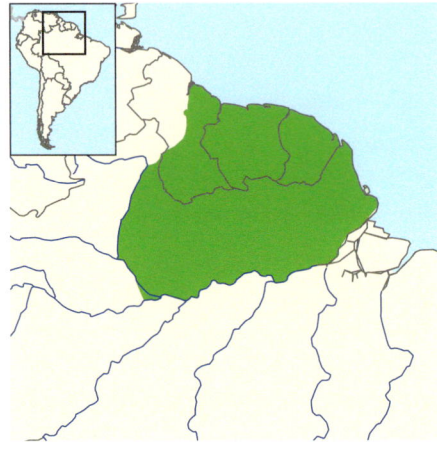

Essequibo River at Hosororo Hill, close to the Venezuelan border.

NT Martins's Bare-faced Tamarin
Saguinus martinsi

Description: Two subspecies are recognized. SSP. *martinsi* is uniformly brown above from the crown to the base of the tail, being slightly paler on the flanks and shoulders, with buff limbs. The underparts are orange and the face is largely naked. SSP. *ochraceus* is paler yellowish-brown above and golden-orange below, with yellowish tones to the ruff and nape and a yellowish-gray nape and base of the mantle.

Similar species: None in range.

Habitat: Lowland tropical primary and secondary forest with a dense understory.

Distribution: Brazil, where SSP. *martinsi* occurs from the west bank of the Nhamundá River east to the Erepecurú River, north of the Amazon River.

SSP. *ochraceus* is thought to occur on the west bank of the Nhamundá River, possibly extending west to the Uatumã River, north of the Amazon.

HB:	21–28 cm
Tail:	34–42 cm
Wt:	400–600 g

SSP. *martinsi* (both photos)

CR Pied Tamarin

Saguinus bicolor

Description: A distinctive tamarin with a bare black head contrasting strongly with the white or yellowish-white throat, chest, crown, neck, shoulders, arms and mantle that are also clearly demarcated from the buffy grayish-brown lower back and rump. The belly and the inner sides of the hind limbs are reddish-orange, and the tail is dark brown above and orange below, normally with a pale tip. Feeds in the forest canopy.

Similar species: None in range.

Habitat: Mature primary forest and young secondary lowland forest, generally on terra firme but also in seasonally flooded forests. Also disturbed forest fragments in urban areas, including Manaus, where in secondary forest.

Distribution: Brazil, where found north of the Amazon River and east of the Negro River, in the vicinity of Manaus. Occurs at up to 30–45 km to the north of Manaus and 100 km to the east as far as the town of Itacoatiara.

HB:	23–33 cm
Tail:	34–42 cm
Wt:	480–600 g

White-footed (Silvery-brown) Tamarin *Oedipomidas leucopus*

Description: Pale silvery gray-brown above, with long whitish tips to the hair creating a streaked appearance. The underparts are rusty-orange. The facial skin appears blackish but is covered by short silvery hair, and the animal has white eyebrows. The cheeks and crown are silvery-brown, and it has long dark brown hairs on the back of the neck. The limbs, including the hands and feet, are whitish and the tail is brown or blackish with a contrasting pale tip. Populations in the southern part of its range are generally darker, with dark stripes on the forearms and legs.

Similar species: None in range.

Habitat: Up to 1,500 m in tropical dry forest, tropical humid forest, and very humid sub-montane forest in the lower and mid strata of both primary and secondary forests, as well as in gallery forest, and in some urban areas. Appears to prefer forest-edge habitats.

Distribution: Colombia, its range bordered by the east bank of the lower Cauca River, the

HB:	22–29 cm
Tail:	35–42 cm
Wt:	approx. 460 g

west bank of the middle Magdalena River and the foot of the Cordillera Central.

CR Cotton-top Tamarin

Oedipomidas oedipus

Description: Readily identified by the large shock of white fur on its head which gives the species its name. The face and ears are black, with a pale gray or whitish supraorbital band and a grayish fringe across the muzzle to each corner of the mouth. The temples and sides of the head are covered with short silvery hairs. There is a wedge-shaped mid-frontal white crest. The dorsal surface of the body is primarily gray-brown, while the underparts of the body, arms and legs are predominantly white. Has variable reddish patches on the thighs and rump, the upper third of the tail is red and the remainder is dark brown.

Similar species: Does not overlap in range with Geoffroy's Tamarin but the ranges of the two species are adjacent: Geoffroy's Tamarin lacks the distinct crest of Cotton-top Tamarin.

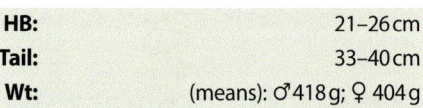

HB:	21–26 cm
Tail:	33–40 cm
Wt:	(means): ♂ 418 g; ♀ 404 g

Habitat: Up to 1,500 m in primary and secondary humid tropical forest and in dry deciduous forest with low seasonal rainfall, scrubland and forest edge. Usually in the middle and lower canopy.

Distribution: Restricted to NW Colombia in the departments of Atlántico, Sucre, Córdoba, western Bolívar, north-west Antioquia, and north-east Chocó. To the south from the Cauca River in the east, extending west to cross the Serranía de Ayapel and the Serranía de San Jeronimo. Introduced populations occur in mangrove swamps near Santa Marta and in the department of Sucre.

Additional photo *p. 239*

Geoffroy's Tamarin

Oedipomidas geoffroyi

Description: The upperparts are black and with buff-white mottling or streaking. The bases and tips of the hairs are blackish. The nape is dark orange-red, and there is a stripe of white fur on the crown of the otherwise dark gray head which contrasts with the pale chest. The chest, underside of the body and outer edge of the upper arms are creamy white. The tail is predominantly black, with a dark red base.

Similar species: Does not overlap in range with **Cotton-top Tamarin** but the ranges of the two species are adjacent: Cotton-top Tamarin has a distinct crest.

HB:	20–29 cm
Tail:	32–42 cm
Wt:	400–680 g

Habitat: Up to 1,000 m in primary and secondary rainforests, in moist seasonal dry forests, and secondary forests and scrub.

Distribution: Colombia along the Pacific coast, south as far as the San Juan River. In the east its range extends to the west of the upper Cauca River and just east of the upper Atrato River. Range extends into E Panama.

265

Lion Tamarins

FAMILY | **Callitrichidae**

Lion tamarins are small monkeys with a lion-like mane around their face. They are unlikely to be mistaken for any other species of monkey. All four species are Endangered and endemic to Brazil. They are diurnal, and forage for food in groups of 2–11 individuals in the lower and middle canopy. They feed on fruits, flowers, nectar, gums and animal prey, particularly spiders and insects, but also including frogs, lizards and snails, and birds' eggs.

EN Golden Lion Tamarin

Leontopithecus rosalia

Description: Distinctive golden-orange, with a naked pale purplish-black face surrounded by a golden-orange mane. Sometimes shows orange-yellow, brown or black coloration on the tail and front feet.

HB:	26–33 cm
Tail:	32–40 cm
Wt:	710–795 g

Similar species: None in range.

Habitat: Primary and secondary Atlantic lowland rainforest, including degraded and secondary forests with sufficient year-round food, and holes in trees which it uses as sleeping sites.

Distribution: Restricted to Rio de Janeiro State in SE Brazil, where it is found in a small area (104.5 km²) of forests in the Centro Hípico de Cabo Frio, Campos Novos, the Poço das Antas Biological Reserve and adjacent forests in the Serra do Mar, and a further ten isolated localities.

Golden-headed Lion Tamarin *Leontopithecus chrysomelas*

Description: Largely glossy black, with a bare purplish-black face surrounded by a striking reddish-orange and golden-red mane. The limbs and the upper side of the base of the tail are yellowish-orange to golden-orange.

Similar species: None in range.

Habitat: Primary and secondary Atlantic Forest, restinga, coastal white-sand forest, and cabruca-cacao plantations interspersed with native trees. Found also in secondary-growth forest in abandoned rubber plantations, but also requires old-growth forest with tree holes for sleeping and epiphytic bromeliads, which are crucial feeding sites.

Distribution: Forest fragments in the south-east of Bahia, in NE Brazil. Introduced into the suburbs of Niteroi, in Rio de Janeiro State.

HB:	22–26 cm
Tail:	33–39 cm
Wt:	480–700 g

EN Black Lion Tamarin

Leontopithecus chrysopygus

Description: Mainly glossy black, with varying amounts of yellowish-brown to golden-red fur on the rump, thighs and base of the tail. The bare blackish face is surrounded by a long black mane. Males are larger than females.

Similar species: None in range.

Habitat: Inland Atlantic Forest at up to 900 m, normally foraging in the lower and middle layers of the canopy, and prefers swampy forests. Found in forests lacking the bromeliads preferred by other lion tamarins.

Distribution: Restricted to 11 widely separated forest patches in São Paulo State, in S Brazil.

HB:	25–30 cm
Tail:	36–41 cm
Wt:	540–690 g

Black-faced Lion Tamarin
Leontopithecus caissara

Description: Golden-orange, with a bare blackish face, and a black mane, chest, forearms, feet and tail.

HB:	approx. 34 cm
Tail:	approx. 40 cm
Wt:	540–710 g

Similar species: None in range.

Habitat: Mature, lowland, coastal rainforest, where it forages largely in swampy and inundated forests but also in terra firme forest and secondary forest.

Distribution: Restricted to an area of 300 km² on Superagui Island, and in parts of the valleys of the Sebuí and Patos Rivers, in the southern Brazilian states of Paraná and São Paulo.

Squirrel Monkeys

FAMILY | **Cebidae**

IUCN Primate Specialist Group recognizes nine species of squirrel monkey, eight of which occur in South America. Squirrel monkeys are small, graceful and slender monkeys that feed in groups of from ten to more than 100 individuals. They are extremely active and feed in the lower and middle strata of forest, occasionally descending to the ground. They feed primarily on insects and ripe fruit and less commonly on flowers and nectar. They are often encountered in association with capuchins (*p. 278*).

LC Humboldt's Squirrel Monkey

Saimiri cassiquiarensis

HB:	♂ 25–37 cm; ♀ 28–34 cm
Tail:	♂ 38–45 cm; ♀ 33–44 cm
Wt:	♂ 590 g–1.2 kg; ♀ 650 g–1.38 kg

Description: Short-furred, with olive-gray to golden-olive upperparts and thighs with black tints, and buff to orange underparts. The arms, hands and feet are bright golden to yellowish-orange. The head is rounded and the face blackish, with a white ring around each eye. The chin, throat and sides of the neck behind the ears are white, and the pointed ears are white and hairy. The crown and nape are gray to black. Females have a darker head and cheeks. The non-prehensile tail is olive-yellow, with a blackish upper side and a paler underside towards the tip.

Similar species: None in range.

Habitat: Humid tropical and upper tropical forest in the Amazonian lowlands and foothills from 200 m to 1,200 m, but normally below 500 m. Primary, secondary and disturbed forests, but most common in inundated

lowland forest and scarce in terra firme forest, hill forest and forest well away from water.

Distribution: Found in the upper Amazon and Orinoco Basins of Brazil, Colombia and Venezuela.

VU Colombian Squirrel Monkey *Saimiri albigena*

Description: Very similar to Humboldt's Squirrel Monkey. Short-furred, with olive-gray to golden-olive upperparts and thighs with black tints, and buff to orange underparts. The back can be orange-toned. The forearms and hands are grizzled gray. The head is rounded and the face blackish, with a white ring around each eye. The chin, throat and sides of the neck behind the ears are white, and the pointed ears are white and hairy. The crown and nape are gray to black. Females have a darker head and cheeks than males. The non-prehensile tail is olive-yellow, with a blackish upper side and a paler underside towards the tip.

HB:	♂ 25–37 cm; ♀ 28–34 cm
Tail:	♂ 38–45 cm; ♀ 33–44 cm
Wt:	♂ 590 g–1.2 kg; ♀ 650 g–1.38 kg

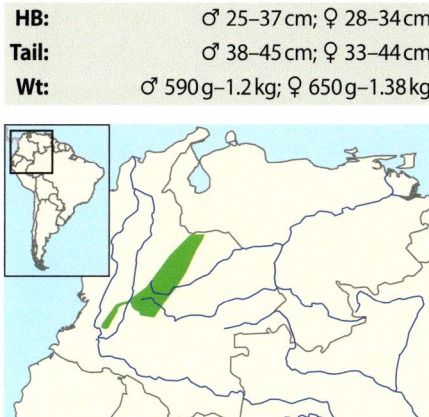

Similar species: None in range.

Habitat: Humid tropical and upper tropical forest in the Amazonian lowlands and foothills from 200 m to 1,200 m, but normally below 500 m. Primary, secondary and disturbed forests, but most common in inundated lowland forest and scarce in terra firme forest, hill forest and forest well away from water.

Distribution: Central Colombia between the east slopes of the Cordillera Oriental and the eastern Llanos.

MALE

271

LC **Ecuadorian Squirrel Monkey** *Saimiri macrodon*

HB:	♂ 25–37 cm; ♀ 28–34 cm
Tail:	♂ 38–45 cm; ♀ 33–44 cm
Wt:	♂ 590 g–1.2 kg; ♀ 650 g–1.38 kg

Description: Very similar to Humboldt's Squirrel Monkey (*p. 270*). Short-furred, with olive-gray to golden-olive upperparts with black tints. The underparts are yellowish-white. The arms, hands and feet are bright gold to yellowish orange. The head is rounded and the face blackish, with a white mask surrounding the eyes. The chin, throat and sides of the neck behind the ears are white, and the pointebld ears are white and hairy. The crown and nape are gray to black. Females tend to have a darker head and cheeks than males. The non-prehensile tail is olive-yellow, with a blackish upper side and a paler underside towards the tip.

Similar species: Black-capped Squirrel Monkey (*p. 276*) appears much darker.

Habitat: Humid tropical and upper tropical forest in the Amazonian lowlands and foothills from 200 m to 1,200 m, but normally below 500 m. Primary, secondary and disturbed forests, but most common in inundated lowland forest and scarce in terra firme forest, hill forest and forest well away from water.

Distribution: Brazil in the state of Amazonas between the Juruá and Japurá Rivers, in

Colombia from the Apaporis River south into E Ecuador, throughout the Ecuadorian Amazon east of the Andes to 1,200 m, and into the departments of San Martín and Loreto in Peru to the north bank of the Marañón-Amazon rivers.

Golden-backed Squirrel Monkey

Saimiri ustus

Description: The crown is dark gray-brown, females having some black markings in and bordering the crown. The back is golden, the sides of the thighs grayish, and the forearms, hands and feet are yellowish or orange. Animals in the south of their range have bright orange forearms, while those from the lower Tapajós River have a yellowish crown and forearms.

HB:	23–43 cm
Tail:	31–45 cm
Wt:	620 g–1.2 kg

Similar species: Guianan Squirrel Monkey (*p. 274*), with which it has been recorded hybridizing, and **Collins's Squirrel Monkey** (*p. 275*) do not have an obvious golden back. Both those species and Golden-backed Squirrel Monkey have a high pointed white arch over each eye, while **Black-capped** (*p. 276*) and **Black-headed** (*p. 277*) **Squirrel Monkeys** have lower, more rounded arches.

Habitat: Seasonally inundated forests, river-edge forest, floodplain and secondary forests, but will also use terra firme forest when food is scarce.

Distribution: South of the Amazon River in the Brazilian states of Amazonas, Pará, Mato Grosso and Rondônia. Occurs from the Tefe River east to the Tapajós River and south to the Guaporé River and the headwaters of the Juruena River.

273

LC Guianan Squirrel Monkey

Saimiri sciureus

Description: Grayish or greenish to reddish-agouti with a grayish-agouti crown, females sometimes exhibiting a blackish border. High white arches over the eyes and white ear-tufts. The face is pink, with black around the muzzle. Yellowish-orange forearms, hands and feet and white underparts. Mercês *et al.* (2015) found considerable variation in a study of specimens with differing amounts of black in the shoulder region and some melanism in the medial region of the back: the hip and saddle were predominantly pure chestnut, or, in some individuals, blackish-chestnut or with patches of black and light chestnut-brown; the hands and forelimbs ranged from burnt orange to reddish, the forearms with some black hairs; the hind limbs varied in color from burnt yellow to reddish, and this coloration extended from the foot past the ankles. Animals forage and travel mainly in the lower canopy and understory.

HB:	25–37 cm
Tail:	36–47 cm
Wt:	550 g–1.4 kg

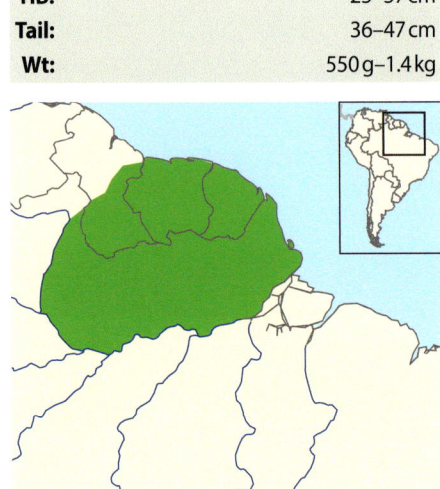

Similar species: Golden-backed Squirrel Monkey (*p. 273*), with which it has been recorded hybridizing, has a conspicuous golden back. **Collins's Squirrel Monkey** – see *opposite* for differences from that species.

Habitat: A wide range of forested habitats below 100 m, including both primary and secondary forests, seasonally inundated forests, river-edge and gallery forests, and swamp forests and mangroves.

Distribution: Widespread across the north-eastern and eastern Amazon region north of the Amazon River, extending from the Negro and Branco Rivers to Amapá, Brazil, and reaching the forests of Guyana, French Guiana and Suriname. Specimens have been collected from the region of Juruti, immediately south of Amazon River [not mapped], while specimens of Collins's Squirrel Monkey have been collected near Faro, to the north of the Amazon River, suggesting that these are areas where both species occur.

LC Collins's Squirrel Monkey

Saimiri collinsi

Description: Split from Guianan Squirrel Monkey on the basis of crown color, morphometric measurements and molecular studies (Mercês *et al.,* 2015). Closely resembles that species, and distinguished in the field by its yellow rather than gray crown and the slightly yellower back. The head and the upper back are grayish. The shoulders are orange-yellow, washed with gray. The orange coloration is brightest in specimens from the west of its range. The hip and saddle are most commonly orange-chestnut with black patches or, less frequently, pure chestnut-brown. The yellow on the hands and feet extends beyond the wrists and the ankles, although Mittermeier *et al.* (2013) described the hands and feet as being dark, rich tawny.

Similar species: Golden-backed Squirrel Monkey (*p. 273*) has an obvious golden back.

Habitat: Primary and secondary forest, seasonally inundated forests, river-edge and gallery forests, and swamp forests and mangroves.

Distribution: South of the Amazon River in Brazil, from the region of the Tapajós River to the state of Maranhão, and in the Marajó

HB:	approx. 25 cm
Tail:	approx. 41 cm
Wt:	550 g–1.4 kg

Archipelago. There are also two specimens from the Faro region, on the north bank of the Amazon River [not mapped], suggesting that there are areas where this species and Guianan Squirrel Monkey both occur.

LC Black-capped (Bolivian) Squirrel Monkey *Saimiri boliviensis*

Description: Males and females both have a blackish crown and golden-yellow forearms, hands and feet but, while males tend to be gray, females are blackish. Two subspecies are recognized. Males of ssp. *peruviensis* have a grayer crown than the females. ssp. *boliviensis* tends to be darker than ssp. *peruviensis* and with a darker blackish tail, while the latter has a grayer tail with a black tip. Both subspecies have rounded white arches over the eyes, including narrow white brows.

Similar species: Much darker-crowned than other squirrel monkeys in its range.

Habitat: Humid forest, including terra firme forest. Most frequently found along the edges of rivers and in seasonally flooded forests, where it feeds in both the upper canopy of tall trees and emergent vegetation.

Distribution: ssp. *boliviensis*: From 50 m to 500 m in the upper Brazilian Amazon, south of the Solimões–Amazon Rivers, between the Juruá and Tefe Rivers in Amazonas and Acre, south into N Bolivia west of the Guaporé River, and SE Peru south of the Abujao River.

HB:	♂ 27.3–32.0 cm; ♀ 26.5–28.5 cm
Tail:	♂ 37.7–43.5 cm; ♀ 37.4–43.5 cm
Wt:	700 g–1 kg

ssp. *peruviensis*: From 90 m to 800 m in Peru south of the Amazon and Marañón Rivers, between the Huallaga and Ucayali Rivers. It is sympatric with Guianan Squirrel Monkey (*p. 274*) east of the Ucayali River from about 04°S southward to the Abujao River.

ssp. *boliviensis*

Black-headed Squirrel Monkey

Saimiri vanzolinii

Description: Both sexes are small, with a black crown and short dense fur. The back is dark, with a broad black band from the crown to the tip of the tail. The forearms, hands and feet are orange-yellow and the shoulders are gray. The pale arches over the eyes are low and rounded. This is the smallest and darkest of the squirrel monkeys, usually with an all-dark tail.

HB:	♂ 26–30 cm; ♀ 22–26 cm
Tail:	♂ 35–40 cm; ♀ 37–41 cm
Wt:	♂ approx. 950 g; ♀ approx. 650 g

Similar species: None in range.

Habitat: Seasonally flooded forests (várzea), preferring swampy ground with low and shrubby vegetation inundated for 6–8 months of the year (chavascal). Enters várzea mainly when certain trees are in fruit.

Distribution: Amazonian Brazil, where confined to the Mamirauá State Sustainable Development Reserve in an area bounded by the Paraná de Jarauá and Paraná do Aiucá in the west and the Japurá River in the east.

Capuchins

The IUCN Primate Specialist Group recognizes 24 species of capuchin, 23 of which occur in South America. Black-horned Capuchin has been split into two, Northern Black-horned Capuchin *Sapajus nigritus* and Southern Black-horned Capuchin *S. cucullatus*, in line with Mittermeier *et al*. (2022). Martins-Junior *et al*. (2018) proposed that Large-headed Capuchin *S. macrocephalus* and Hooded Capuchin *S. cay* should be treated as junior synonyms of Guianan Brown Capuchin *S. apella*. The former is now generally lumped with that species, but the latter remains split pending further studies. This treatment is followed here.

Capuchins are medium to large-sized monkeys that have a prehensile tail. They feed on a wide variety of fruits, seeds and arthropods, frogs, nestlings and even small mammals, supplemented by stems, flowers and leaves. They are largely arboreal and occur at all levels, but are most commonly found in the lower to middle canopy and the understory. They will also descend to the ground to feed. Groups generally range in size from 6 to 20 individuals, with the numbers of females exceeding the numbers of males, although groups of up to 35 have been recorded for some species.

Colombian White-fronted Capuchin

Northern Black-horned Capuchin   *Sapajus nigritus*

Description: A large capuchin with conspicuous horn-like tufts of fur on each side of the head. There is a dark triangular patch on the forehead with the point directed towards the front and reaching the base of the nose. The beard is long and thick. The fur is long and silky. There is considerable geographical variation in coloration, but normally dark brown or black above, with a shiny black crown which contrasts with a paler face, and a black nape. The dorsal stripe is inconspicuous. The flanks are blackish-brown. The shoulders and the fronts of the forelimbs are black, the throat is yellowish-white, the remainder of the underparts reddish, and the outer arms are blackish-brown. The hands and feet are black, contrasting slightly with the wrists. The upper surfaces of the limbs are blackish-brown. There is a contrast between the black upper surface and the blackish-brown lower surface of two-thirds of the tail.

HB:	42–56 cm
Tail:	43–56 cm
Wt:	2.3–4.8 kg

Continued on next page...

MALE

Northern Black-horned Capuchin (*continued*)

Similar species: Southern Black-horned Capuchin tends to have fused horns, a completely black forehead rather than a dark triangular patch, blackish-brown rather than black shoulders and the fronts to the forelimbs, and a blackish-brown throat and reddish underparts rather than a yellowish-white throat, and reddish underparts. **Crested Capuchin** (*p. 283*) has distinctive very dark reddish-brown or blackish long, silky fur. It has a yellowish-brown nape. The underparts are more reddish-brown than the upperparts. The face is dark gray, with black patches in front of the ears that connect below the chin, and it has a conical-shaped blackish-brown crown with two tufts which converge into a single conspicuous central tuft. A black strip in front of the ears connects with the cap. The outer arms and hind limbs are reddish-brown rather than blackish-brown.

Habitat: Lowland, sub-montane, montane tropical and subtropical forests, including gallery and secondary forests, in the Atlantic Forest.

Distribution: Groves (2001) described the distribution as Atlantic Forest in a coastal strip in Brazil from about 20°S to the Rio de Janeiro District. Now believed to extend south into the state of São Paulo although it remains unclear whether Northern or Southern Black-horned Capuchin occurs in the southern Brazilian highlands. More information is needed to delimit the distributions of the two species. To the north, Northern Black-horned Capuchin occurs between the Piracicaba River and the Santo Antônio River. North of here there is evidence of hybridization with Crested Capuchin between the Santo Antônio and Suaçui Grande Rivers, and also in Minas Gerais.

FEMALE

Southern Black-horned Capuchin   *Sapajus cucullatus*

Description: A large capuchin with conspicuous horn-like tufts of fur on each side of the head. The horns appear fused with the hairs projecting to the front and sides. The forehead is completely black and linked to the crown. The beard is long and thick. The fur is long and silky. There is considerable geographical variation in coloration, but normally dark brown or black above, with a shiny black crown which contrasts with a paler face, and a black nape. The dorsal stripe is inconspicuous. The flanks are blackish-brown. The hands and feet are black, contrasting slightly with the wrists and ankles. The upper surfaces of the limbs are blackish-brown. The shoulders and the fronts of the forelimbs are blackish-brown, the throat and the rest of the underparts are blackish-brown, and the outer arms are black. There is a contrast between the black upper surface and the blackish-brown lower surface of two-thirds of the tail.

Similar species: Northern Black-horned Capuchin (*p. 279*) tends to have upright horns, a dark triangular patch on the forehead, black shoulders and the fronts to the forelimbs, and a yellowish-white throat, and reddish underparts.

Habitat: Lowland, sub-montane, montane tropical and subtropical forests in the Atlantic Forest, including gallery and secondary forests.

HB:	42–56 cm
Tail:	43–56 cm
Wt:	2.3–4.8 kg

Distribution: Occurs in the Tietê River valley, along the banks of the Paraná River in São Paulo State and along the Serra do Mar from São Paulo through northern Santa Catarina State in Brazil. Occurs also in S Brazil and into Misiones Province in N Argentina. It remains unclear whether Northern or Southern Black-horned Capuchin occurs in the southern Brazilian highlands.

FEMALE

 Hooded (Azara's) Capuchin

Sapajus cay

HB:	39.5–45.0 cm
Tail:	41–47 cm
Wt:	♂ 1.35–4.80 kg; ♀ 1.76–3.40 kg

Description: Small and relatively short-limbed, short-furred and usually quite pale, with a prominent dark dorsal stripe. The crown is blackish-brown with two very small horn-like ear-tufts. The face has a pale rounded arc with no dividing line on the forehead. It has a small whitish or pale orange-yellow beard, and a black line runs along the sides of the face from the chin to the ears. The eyes, nose and mouth are surrounded with white hairs. The upperparts, including the flanks, shoulders, forearms and outer thighs, are pale grayish-brown with a diffuse dorsal line, and the nape and most of the tail are warmer brown or burnt yellow. The underparts, including the inner arms and thighs, are orange-yellow, being darkest on the throat and chest and paler on the abdomen. The lower limbs, including the forearms and feet, are blackish. The upper two-thirds of the tail are dark warm yellow above and yellow-brown below, and the remainder of the tail is blackish-brown.

Similar species: None in range.

Habitat: Subtropical humid and semi-deciduous forest in the Yungas, the dry deciduous forests of the northern Bolivian Chaco, eastern Paraguay (except for the Paraguayan Chaco) and the Pantanal.

Distribution: Mato Grosso, Mato Grosso do Sul, and extreme SE Goiás in Brazil into Paraguay, east of the Paraguay River. A separate population in SE Bolivia and N Argentina, including the provinces of Salta, Formosa, Chaco and in the extreme south-east of Jujuy, although absent from the Chaco region of the provinces of Formosa and Chaco.

EN Crested Capuchin

Sapajus robustus

Description: Distinctive, with very dark reddish-brown or blackish long, silky fur. Has a yellowish-brown nape and some individuals have a faint dorsal stripe. The underparts are more reddish-brown than the upperparts. The face is dark gray, with black patches in front of the ears that connect below the chin, and it has a long, thick beard and a conical-shaped blackish-brown crown with two tufts which converge into a single conspicuous central tuft. A black strip in front of the ears connects with the cap. The outer arms and hind limbs are reddish-brown and darker towards the wrists, the hands and feet are black, and the tail is black.

Similar species: Northern Black-horned Capuchin (*p. 279*) is normally dark brown or black above, has separate upright ear tufts (not converging), lacks a black strip in front of the ears, and has blackish-brown outer arms and hind limbs.

Habitat: Tropical lowland and sub-montane Atlantic Forest, and in semi-deciduous forest patches in Minas Gerais.

Distribution: Brazil from southern Bahia and northern Minas Gerais, south through Espírito Santo to the Suaçuí Grande River

HB:	♂ 42–56 cm; ♀ 33–44 cm
Tail:	43–56 cm
Wt:	1.3–4.8 kg

and the Serra do Espinhaço in the south-west, and to the Doce River in Minas Gerais and Espírito Santo in the south-east. Hybridizes with Northern Black-horned Capuchin in the region of the Santo Antônio and Suaçuí Grande Rivers.

NT **Bearded Capuchin** *Sapajus libidinosus*

Description: A small capuchin with a black crown and two small, rounded black tufts. The pale forehead creates a rounded arc, but it lacks obvious eyebrows. The fur is short and silky, and there is diagnostic rusty-red hair on the back of the neck, and dark brown stripes running down the sides of the face and sometimes on to the beard. The throat, chest, upperparts, flanks, and the lower two-thirds of the tail are orange-yellow with no contrast between the upper and lower sides of the tail. The saddle, rump and outer thighs are grayish-brown with some reddish hairs. The outer arms are yellowish-brown and blackish-brown, the two colors merging where they meet.

Similar species: None in range.

Habitat: Dry deciduous forest, forest and scrub of the Caatinga in north-east Brazil, and gallery and dry forests in the Cerrado of central Brazil.

Distribution: NE Brazil in the east of the state of Maranhão, from the Mearim and Itapecuru Rivers, through Piaui and Ceará into Rio Grande do Norte, Pernambuco, Paraíba and Apagoas. West of the São

HB:	33–44 cm
Tail:	38–49 cm
Wt:	1.3–4.8 kg

Francisco River, through the Cerrado in Tocantins, Goiás, western Minas Gerais and part of western Bahia, and north-east Mato Grosso. West to the eastern bank of the Araguaia River and south to the Rio Grande in western Minas Gerais.

CR Yellow-breasted Capuchin *Sapajus xanthosternos*

Description: Grizzled reddish-brown above, with contrasting golden-red underparts. The cap is black and the face and forehead are pale buff. The tail, hind limbs and forearms are black. Individuals in the south-west of its range are darker overall, while those in the north of its range tend to be much paler. Normally appears to lack ear-tufts, but when seen well it does have small backward-pointing tufts.

Similar species: None in range.

Habitat: Humid tropical lowland and sub-montane forest, and also in deciduous and dry semi-deciduous forest patches in the western part of its range in Bahia. Restricted to hills and mountain ranges within the Caatinga.

HB:	♂ 39–42 cm; ♀ 35.1–39.0 cm
Tail:	♂ 35–41 cm; ♀ 38–45 cm
Wt:	♂ 2.0–4.8 kg; ♀ 1.3–3.4 kg

Distribution: Atlantic Forest of southern Bahia, Brazil, where it occurs north of the Jequitinhonha River, with the northern limit of its range probably the Paraguaçú River near Salvador.

LC **Guianan Brown** (Black-capped) **Capuchin** *Sapajus apella*

Description: Identified by the characteristic head coloration. The nominate ssp. *apella* has a broad black or dark brown crown with thick, black sideburns which on many individuals meet below the chin and contrast strongly with the rest of the body, giving it a hooded appearance. Small ear-tufts appear as horns or as a single tuft in the center of the crown. The face ranges in color from light gray-brown to pink with a pale, often whitish, surround. A darker area above the eyes creates an eye-browed appearance. There is significant variation in face color, even among members of the same group. The fur

HB:	38–46 cm
Tail:	38–49 cm
Wt:	♂ 2.3–4.8 kg; ♀ 1.3–3.4 kg

is long and coarse and ranges in color from light gray-brown to dark brown, with a dark dorsal stripe. The shoulders are paler than the back. The underparts tend to be slightly more reddish or yellowish-toned. The lower limbs and the tail are black. In the form previously split as **Large-headed Capuchin**, ssp. *macrocephalus*, black fur extends from the crown down the center of the forehead, creating a distinct black triangle with the sides of the forehead whitish. The beard is poorly developed. The fur is very short and silky. The upperparts are gray-brown or gray-ochre to dark reddish-brown, with a blackish-brown dorsal line. The shoulders are yellowish-brown to brown and slightly paler than the back. The reddish-brown underparts have a slightly yellowish tone to the chest. The upper surfaces of the arms are reddish-brown, the outer thighs reddish-brown and the inner thighs the same reddish tone as the underparts. The hands, hind legs and feet are black. The tail is long and thick, and the upper two-thirds are reddish-brown, with no difference between the upper and lower surfaces, while the lower third is black and contrasts with the base of the tail.

PRESUMED SSP. *macrocephalus* [Large-headed Capuchin]

Similar species: Marañón White-fronted Capuchin (*p. 290*) has a more yellowish-cream appearance, including the hands and feet.

Habitat: Tropical lowland, sub-montane and montane rainforest, seasonally inundated forest, open swamp forest, mangroves along the coasts, and savanna forests including white-sand forests and savanna scrub. 'Large-headed Capuchin' occurs in a wide range of habitats, including Amazonian lowland, sub-montane, deciduous gallery and humid evergreen forests. It also occurs in secondary growth, but appears to avoid flooded forest. ssp. *margaritae* (**Margarita Island Capuchin**) occurs in isolated cloud forest roughly 600 m above sea level.

Distribution: A wide-ranging species in the lower Amazon of Bolivia, Brazil, Colombia, Ecuador, French Guiana, Guyana, Peru, Suriname and Venezuela. The full extent of its range is unclear owing to uncertainty of the taxonomic status and the ranges of the subspecies currently attributed to other species of *Sapajus* capuchins. Margarita Island Capuchin occurs on the east side of Margarita Island, Venezuela.

ssp. *macrocephalus* [**Large-headed Capuchin**] (TOP); ssp. *apella* (BOTTOM)

 Blond Capuchin

Sapajus flavius

Description: A small and distinctive golden-yellow capuchin with a rounded head that lacks tufts. The underparts are slightly darker, and the hands and feet are black, although the limbs do not contrast so strongly with the dorsal parts as they do in Yellow-breasted (*p. 285*) and Bearded Capuchins (*p. 284*), which have adjacent ranges to that of Blond Capuchin. The tail is golden-blond, its upper side appearing darker than the rest of the upperparts. A yellow-buff cap on the front of the head extends to just above the ears. The face is pinkish, and the forehead pale yellowish-buff and not contrasting with the cap.

Similar species: None in range.

Habitat: Fragmented patches of lowland Atlantic coastal forest, swamps and secondary and semi-deciduous seasonal forest. Also low, xerophytic, spiny scrub (Caatinga), sand dunes, mangroves, and in sugar-cane and corn plantations bordering forest.

HB:	35.1–40.0 cm
Tail:	37.8–42.0 cm
Wt:	1.8–3.0 kg

Distribution: Coastal NE Brazil in the states of Alagoas and Paraíba, and north-east Pernambuco.

NT Shock-headed Capuchin

Cebus cuscinus

Description: Usually light brownish-yellow, becoming tawny on the lower back, and often with reddish hues, particularly on the legs. The underparts are yellowish-orange and silvery, the chest being a warmer buff color. The fronts of the shoulders are whitish, and the upper arms are brown on the outside and white on the inside. The lower arms are orange-rufous. The tail is cinnamon-brown with a dark dorsal stripe above, and is brown below. The crown contrasts with the body, being dark brown to black and forming a large cap that ends abruptly just behind the eyebrows. The cap is more clearly defined on males, which have a broad pale forehead, while females have a dark forehead.

HB:	39–46 cm
Tail:	39.0–47.5 cm
Wt:	2.8–3.0 kg

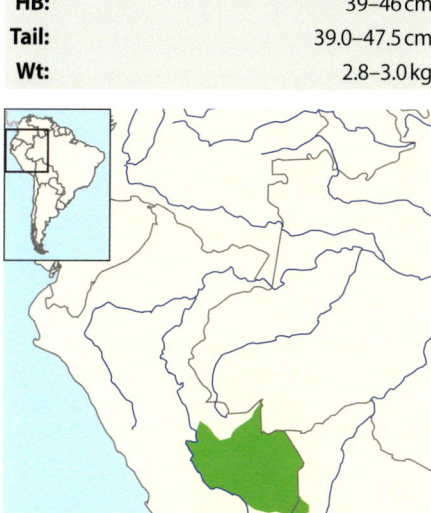

Similar species: Spix's White-fronted Capuchin (*p. 291*) is bright ochraceous-brown to grayish-brown and has shorter, less silky fur.

Habitat: Lowland terra firme and seasonally inundated forests in the western Amazon Basin to the western slopes of the Andes, in montane forests up to 1,800 m.

Distribution: Poorly known. Occurs in SE Peru from the upper reaches of the Purús River, west into the Urubamba River Valley, including the upper reaches of the Madre de Dios River. The range extends south and east to the Tambopata Basin and into NW Bolivia. May occur also in the south of the state of Acre in Brazil.

NT Marañón White-fronted Capuchin *Cebus yuracus*

Description: Largely yellowish-brown to dark grayish-brown, often with reddish hues on the upper thighs, similar to Ecuadorian White-fronted Capuchin (*p. 302*). The crown has a dark brown forward-pointing wedge-shaped patch, while the face is pinkish with a silvery-white border but can be grayish on older individuals. The silvery-white to cream-colored tail is palest at the tip, which is often coiled. The upper side of the tail can be dark in some individuals. The hands and feet tend to be yellower and brighter than the rest of the body.

Similar species: Guianan Brown Capuchin (*p.286*) is much darker, has dark brown hands, feet and tail, and has obvious fringes of black hairs in front of the ears.

Habitat: Primary, secondary and disturbed humid tropical, subtropical and temperate forest from 200 m to 2,515 m, but generally below 900 m in the Amazonian lowlands and on the lower eastern slopes of the Andes.

Distribution: Occurs in S Colombia, E Ecuador and NE Peru north of the Amazon

HB:	37–43 cm
Tail:	45–47 cm
Wt:	2.0–4.7 kg

River as far north as the Putumayo River. The eastern limit of the range is the Ucayali River and the southern limit is the Pachitea River.

VU Spix's White-fronted Capuchin

Cebus unicolor

Description: Largely uniformly colored, with a black crown ending just above the eyebrows. Generally bright ochraceous-brown to grayish-brown, the flanks being grayer than the darker brown middle back. There is no white on the front of the shoulders, the pale areas being yellowish or creamy-fawn. The limbs and tail are reddish-yellow or reddish. The face is uniformly pale.

Similar species: Shock-headed Capuchin (*p. 289*) has less brightly colored, longer, silkier fur.

Habitat: Lowland terra firme and seasonally inundated forest, including forest patches within savanna.

Distribution: South of the Amazon–Solimões Rivers in the upper Brazilian Amazon Basin from the Tapajós River west through the northern parts of the states of Mato Grosso and Rondônia (south to at least 10°S, and the basins of the Madeira, Purús, Juruá and Javari Rivers), to the Ucayali River in E Peru. May occur also in N Bolivia. It is unclear

HB:	36.5–37.5 cm
Tail:	42–46 cm
Wt:	No info. available

where its range meets that of Shock-headed Capuchin in SE Peru.

LC Humboldt's White-fronted Capuchin *Cebus albifrons*

Description: Ranges in color from light to dark brown, some individuals exhibiting a reddish cast. The front of each shoulder is whitish. The limbs are darker than the body, and the hands and feet are yellowish-brown. The tail is ashy gray above and whitish below, the tip of the tail being light on some individuals and brownish-black on others. The wedge-shaped cap on the head is dark brown, the forehead is white, and the face is naked and pink.

Similar species: None in range.

Habitat: Moist, undisturbed primary and secondary deciduous forest, gallery and flooded forests, and high-elevation forests at up to 2,000 m.

Distribution: Found throughout the Colombian Amazon north of the Putumayo and Amazon Rivers. Occurs also in the Venezuelan Amazon, south into the Brazilian states of Amazonas and Roraima as far as the Solimões River and the western banks of the Branco and Negro Rivers. In the west

HB:	♂ 36.5–38.5 cm; ♀ 36.5–37.5 cm
Tail:	♂ 42.5–43.0 cm; ♀ 41–46 cm
Wt:	♂ 2.1–2.6 kg; ♀ 2–3 kg

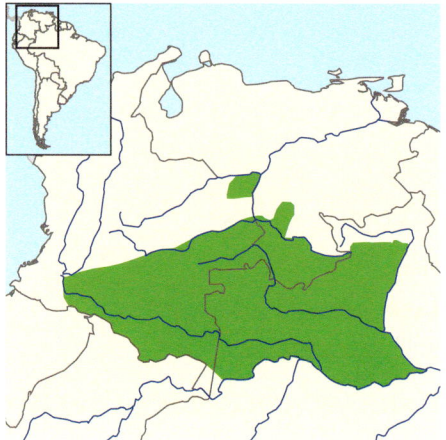

and south, the limits of the range are the boundaries of the ranges of Marañón (*p. 290*) and Spix's (*p. 291*) White-fronted Capuchins.

Ka'apor Capuchin

Cebus kaapori

Description: Long-bodied capuchin with long, silky, grayish-brown fur with a dark dorsal stripe. The flanks are paler, and the face, shoulders, underparts, mantle and tip of the tail are silvery-gray. The limbs are yellowish-brown or grayish-brown with dark brown or black hands and feet. A triangular black cap extends from the crown as a dark stripe down to the nose. The nape is blackish-brown. Lacks tufts but does have a line of erect hairs along the medial line. The upper two-thirds of the tail are grayish-brown and the remainder is silvery-gray.

Similar species: None in range.

Habitat: Undisturbed and moderately disturbed dense lowland terra firme Amazonian Forest at altitudes below 200 m. Found also in forest edge bordering cocoa plantations.

Distribution: Brazil in the west of the state of Maranhão and the eastern part of the state of Pará from east of the lower Tocantins River to the right bank of the Grajaú River, in Maranhão. The precise limits of its range are unclear.

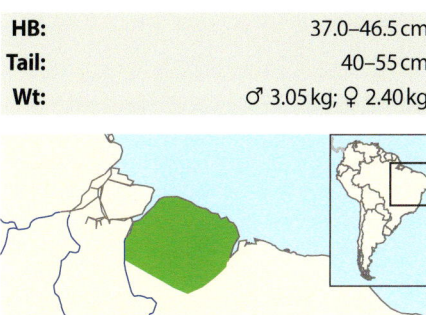

HB:	37.0–46.5 cm
Tail:	40–55 cm
Wt:	♂ 3.05 kg; ♀ 2.40 kg

LC Weeper Capuchin

Cebus olivaceus

Description: Large and heavy-set, with long limbs and long, coarse, shaggy fur. Brown with blackish-agouti banding on the flanks, limbs and tail and blackish hands, feet and the tip of the tail. The underparts appear dark brown to blackish. Has a dark brown, wedge-shaped cap, with a dark line running down the light gray-brown forehead and face from the point of the cap to the nose. The face is naked and usually pinkish. The head lacks tufts. Formerly treated as conspecific with Chestnut Weeper Capuchin.

Similar species: None in range.

Habitat: A range of forested habitats at up to 2,000 m, including primary rainforest, evergreen and gallery forests and, less commonly, dry deciduous forest and shrub woodland.

Distribution: Venezuelan Amazon Basin from the upper Orinoco River through the Orinoco savanna as far north and west as Sierra de Perijá and the Venezuelan coastal ranges, and east to the west bank of the Essequibo River

HB:	37–46 cm
Tail:	45–55 cm
Wt:	♂ 3.0–4.2 kg; ♀ 2.3–3.0 kg

in W Guyana. In Brazil, it is restricted to the Branco River/Aracá River interfluvial area in the states of Roraima and Amazonas.

Chestnut Weeper Capuchin

Cebus castaneus

Description: A large, heavy-set capuchin with long limbs and long, coarse, shaggy fur. It has a yellowish-white head lacking tufts, and a narrow dark brown wedge-shaped cap on the crown with a dark line running down the center of the face to the nose. The areas above the ears and nape are reddish-chestnut, as are the upperparts of the body and limbs. The head lacks tufts. The shoulders and the front of each upper arm above the elbow are pale yellow. The hands and feet are blackish, and the tail is blackish-brown, with gray-tipped hairs on the upper surface. Formerly treated as conspecific with Weeper Capuchin.

Similar species: None in range.

Habitat: Undisturbed high terra firme rainforest, but avoids liana forest and mountain savanna forest.

Distribution: Poorly known but believed to occur north of the lower Amazon River on the upland Guiana Shield of Guyana (east of the Essequibo River), Suriname and French Guiana, and N Brazil between the Negro and Branco Rivers in the west, the Amazon River in the south and the Atlantic coast in the east. It is not found along the alluvial coastal plains of Guyana, Suriname, French Guiana and the state of Amapá, Brazil.

HB:	37–46 cm
Tail:	45–55 cm
Wt:	♂ 3.0–4.2 kg; ♀ 2.3–3.0 kg

EN Venezuelan Brown Capuchin

Cebus brunneus

Description: The thick fur is brown, with a dorsal band that is darker than the sides of the back. The head and face are pale yellowish-gray. A 'V'-shaped cap on the crown comes to a point on the forehead, with a thin black line running down the face on to the nose. The cheeks and chin are whitish, and the throat is paler than the blackish-brown chest and belly. The upper arms are pale yellowish as far as the elbows. The legs and the tail are brown, the outside of the thighs being slightly paler. The hands and feet are almost black.

Similar species: None in range, but introduced to Trinidad, where **Trinidad White-fronted Capuchin** (*p. 301*) is native.

HB:	approx. 42 cm
Tail:	approx. 44 cm
Wt:	No info. available

Habitat: Moist lowland, montane and sub-montane forest in the coastal ranges, and dry semi-deciduous and gallery forests in the western Llanos.

Distribution: Coastal ranges of Venezuela east of the Sierra de Perijá and in the western Llanos of Venezuela. Introduced to Trinidad.

Sierra de Perijá White-fronted Capuchin *Cebus leucocephalus*

Description: The darkest of the white-fronted capuchins. The cap is brownish-yellow to cinnamon-black. The back is cinnamon-brown, interspersed with tawny on the upper back and russet on the lower back. The flanks are paler grayish to yellowish-brown, and the upper sides of the shoulders and the outer surfaces of the upper arms are cinnamon-brown. The outer sides of the forearms are deep reddish-brown, grading into paler reddish-brown. The outer surfaces of the thighs are cinnamon-brown, and the forelegs and front of the thighs are deep reddish-brown, as are the wrists, ankles and upper surface of the hands. The belly and lower chest are deep reddish-brown, becoming orange-rufous on the upper chest and on the inner sides of the upper arms. The upper tail is cinnamon-brown, becoming paler towards the tip, and the tail is paler buff below.

HB:	37.0–40.7 cm
Tail:	39.2–49.9 cm
Wt:	No info. available

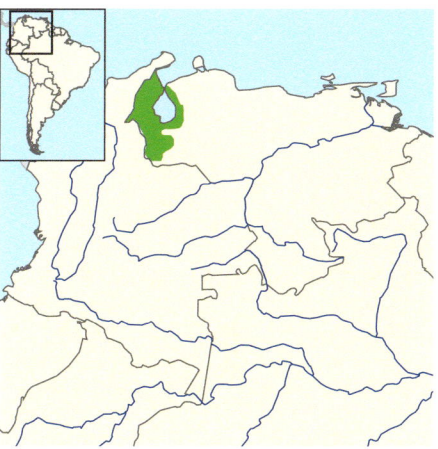

Similar species: None in range.

Habitat: Moist lowland forests, semi-deciduous dry forests and mangroves.

Distribution: NW Venezuela (Zulia State), through the basins of the Zulia and Catatumbo Rivers to the western slope of the Cordillera Oriental in the north of Santander Department, in N Colombia. The geographical limits of the ranges of Sierra de Perijá and Varied (*p. 298*) White-fronted Capuchins in Colombia are unclear.

EN Varied White-fronted Capuchin

Cebus versicolor

Description: Rather pale, with red tones on the back, forearms and forelegs which contrast with the rest of the body. The crown is dark brown, rather than black as in Colombian White-fronted Capuchin (*p. 303*), and the temple and forehead are creamy yellow. The nape and the undersides of the hands are dark brown.

Similar species: None in range.

Habitat: Moist lowland forest and palm swamps.

Distribution: Found on both banks of the Magdalena River in northern Colombia between the Central and Eastern Andes Mountains, excluding the western slope of the Eastern Andes. The geographical limits of the ranges of Varied, Río Cesar and Sierra de Perijá (*p. 297*) White-fronted Capuchins in Colombia are unclear.

HB:	45–50 cm
Tail:	42–45 cm
Wt:	No info. available

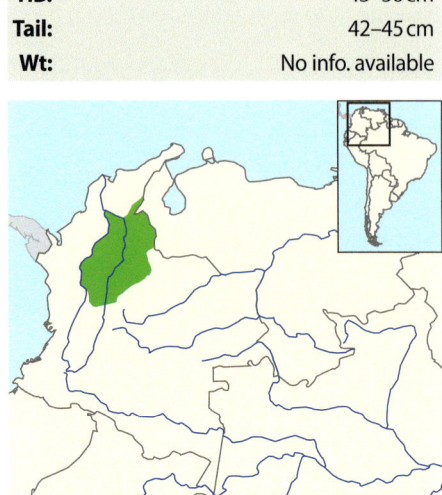

Río Cesar White-fronted Capuchin

Cebus cesarae

Description: The palest of the white-fronted capuchins in N Colombia; smaller than Varied White-fronted Capuchin, and with a longer tail than that species. The upperparts are buff-brown, although the cap is cinnamon or orangey-brown, and the middle of the back and the forearms and forelegs are orange, contrasting with the sides of the body. The upper tail is cinnamon-brown. The belly and chest are pale buff and silvery, and contrast with the pale upper surfaces of the shoulders and the inner sides of the upper arms. The frosted tail is cinnamon-brown.

Similar species: None in range.

Habitat: Found at up to 500 m in mangroves, gallery forest and dry semi-deciduous forest patches.

HB:	35–41 cm
Tail:	42–50 cm
Wt:	No info. available

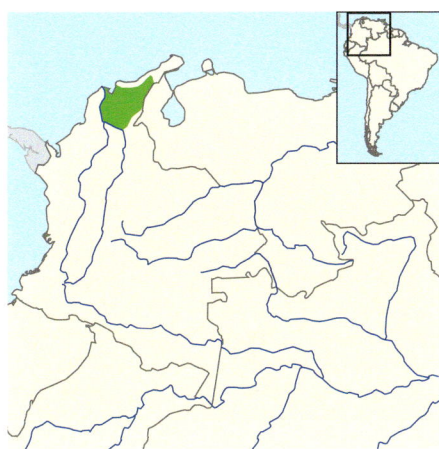

Distribution: Restricted to the Cesar River valley and west on to the southern and eastern slopes of the Sierra Nevada de Santa Marta, in N Colombia, north of the range of Varied White-fronted Capuchin, and south of the range of Santa Marta White-fronted Capuchin (*p. 300*).

EN Santa Marta White-fronted Capuchin *Cebus malitiosus*

Description: Pale brown, with relatively uniform cinnamon-brown upperparts and limbs. The belly and chest appear cinnamon-brown or ochraceous-tawny with silvery hairs. Has a contrasting pale area extending across the upper surface of the shoulders and the inner side of the upper arms. The crown has a brown cap. The ears and face, including the skin around the eyes, are pale and flesh-colored, except for the dark brown lips and the tip of the nose.

Similar species: None in range.

Habitat: Lowland, sub-montane and montane forests.

Distribution: Recorded only from the north-west base of the Sierra Nevada de Santa Marta, in N Colombia, although likely to occur more widely on the northern and western slopes of the range.

HB:	approx. 46 cm
Tail:	approx. 43 cm
Wt:	No info. available

CR Trinidad White-fronted Capuchin *Cebus trinitatis*

Description: The only *Cebus* capuchin native to Trinidad. A pale yellowish capuchin with a poorly distinguished brownish-yellow crown patch, and a clearly defined silvery forehead. The nape is ochraceous-buff and the rest of the back is buffy-brown, the sides being paler pinkish-buff. The upper surfaces of the limbs are ochraceous-buff, the undersides paler and silvery. The hands and feet are pinkish-buff to cinnamon-buff. The tail is brown above and ochraceous-buff below, becoming silvery towards the tip.

Similar species: None in range, but see **Venezuelan Brown Capuchin** (*p. 296*), which has been introduced to the island.

Habitat: Forests, particularly on the northern slopes of Trinity Hills.

Distribution: Endemic to Trinidad.

No information on measurements and weight available

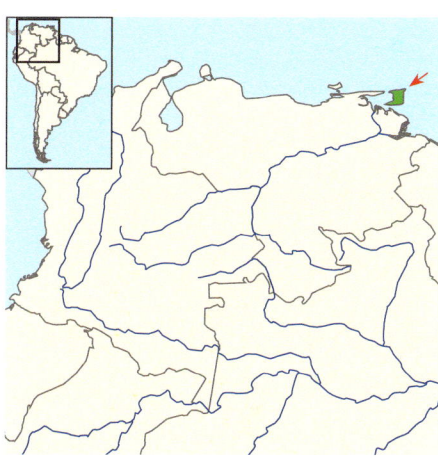

301

CR **Ecuadorian White-fronted Capuchin** *Cebus aequatorialis*

Description: A slender, medium-sized capuchin with pale cinnamon-rufous upperparts and a darker dorsal line. The crown is dark brown, with a narrow black line extending down to the nose. The front and sides of the head are yellowish-white. The face is either unpigmented or pinkish, with a silvery-white border. The shoulders, chest, upper arms and thighs are creamy, while the remainder of the underparts is reddish-brown. The tail is silvery-yellow to cream-colored, with the tip usually paler than the base. The hands and feet are normally brighter and yellower than the rest of the body.

Similar species: None in range.

Habitat: Prefers tall terra firme forest up to 2,040 m, but also known in degraded secondary forest and cultivated areas. In both dry forests and wet sub-montane Andean forests.

Distribution: The Ecuadorian lowlands and extreme NW Peru, west of the Andes – from

HB:	35–51 cm
Tail:	40–50 cm
Wt:	♂ 1.7–3.6 kg; ♀ 1.2–2.2 kg

the Esmeraldas–Guayllabamba River in the north to Cerros de Amotape National Park in the south.

Colombian White-fronted Capuchin

Cebus capucinus

Description: Largely black, with a white to yellowish-white face, neck, chest, shoulders and upper arms. The yellowish-white head has a black 'V'-shaped patch on the crown which points backwards to connect with the black back. The nose is pinkish with whitish hairs. The tail is black, but can appear brown on the underside. Unlikely to be confused with any other species in its range.

Similar species: None in range.

Habitat: A variety of forest types, but prefers terra firme primary forests and older secondary forests. Also in seasonally inundated forests, degraded forest fragments, mangroves, and dry deciduous forests. Up to 1,800–2,100 m in the western Andes.

Distribution: The Pacific coastal region of W Colombia and NW Ecuador as far south as the Esmeraldas–Guayllabamba River. Found also on Gorgona Island, off the coast of Colombia. Range extends into E Panama.

HB:	33.5–45.3 cm
Tail:	35.0–55.1 cm
Wt:	♂ 3–4 kg; ♀ 1.4–3.0 kg

Additional photo p. 278

Night Monkeys

FAMILY | **Aotidae**

The taxonomic status of many of the forms of night monkey is subject to debate, and a further review of their taxonomy is highly likely. Defler & Bueno (2007) is an essential reference on night monkey taxonomy. At the time of writing, there are 11 species, seven of which are gray-necked, the remaining four being red-necked.

Night monkeys are small to medium-sized monkeys with a small, rounded head, large eyes reflecting their normally nocturnal lifestyle, and small ears, and they often have a complex head pattern with whitish hairs around or above the eyes and around the muzzle. They have black stripes on the head which sometimes meet on the back of the head, and they are long-tailed, although the tail is not prehensile. Some species have an interscapular whorl (a circular pattern of fur in the area between the shoulder blades). They are highly variable in color, even within the same family group.

Most are primarily nocturnal and active around dawn and dusk, although they can often be found at the entrance of tree hollows during the day; Azara's Night Monkey is cathermal (active by day and night) in the Chaco of S Bolivia, Paraguay and Argentina.

Night monkeys feed on fruit, nectar, flowers, leaves, and small animal prey such as insects. They live in small territorial groups of up to six individuals.

Brumback's (LEFT); **Gray-legged** (TOP RIGHT); **Black-headed** (BOTTOM RIGHT) **Night Monkeys**

Lemurine Night Monkey

Aotus lemurinus

Description: Gray-necked, with two color forms: a light grayish form and a darker, reddish-brown form, with variations in between. Both forms can show a brownish medial dorsal band and frequently occur in the same family group. The underparts are always a rather dull yellow or pale orange, some individuals being indistinguishable from Gray-legged and Panamanian Night Monkeys. The tail is black, the tip being thicker than the base. The face pattern is individually variable, although the eyebrows are normally white and thick and outlined with black fur. The fur is extremely long and soft.

HB:	♂ approx. 31 cm; ♀ approx. 33 cm
Tail:	approx. 34 cm
Wt:	800–1,050 g

Similar species: The most important differences for separating **Lemurine**, **Gray-legged** (*p. 306*) and **Panamanian** (*p. 307*) **Night Monkeys** are the black or brown hands and feet and very long fur of Lemurine Night Monkey, the generally brownish feet and hands and short fur of **Gray-legged Night Monkey**, and the generally blackish hands and feet and short fur of **Panamanian Night Monkey**.

Habitat: Tropical and high-elevation forest from 1,000 m up to the treeline at 3,000–3,200 m.

Distribution: Ecuador, Colombia and extreme W Venezuela. Tirira (2017) considers this to be the species occurring in the subtropical humid forest along the Cordillera Oriental of the Andes in Ecuador at altitudes of 940–1,800 m, although this has yet to be confirmed. In the north of its range it occurs in highlands, and its range is divided by the lowlands of the middle Magdalena Valley where Gray-legged Night Monkey occurs.

Gray-legged Night Monkey

Aotus griseimembra

Description: Gray-necked with a light grayish form and a darker, reddish-brown form; there are also variations in between. The fur is short and dense. Both forms can show an indistinct brownish medial dorsal band and frequently appear in the same family group. The underparts are always a rather dull yellow.

HB:	approx. 48 cm
Tail:	approx. 42 cm
Wt:	800–1,000 g

Similar species: See **Lemurine Night Monkey** (*p. 305*) for details of further differences between **Lemurine**, **Gray-legged** and **Panamanian Night Monkeys**.

Habitat: Primary and secondary lowland tropical forest, with some records from coffee plantations.

Distribution: Along the Magdalena River and the valleys of the Cauca and São Jorge Rivers, in N Colombia, north to the Sierra Nevada de Santa Marta, up the Guajira Peninsula and east across the Sierra de Perijá to Lake Maracaibo, in NW Venezuela.

Additional photo *p. 304*

Panamanian Night Monkey

Aotus zonalis

Description: Gray-necked, with short, compact fur and two color forms, a light grayish form and a darker reddish-brown form, with variations in between. Both forms often occur in the same family group. The hands and feet are dark brown or blackish. Feeds in the forest canopy 7–30 m above the ground.

HB:	approx. 30 cm
Tail:	approx. 36 cm
Wt:	880–916 g

Similar species: See **Lemurine Night Monkey** (*p. 305*) for details of further differences between **Lemurine**, **Gray-legged** and **Panamanian Night Monkeys**.

Habitat: Lowland rainforest at up to 650 m.

Distribution: The Chocó region of Colombia, extending southward, west of the Andes, to the Raposo River just south of Buenaventura, and including the Urabá region and eastward to the Sinú River, and possibly including the upper San Jorge Valley to the region of Puerto Valdivia, in northern Antioquia. It is unclear whether the ranges of Panamanian and Gray-legged Night Monkeys overlap, and if so where.

VU Brumback's Night Monkey

Aotus brumbacki

Description: Short-haired, with two temporal head stripes which continue as shadowy stripes, appearing to merge into a generally dark blotch on top of and behind the crown. The upperparts are usually grayish-buff and lack an obvious dorsal stripe. The underparts are pale orange. A pale band between the buff supraorbital and subocular patches is interrupted by an extension of the blackish temporal stripe to the outer corner of the eye; a blackish malar stripe is also present. The white above the eyes has a yellow tinge and the white on the face extends down on to the chin.

Similar species: None in range.

No information on measurements and weight available

Habitat: Mainly in gallery and closed-canopy forest from 467 m to 1,543 m.

Distribution: Known only from western Meta and Boyaca Departments in Colombia, although the night monkeys of eastern central Colombia north of the Guaviare River may also be of this species.

Additional photo p. 304

Humboldt's (Northern) **Night Monkey** *Aotus trivirgatus*

Description: The most distinct of the group of northern gray-necked night monkeys. The dorsum is usually grayish, sometimes with buffy gray-brown tones, and with a narrow and strongly contrasting orange mid-dorsal band. The neck and outer sides of the limbs are grayish, and the chest, belly and inner sides of the limbs are orange-buff to dark yellow. The head has no whorls, crests or tufts, but does have two parallel black stripes that extend from the top of the head to the lining of the eye but do not converge posteriorly. The ears are more prominent than on other night monkeys. The face is rather grayer than that of most night monkeys, which have whitish faces. The tail is gray at its base, and becomes dark gray to black halfway along its length. The feet and hands are dark brown.

HB:	30–38 cm
Tail:	33–40 cm
Wt:	736–813 g

Similar species: None in range.

Habitat: Tropical and humid forests, including dry forests.

Distribution: Eastern Colombia east of the Andes (east of the Negro, lower Guainía and Atabapo Rivers), southern Venezuela as far north as the middle reaches of the Orinoco River, and parts of N Brazil to the north of the Negro and Amazon Rivers and including the north of Amazonas, Roraima and the north-west of Pará States. There are unconfirmed reports from Guyana.

LC Spix's Night Monkey   *Aotus vociferans*

Description: Gray-necked and short-haired, with a white face, a light brown to gray back, a light yellow belly and this color extending on to the chin and limbs, the hands and feet being black. Two thick brownish crown stripes converge at the back of the head, and there is a dark brown stripe with light brown edges at the center of the head. It has an interscapular whorl and a small whitish patch above each eye. The upper third of the tail is reddish or gray and the lower two-thirds are blackish.

HB:	33–45 cm
Tail:	31–47 cm
Wt:	698–708 g

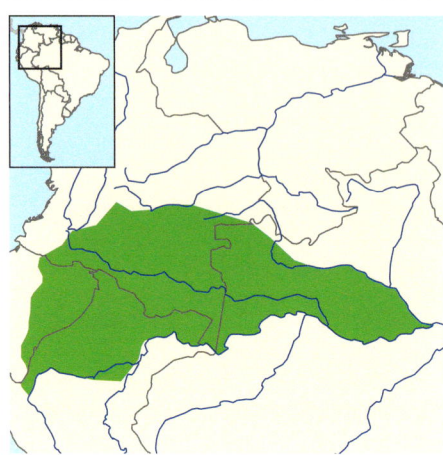

Similar species: In the small area of overlap with **Ma's Night Monkey** (*p. 313*), readily identified by its gray, rather than red or orange, neck.

Habitat: Mainly at 200–900 m in tropical lowland forest, including seasonally inundated forests, swamp forests and terra firme forests, although recorded at up to 1,550 m in Cordillera del Condor in Ecuador and Peru.

Distribution: Widespread in Colombia, N Brazil, N Peru and E Ecuador. In Brazil north of the Amazon–Solimões Rivers, west from the Negro River; south of the Solimões River on each side of the lowermost reaches of the Purús River. In the west it extends into Peru north of the Amazon and Marañón Rivers, east of the Chinchipe River to the borders with Ecuador and Colombia. It extends north through the Ecuadorian Amazon into Colombia to the Guaviare River, probably extending east as far as the Negro, Atabapo and Orinoco Rivers.

Hernández-Camacho's Night Monkey *Aotus jorgehernandezi*

Description: A poorly known species. Gray-necked, with the face having two discrete supraocular white patches separated by a broad black frontal stripe. Subocular white bands of fur are separated by a thin black malar stripe on each side of the head. The ventral part of the arms from the wrists running up into the chest and belly are white and thickly furred.

Similar species: None in range.

Habitat: The type locality is in sub-montane tropical forest.

No information on measurements and weight available

Distribution: Described from a specimen in captivity in Quindío Department, Colombia, said to be from the Parque de los Nevados on the border between Quindío and Riseralda. This locality, however, is within the range currently considered to be that of Panamanian Night Monkey (*p. 307*).

EN Andean Night Monkey *Aotus miconax*

Description: A red-necked brownish to buff-gray night monkey, the upperparts being light gray with a brownish or reddish-brown tint. The underparts, including the inner sides of the limbs, are pale orange. The tail is blackish above, reddish-orange below, and bushy. The facial pattern is relatively inconspicuous. Lacks an interscapular whorl.

Similar species: None in range (separated altitudinally from **Ma's Night Monkey**).

Habitat: Reported from 800–3,100 m in primary and secondary humid lower montane cloud forest, including small forest patches close to houses.

Distribution: Restricted to the Peruvian Andes south and east of the Marañón River and west of the Huallaga River. In the Cordillera de Colán, south of the Chiriaco River, from 1,730 m to 2,400 m above sea level.

HB:	39.4 cm
Tail:	22 cm
Wt:	No info. available

Ma's Night Monkey
Aotus nancymaae

Description: Red-necked, with a gray-brown back with a slightly darker mid-dorsal band. The underparts, including the sides of the neck and the upper half of the inner limbs, are pale orange. The hands and feet are blackish. The upper third of the upper tail is usually grayish-orange, the lower two-thirds being blackish. The underside of the tail is blackish. The face is grayish-white, with narrow dark brown crown stripes that do not normally meet behind the head. Lacks an interscapular whorl.

Similar species: Overlaps with **Spix's Night Monkey** (*p. 310*) in a small area north of the Amazon and Marañón Rivers in Peru; that species is readily identified by its gray neck.

Habitat: Lowland tropical forest, including seasonally flooded forests, at 60–130 m. Prefers swamp and seasonally inundated forests to terra firme because of the former's greater availability of tree holes for sleeping.

HB:	29–34 cm
Tail:	35–42 cm
Wt:	780–794 g

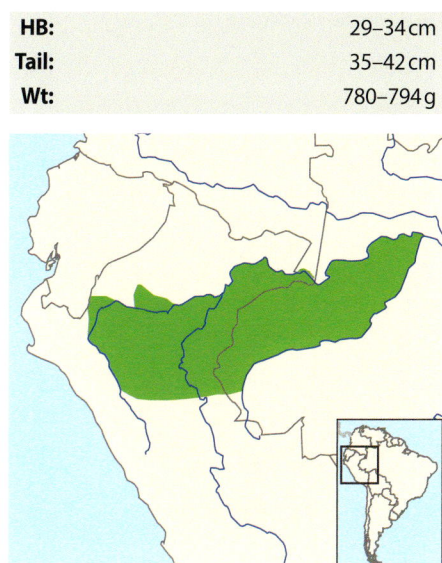

Distribution: Brazil: occurs south of the Amazon and Marañón Rivers, north and west of the Jutaí River; its range extends west to cross the Javari River level with the headwaters of the Tapiche River, and across the Ucayali Basin to the upper Marañón River. Peru: found south of the Amazon and Marañón Rivers between the Huallaga and Yavari Rivers and to the north of the Marañón between the Tigre and Pastaza Rivers; it occurs west to the Andean foothills in San Martín. Colombia: reported from north of the Amazon River west from San Juan de Atacuari to the border with Peru.

LC Black-headed Night Monkey
Aotus nigriceps

Description: Red-necked, with gray upperparts having some brownish coloration on the back. Populations along the right bank of the Juruá River are gray-backed, while those on the left bank have a reddish-brown back. The underparts, including the throat and upper half of the inner limbs, are bright orange. The hands and feet are blackish. The proximal part of the upper tail is gray-brown, with the distal section dark gray to black. The underside of the tail is orange. The top of the head has a lozenge-shaped black patch. The face has conspicuous areas of white around the head patch and on the cheeks, and two broad, black stripes on the sides of the face that converge on the upper part of the head, but there is no interscapular whorl.

Similar species: None in range.

Habitat: Primary tropical forest, seasonally inundated forest and swamp forests.

Distribution: Brazil to the south of the Amazon in south-west, central and eastern Amazonas, eastern Pará, Acre, Rondônia and western Mato Grosso. West of the Tapajós–Juruena Rivers into Peru and south

HB:	35–42 cm
Tail:	35–44 cm
Wt:	875–1,040 g

to the Madre de Dios and Guaporé Rivers. In SE Peru east of the Huallaga River and south of the Cushabatay River. In Bolivia north of the Madre de Dios River in the department of Pando.

FEMALE WITH INFANT

Additional photo *p. 304*

LC Azara's Night Monkey

Aotus azarae

Description: A highly variable red-necked night monkey. SSP. *azarae* is grayish to pale buff-gray above and pale whitish-orange below, with long, thick, shaggy fur. The face stripes are narrow and the proximal half of the tail is orange, the remainder being blackish. SSP. *boliviensis* has relatively short fur and the upperparts are olive-toned, with contrasting grayer limbs. The central face stripe is narrow but expands on the crown, while the other crown stripes are very narrow. The black temporal stripe is poorly defined, and there is often a whitish stripe between the temporal stripe and the eyes. The black malar stripes are indistinct and can be absent. Has a conspicuous interscapular whorl. SSP. *infulatus* resembles SSP. *boliviensis* but has more prominent areas of white on the face, and the well-defined black head stripes are continuous with the malar stripes. The throat can be orange and gray or completely orange, and the tail is reddish apart from a black tip. SSP. *azarae* is the only night monkey active during the day as well as at night.

HB:	approx. 34 cm
Tail:	approx. 31 cm
Wt:	990–1,580 g

Similar species: None in range.

Habitat: A wide range of habitats, including palm forests, forest patches and gallery forests in the Cerrado of central Brazil, deciduous scrub and forest, and mangrove forest in NE Brazil, and deciduous, gallery, riparian and secondary forest in northern Argentina, Bolivia and Paraguay.

Distribution: SSP. *azarae*: The southern and western Chaco region of Paraguay west of the Paraguay River, and across the Pilcomayo River to south of the Bermejo River to the Negro River in the provinces of Formosa and Chaco, in Argentina. May extend to Bañado do Izozog in Bolivia.

SSP. *boliviensis*: South of the Madre de Dios River, as far west as the Inambari River in Peru, extending along the Cordillera Oriental to southern Bolivia. To the east as far as the Guaporé River and the Brazilian border.

SSP. *infulatus*: In Brazil in the south-east of Amapá (Carmo de Macacoari), south-central Pará, Tocantins, Maranhão, west of Goiás and Mato Grosso. Occurs also in the upper Paraguay River Basin in the extreme west of Mato Grosso do Sul.

SSP. *infulatus*

Titi Monkeys

FAMILY | **Pitheciidae**
(SUBFAMILY | Callicebinae)

The IUCN Primate Specialist Group recognizes 35 species of Titi Monkey, which follows Byrne *et al.* (2016), with the subsequent additions of Groves's Titi *Plecturocebus grovesi*, described by Boubli *et al.* (2018), and Aquino's Collared Titi *Cheracebus aquinoi*, described by Rengifo *et al.* (2022). Parecis Titi *Plecturocebus parecis*, described by Gusmão *et al.* (2019), is now believed to be a cline of Ashy Titi *P. cinerascens* (Byrne *et al.*, 2021, who also highlighted the need for further research on the distribution of titi monkeys).

The species are treated as three genera, *Plecturocebus*, *Cheracebus* and *Callicebus* (an example of each of which is shown *below*), based on phylogenetic and morphological studies by Byrne *et al.* (2016). This taxonomy is followed here. The IUCN Primate Specialist Group continues to treat Doubtful Titi *P. dubius* as a valid species rather than as a color form of Chestnut-bellied Titi *P. caligatus* as proposed by Byrne *et al.* (2021) and adopted by the Mammal Diversity Database and Lynx Nature Books (2023).

In general, titi monkeys are relatively stocky, with a longish tail. The descriptions in these accounts draw heavily on van Roosmalen *et al.* (2002), an important reference for this group. Titi monkeys occur in a wide range of habitats and feed mainly on fruit pulp, leaves, insects and seeds. They form small territorial groups, usually of up to seven individuals. They are inconspicuous for much of the time, feeding quietly in thick vegetation, and are easily overlooked. They advertise their presence, however, by loud vocalizations in the morning that probably serve to define the boundaries of their territory.

Hoffmann's Titi
Plecturocebus hoffmannsi

Aquino's Collared Titi
Cheracebus aquinoi

Black-fronted Titi
Callicebus nigrifrons

Plecturocebus titi monkeys

There are 25 species of titi monkeys in the genus *Plecturocebus*. Species within the genus occur in Brazil, Colombia, Ecuador, Peru, Bolivia and Paraguay. All are illustrated on this and the next page for comparative purposes.

Rio Beni Titi
Plecturocebus modestus

Rio Beni Titi
Plecturocebus modestus
(*p. 319*)

White-eared Titi
Plecturocebus donacophilus
(*p. 320*)

White-coated Titi
Plecturocebus pallescens
(*p. 321*)

Olalla Brothers' Titi
Plecturocebus olallae
(*p. 322*)

Rio Mayo Titi
Plecturocebus oenanthe
(*p. 323*)

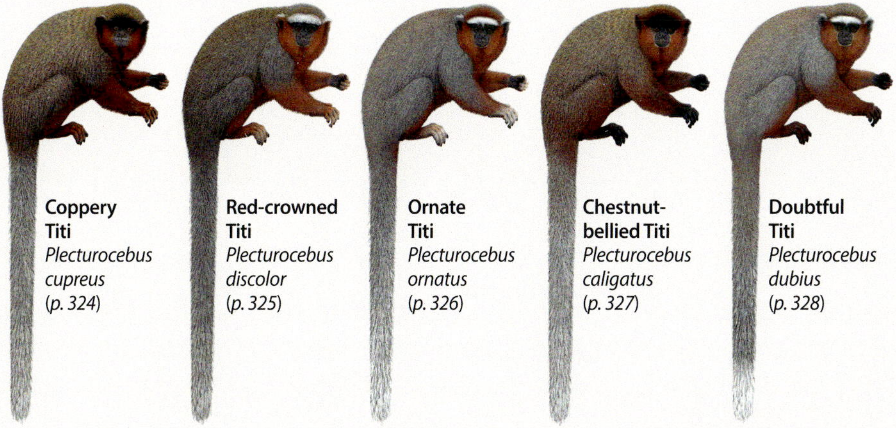

Coppery Titi
Plecturocebus cupreus
(*p. 324*)

Red-crowned Titi
Plecturocebus discolor
(*p. 325*)

Ornate Titi
Plecturocebus ornatus
(*p. 326*)

Chestnut-bellied Titi
Plecturocebus caligatus
(*p. 327*)

Doubtful Titi
Plecturocebus dubius
(*p. 328*)

Continued on next page...

Plecturocebus titi monkeys (continued)

Stephen Nash's Titi
Plecturocebus stephennashi
(p. 329)

Caquetá Titi
Plecturocebus caquetensis
(p. 330)

Ashy Titi
Plecturocebus cinerascens
(ashy form)
(p. 331)

Ashy Titi
Plecturocebus cinerascens
('Parecis' form)
(p. 331)

Hoffmann's Titi
Plecturocebus hoffmannsi
(p. 332)

Lake Baptista Titi
Plecturocebus baptista
(p. 333)

Red-bellied Titi
Plecturocebus moloch
(p. 334)

Vieira's Titi
Plecturocebus vieirai
(p. 335)

Brown Titi
Plecturocebus brunneus
(p. 336)

Prince Bernhard's Titi
Plecturocebus bernhardi
(p. 337)

Madidi Titi
Plecturocebus aureipalatii
(p. 338)

Milton's Titi
Plecturocebus miltoni
(p. 339)

Toppin's Titi
Plecturocebus toppini
(p. 340)

Urubamba Brown Titi
Plecturocebus urubambensis
(p. 341)

Groves's Titi
Plecturocebus grovesi
(p. 342)

Plecturocebus titi monkeys compared *pp. 317–318*

EN Rio Beni Titi

Plecturocebus modestus

HB:	♂ approx. 32 cm; ♀ No info. available
Tail:	♂ approx. 40 cm; ♀ No info. available
Wt:	♂ approx. 800 g; ♀ No info. available

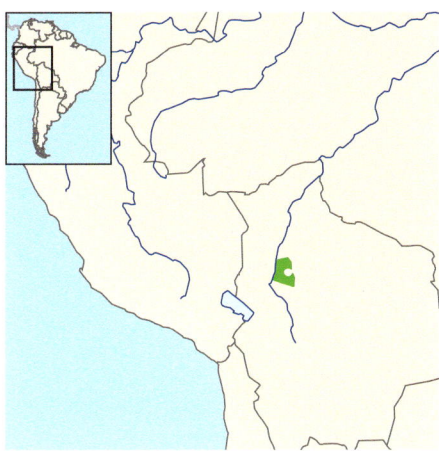

Description: A largely light brown or reddish-agouti colored titi monkey. The ear-tufts are whitish, the forehead, crown and sideburns are reddish-brown agouti and there is a thin black supraocular fringe. The outer surfaces of the limbs are reddish-brown agouti and the hands and feet are blackish sometimes mixed with reddish. The tail is blackish-agouti.

Similar species: **Olalla Brothers' Titi** (*p. 322*) is orange, and the face is fringed with blackish fur. **White-eared Titi** (*p. 320*) is largely buff to orange-agouti and its tail, hands and feet are buff mixed with blackish (rather than being predominantly black). **Madidi Titi** (*p. 338*) has a bright orange crown and orange underparts which contrast with the browner upperparts.

Habitat: Relatively dry forest patches within a forest–savanna mosaic.

Distribution: Bolivia, in the upper Beni River Basin. Occurs east of the Beni River and west to the Manique River in the south-west of Beni Department. The range overlaps with that of Olalla Brothers' Titi, and although the transitional zones between the two species are unclear they appear to occupy different habitats.

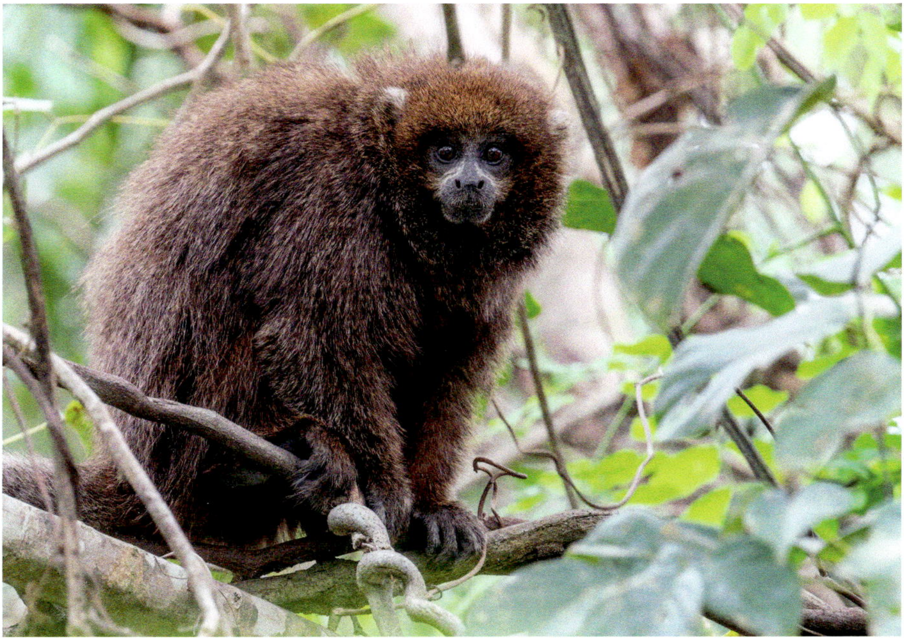

Additional photo *p. 317*

LC **White-eared Titi**

Plecturocebus donacophilus

HB:	♂ 27.8–33.0 cm; ♀ 30.5–42.0 cm
Tail:	♂ 37.2–44.5 cm; ♀ 41–46 cm
Wt:	approx. 800 g

Description: The upper and outer parts of the head and body, and the limbs, are buff or grayish-agouti to orange-agouti. Lacks distinct sideburns and a blackish supraocular fringe. The forehead and crown are uniformly colored. The chest and belly are largely uniform orange. The upper surfaces of the hands and feet are buff or buff-agouti, and paler than the forearms. The tail is buff and mixed with blackish hairs, and is palest at the base.

Similar species: Although their ranges do not overlap, most closely resembles **White-coated Titi**, which is distinguished by its conspicuous whitish ear-tufts and shaggy appearance. **Rio Beni Titi** (*p. 319*), is light brownish to reddish-agouti in color.

Habitat: Tropical humid forests, but apparently restricted to the slightly drier forests of southern Amazonia and seemingly absent from more humid forests in northern Beni Department. Appears to occur in disturbed habitats.

Distribution: East from the Manique River in Beni, Bolivia, and north into Brazil in southern Rondônia, and possibly as far north as the Serra dos Pacaás Novos. In Bolivia, occurs south to the forests around the city of Santa Cruz.

LC **White-coated** (Pale) **Titi**   *Plecturocebus pallescens*

Description: Shaggy-looking with extremely long fur. The upper and outer parts of the head and body, the outer sides of the limbs, hands and feet, and the tail are uniformly pale buff-agouti. The facial skin is largely concealed by facial hairs, and it has conspicuous whitish ear-tufts, a distinct malar stripe and an inconspicuous blackish supraocular fringe.

HB:	♂ approx. 32 cm; ♀ approx. 36 cm
Tail:	♂ approx. 42 cm; ♀ approx. 39 cm
Wt:	approx. 800 g

Similar species: None in range but most closely resembles **White-eared Titi**, which has a less shaggy appearance, lacks conspicuous whitish ear-tufts and appears darker overall.

Habitat: Most common in more humid forest types, particularly riverine forests within the Chaco region. Appears to be restricted to the Chaco and Pantanal, although it may occur also in the neighboring southern portions of the Chiquitania dry forest in Santa Cruz Department, Bolivia. In Paraguay found in continuous xeric forest in the north, and swampland, gallery and palm savanna with forest patches in the south.

Distribution: A few scattered locations in the Chaco ecosystem of S Bolivia, the Gran Chaco of Paraguay south from the border with Bolivia to about 23°S and west from the Paraguay River to about 61°30'W, and in the Pantanal of Mato Grosso do Sul, Brazil.

Plecturocebus titi monkeys compared *pp. 317–318*

CR Olalla Brothers' Titi

Plecturocebus olallae

Description: The sideburns, beard and forehead are blackish, contrasting with a reddish-brown agouti crown. It has conspicuous blackish whiskers and weakly developed whitish ear-tufts. The back and inner surfaces of the limbs are orange, the outer surfaces of the limbs are reddish-brown, and the hands and feet are normally blackish. The tail is entirely dark agouti, contrasting sharply with the orange back.

Similar species: Only likely to be confused with **Rio Beni Titi** (*p. 319*), which differs in being light brownish or reddish-agouti colored and lacks a blackish facial fringe; that species and Olalla Brothers' Titi appear to occupy different habitats.

Habitat: Mainly up to 400 m in patches of gallery forest within forest and savanna mosaics.

Distribution: Bolivia, where occurs in the south-west of Beni Department, with a very restricted range to the east of the Beni River within the gallery forest and adjacent forest islands of the Yacuma River, and with a single additional locality in gallery forest of the Manique River.

HB:	♂ approx. 33 cm; ♀ No info. available
Tail:	♂ approx. 43 cm; ♀ No info. available
Wt:	approx. 800 g

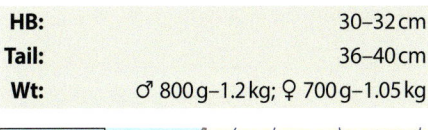 Rio Mayo (San Martín) Titi　　*Plecturocebus oenanthe*

Description: The upperparts are grizzled buffy-grayish to light brown agouti, paler than any other titi other than White-coated Titi (*p. 321*), although the general coloration varies across its range. It has a distinct buff or whitish frontal blaze continuous with long whitish hairs bordering the face. It has whitish malar stripes and narrow black patches around the eyes. The sideburns, crown, outer surfaces of the limbs, the hands and feet and the tail are predominantly

HB:	30–32 cm
Tail:	36–40 cm
Wt:	♂ 800 g–1.2 kg; ♀ 700 g–1.05 kg

buff to light brown agouti (on females the sideburns are white). The inner surfaces of the limbs, chest and belly are orange. Many specimens, especially in the south of the range, are darker and lack the whitish supraorbital blaze and mask. Mixed groups of animals with and without the white mask have been observed.

Similar species: None in range.

Habitat: At 252–1,053 m in low secondary forests and remnant forest along with bamboo stands, vine thickets, fruit crops, and semi-flooded forests dominated by palms. Has been recorded also in riparian forests and in drier zones, and is able to survive in fragmented forest where hunting pressure is low.

Distribution: Upper Mayo River Valley in Peru and farther to the south in the department of San Martín, extending into the Bajo Mayo and Huallaga Central, and almost reaching the Huallabamba River.

Plecturocebus titi monkeys compared *pp. 317–318*

LC Coppery Titi

Plecturocebus cupreus

Description: The sideburns, sides of neck, throat, inner surfaces of the limbs, and the underparts are uniformly reddish, contrasting sharply with the buff-brown agouti crown, dorsum, sides of the body and base of the predominantly buff to white tail. The forehead is reddish-brown agouti.

Similar species: Doubtful Titi (*p. 328*), with which it is parapatric along the Purús River, and **Red-crowned Titi** have a pale blaze on the forehead. **Chestnut-bellied Titi** (*p. 327*) has a broad black frontal blaze and **Stephen Nash's Titi** (*p. 329*) has a black frontal blaze that contrasts with a silvery crown, and has a buff to white tail for three-quarters of its length. **Toppin's Titi** (*p. 340*) has a largely dark tail with a pale tip.

Habitat: Lower and middle strata of primary and secondary forests, including disturbed forests, from 100 m to 300 m. May be restricted to terra firme forests in some areas, where it occurs alongside White-collared (*p. 345*) and Juruá Collared (*p. 348*) Titis.

Distribution: South of the Marañón/Solimões Rivers as far as the east bank of the Ucayali River, in Loreto and northern Ucayali, in Peru. To the south it extends into Acre in Brazil as

HB:	♂ 29–34 cm; ♀ 27–41 cm
Tail:	♂ 39.5–48.0 cm; ♀ 40.5–47.0 cm
Wt:	♂ 1.0–1.2 kg; ♀ No info. available

far as the headwaters of the Juruá and Purús Rivers, and north-east as far as the Juruá River and extending into the Juruá/Purús interfluvium south (and possibly also north) of the Tapauá River. To the east it ranges as far as the west bank of the Purús River.

^{LC} Red-crowned (White-browed) Titi *Plecturocebus discolor*

Description: Distinctive, having a white forehead with a white or buff tuft, contrasting with a dark brown transverse band, or blaze, above the white, and normally contrasting with the reddish crown and sideburns. May have small white patches along the lower jaw which contrast with the reddish sideburns. The sideburns, crown, side of the neck, forearms, lower legs, hands and feet, chest and belly are reddish, contrasting sharply with the agouti-colored back, sides of body and tail. The tail is a mix of brownish and buff-agouti hairs, the last third to three-quarters being mainly buff or white.

Similar species: Most likely to be confused with **Coppery Titi**, which has an adjacent range but is more buffy-brown above and lacks the distinctive pale band across the forehead. **Caquetá Titi** (*p. 330*) has a grey rather than white band across the forehead.

HB:	♂ 30.0–33.5 cm; ♀ 28.5–33.9 cm
Tail:	♂ 39.2–47.0 cm; ♀ 38.2–51.0 cm
Wt:	♂ 850 g–1.01 kg; ♀ approx. 1 kg

Habitat: Dense vegetation in primary and secondary forests, swamp forests and terra firme forest within lowland Amazonian rainforest from 200 m to 980 m, but normally below 500 m.

Distribution: In Peru north of the Marañón River between the Napo and Santiago Rivers, and south of the Marañón River from the Cordillera Cahuapanas to the Ucalayi River and south to the Tambo River. In Ecuador from the Andean foothills east to the Napo/Aguaric interfluvium and north to the Putumayo River, while in Colombia it is found between the Guamués River and San Miguel.

VU **Ornate Titi**

Plecturocebus ornatus

HB:	♂ 31.2–36.0 cm; ♀ 30.4–34.7 cm
Tail:	♂ 40.5–45.0 cm; ♀ 38.3–44.4 cm
Wt:	approx. 1.2 kg

Description: The ears, frontal tuft, and transverse band or blaze are whitish, contrasting with a reddish-brown crown. The sideburns, underparts, and inner sides of the limbs are reddish. The outer surfaces of the upper limbs and sides of the body are buff-agouti, while the outer surfaces of the lower limbs are reddish. These contrast with pale to whitish feet and hands and/or toes and digits. The base of the tail is dark reddish-brown, the rest of the tail generally appearing white.

Similar species: None in range, but the similar **Caquetá Titi** (*p. 330*) has a grey rather than white band across the forehead.

Habitat: Tall primary terra firme lowland Amazon rainforest at up to 1,000 m. Found also in disturbed and secondary forest alongside White-chested Titi (*p. 346*), although that

species occupies taller forest, whereas Ornate Titi occurs in dense lower forest.

Distribution: Restricted to the department of Cundinamarca (Medina) in E Colombia north to the lower Upía/Meta Rivers, and south into the department of Meta, along the base of the Cordillera Oriental and the Sierra de la Macarena to the Guayabero/upper Guaviare Rivers.

LC **Chestnut-bellied Titi** *Plecturocebus caligatus*

Description: The forehead and front of the crown are black, while the remainder of the crown, the neck, back and sides of the body are dark reddish-brown. The sideburns, underparts and the inner surfaces of the limbs are reddish to reddish-brown, and the back is reddish-brown agouti. The forearms and lower legs are dark reddish-brown, while the hands and feet, including the wrists and ankles, are blackish-brown to black. The tail is buff with black hairs, being darker with more black hairs towards the base and whiter towards the tip.

Similar species: None in range, but the superficially similar **Doubtful Titi** (*p. 328*) has a white or buff frontal tuft or blaze, **Coppery Titi** (*p. 324*) lacks the distinctive contrastingly black forehead, and **Stephen Nash's Titi** (*p. 329*) has a white or buff tail.

Habitat: Lower levels of lowland Amazonian rainforest at up to 200 m.

Distribution: Brazil in Amazonas State south of the Solimões River, in the interfluvium

HB:	♂ 31–41 cm; ♀ 30–40 cm
Tail:	♂ 37.0–47.6 cm; ♀ 38.5–46.0 cm
Wt:	approx. 800 g

delineated by the lower Purús, Solimões and Madeira Rivers, south as far as the Ipixuna (or Paranapixuna) River.

Plecturocebus titi monkeys compared *pp. 317–318*

LC Doubtful (Hershkowitz's) **Titi**   *Plecturocebus dubius*

Description: Has a white or buff frontal tuft or blaze, and a narrow black line connecting the blackish ears. The crown, nape, back, rump and upper arms are brownish-agouti. The outer surfaces of the forearms and lower legs are reddish. The beard is deep reddish, and the throat, chest, belly and inner sides of the limbs are reddish to reddish-brown and lack any banding. The hands and feet are blackish-agouti with white or pale fingers and toes. The upper third of the tail is brownish-

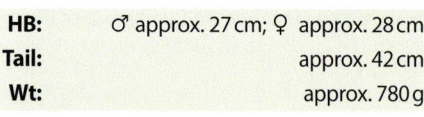

HB:	♂ approx. 27 cm; ♀ approx. 28 cm
Tail:	approx. 42 cm
Wt:	approx. 780 g

agouti like the rump; the remainder of the tail is blackish with a white tip.

Similar species: Coppery Titi (*p. 324*), with which it is parapatric along the Purús River, lacks a white frontal tuft or blaze. **Chestnut-bellied Titi** (*p. 327*) and **Stephen Nash's Titi**, which have adjacent ranges, have a black forehead and front of crown, lacking a white or buff frontal tuft or blaze.

Habitat: Lower levels of lowland Amazonian rainforest from 100 m to 400 m.

Distribution: Western Amazonian Brazil south of the Ituxí River, or possibly the Mucuím River, east-bank tributaries of the Purús River, east as far as the Madeira River south of the town of Humaitá, and west to the Purús River. The southern limits of its range are unknown.

Stephen Nash's Titi   *Plecturocebus stephennashi*

Description: The forehead and front of the crown are black and contrast strongly with the remainder of the crown, nape, dorsum and rump, which are silvery or buff mixed with brownish-agouti or brownish-black. The ears are blackish. The lower arms and legs are normally bright red, but can be dark red. The sideburns, underparts, and inner surfaces of the limbs are bright red and contrast with the upper and outer parts. The upper surfaces of the feet are silvery buff to white, with the

HB:	♂ approx. 27 cm; ♀ approx. 28 cm
Tail:	approx. 42 cm
Wt:	approx. 780 g

lower legs red. The hands are entirely silvery or white. The base of the tail is silvery mixed with brownish agouti to blackish-brown; it becomes white or buff mixed with black, and the final half to two-thirds is entirely white or buff.

Similar species: Doubtful Titi has a white blaze on the forehead, **Coppery Titi** (*p. 324*) lacks the distinctive black forehead, and **Chestnut-bellied Titi** (*p. 327*) has a darker buff tail (all those species have adjacent ranges).

Habitat: Lowland Amazonian rainforest up to 200 m.

Distribution: This species has yet to be found in the wild and its precise type locality is unclear. It has been suggested that it may occur along the right bank of the Purús River between the distributions of Chestnut-bellied and Doubtful Titis in Brazil, but it has also been suggested that it may in fact be a hybrid between these species.

CR Caquetá Titi

Plecturocebus caquetensis

Description: Similar in color and markings to Ornate and Red-crowned Titis, but with a gray-agouti rather than white bar across the forehead. It has agouti patterning above the forehead bar. The crown is light buffy-brown, while the neck, sides, back and tail are mixed grayish-brown and buff-agouti. The final third of the tail is largely white and blackish- agouti. The upperparts are grayish-brown agouti with a slight reddish wash extending from the back on to the upper surfaces of the arms and legs down to the elbows and the knees. The underparts, body, arms, legs and face are sparsely haired chestnut-red, which extends to the upper surfaces of the lower arms and lower legs up to the elbows and knees. The reddish coloration extends on to the underside of the neck and the cheeks, creating a bearded appearance.

HB:	35 cm (one specimen)
Tail:	60 cm (one specimen)
Wt:	No info. available

Similar species: Ornate Titi (*p. 326*) and **Red-crowned Titi** (*p. 325*) most closely resemble Caquetá Titi but both have a white rather than grey forehead band (the three species do not, however, overlap in range).

Habitat: Disturbed humid tropical lowland forest fragments on terra firme from 190 m to 260 m, and in low swampy forest. Occurs also in forest fragments surrounded by pasture.

Distribution: SW Colombia in the south of Caquetá Department between the Orteguaza and Caquetá Rivers near the base of the Cordillera Oriental de los Andes.

Ashy Titi

Plecturocebus cinerascens

Description: The forehead, crown, sides of the body, chest, belly, limbs, and tail are grayish to blackish-agouti, contrasting with tawny or reddish-brown agouti on the middle of the back. The hands and feet are blackish mixed with gray, the tips of the hairs being grayish. The tail is predominantly blackish, mixed with gray, the proximal third being mixed with tawny-agouti. The arms are blackish with grayish-tipped hairs.

The form previously described as **Parecis Titi** is dark grayish-agouti, with a distinctly reddish-chestnut dorsal region. The forehead is light gray with dark gray tips to the fur. The beard and sideburns are grayish-white, as are the hands, feet, and the distal portion of the tail. The remainder of the tail is gray. The ears are grayish-agouti, the chin, throat and ventral parts grayish-white, becoming darker grayish-agouti towards the hind limbs. The inner surfaces of the forelimbs are grayish-white and are continuous with the throat and chin (see illustration on *page 318*).

HB:	32–40 cm
Tail:	39–48 cm
Wt:	780–950 g

Similar species: Hoffmann's Titi (*p. 332*) has pale yellowish sideburns and underparts. **Lake Baptista Titi** (*p. 333*) has uniformly reddish or reddish-brown sideburns, underparts and inner surfaces of limbs. **Milton's Titi** (*p. 339*) has contrastingly dark orangey-brown sideburns and throat, and a predominantly orange tail.

Habitat: Lower levels of primary and secondary Amazon rainforest with dense understorys and vine tangles. Can be in white-sand terra firme forest and riparian forests. Most of the known locations for 'Parecis Titi' are higher-elevation areas in the transition zone between the Amazon rainforest and the Cerrado savanna.

Distribution: Brazil as far east as the west bank of the Abacaxís, Tapajós and Juruena Rivers, in the states of Amazonas and Mato Grosso. Also recorded in the state of Rondônia, extending its distribution to the right bank of the Guaporé River. Likely to occur north to the Paraná do Urariá River, east to the west bank of the Abacaxís River and the west bank of the upper Tapajós River,

Continued on next page...

ASHY FORM

Ashy Titi (*continued*)

west as far as the east bank of the Madeira River, and south to Otoho on the right bank of the Roosevelt River in Mato Grosso.

'Parecis Titi' occurs in southern Rondônia and western Mato Grosso, including part of the Chapada dos Parecis, extending into the Aripuanã/Juruena and Aripuanã/Roosevelt interfluves, including Juruena National Park.

LC # Hoffmann's Titi   *Plecturocebus hoffmannsi*

Description: The head, trunk, and outer surfaces of the limbs are normally grayish-agouti, but some individuals are light gray to almost white. The forehead and crown are grayish. The grayish ears lack white ear-tufts. The sideburns, the underparts and the inner surfaces of the limbs are yellowish to white. The middle of the back is olivaceous-gray. The hands, feet and tail are blackish-agouti.

HB:	♂ 28–36 cm; ♀ 27–35 cm
Tail:	♂ 40–53 cm; ♀ 42–51 cm
Wt:	920 g–1.1 kg

Similar species: Red-bellied Titi (*p. 334*) appears light orange below and has a buff distal half of the tail. **Lake Baptista Titi** has uniformly reddish or reddish-brown sideburns, underparts and inner surfaces of the limbs. **Ashy Titi** (*p. 331*) has grayish to blackish-agouti sideburns and underparts.

Habitat: Swamp forests and terra firme forest in lowland Amazonian rainforest.

Distribution: Brazil south of the Amazon River in the states of Amazonas and Pará. From the right bank of the Abacaxís River to the left bank of the Tapajós River, south to the north bank of Palmares River and north along the south bank of the Paraná do Urariá and Paraná do Ramos Rivers, east along the left bank of the Andirá River and the right bank of the Uíra-Curapá River south of the town of Parintins.

LC Lake Baptista Titi

Plecturocebus baptista

Description: The sideburns, the underparts and the inner sides of the limbs (including hands and feet) are bright to dark reddish or reddish-brown. The head (including forehead and crown), trunk and limbs are grayish to blackish agouti. The forehead does not contrast with the crown, and the ears lack white tufts. The tail is largely blackish agouti to blackish.

Similar species: Hoffmann's Titi has pale yellowish sideburns and underparts, and a black tail. **Ashy Titi** (*p. 331*) has grayish to blackish-agouti underparts.

Habitat: Primary and secondary Amazonian lowland rainforest.

Distribution: Brazil, with a disjunct distribution in the state of Amazonas south of the Amazon River, east of the Madeira River and north of the Paraná do Canumã, Paraná do Urariá and Paraná do Ramos Rivers. May occur also in the lower Uíra–Curupá–Andirá interfluvium. Printes *et al.* (2018) reported its occurrence in Itaituba,

HB:	♂ 30–41 cm; ♀ 33–39 cm
Tail:	♂ 43–49 cm; ♀ 42–50 cm
Wt:	930 g–1.4 kg

Pará, on the west bank of the Tapajós River, extending its distribution and cutting through the supposed range of Hoffmann's Titi from north to south.

Plecturocebus titi monkeys compared *pp. 317–318*

LC Red-bellied Titi

Plecturocebus moloch

Description: The head, trunk, and limbs are buff or grayish to pale brown-agouti. The forehead and crown are normally grayish, but the forehead can sometimes appear distinctly paler. It lacks white ear-tufts. The sideburns, the underparts, and the inner surfaces of the limbs are light orange to buff-orange. The hands and feet are buffy or grey, and the upper half of the tail is blackish-agouti with orange or buff-tipped hairs. The distal half of the tail is buff.

Similar species: Hoffmann's Titi (*p. 332*) has yellowish-white underparts, blackish-agouti upper surfaces to the hands and feet, and a blackish tail. **Vieira's Titi** has a white beard and sideburns, paler orange underparts, and white or light grey upper surfaces and black undersides to the hands and feet. **Groves's Titi** (*p. 342*) has reddish tones to the centre of the back, light ochre underparts, light ochre cheeks, and a black tail with a greyish tip.

HB:	♂ 28.5–37.5 cm; ♀ 27.2–43.4 cm
Tail:	♂ 41–51 cm; ♀ 35.0–54.6 cm
Wt:	♂ 850 g–1.2 kg; ♀ 700 g–1.02 kg

Habitat: Lower levels of lowland Amazonian rainforest, including forest fragments, at up to 300 m.

Distribution: Brazil south of the Amazon River. In Pará from the west bank of Tocantins/Araguaia Rivers west as far as the east bank of Tapajós River and south as far as Ilha do Bananal, north of the confluence of the Mortes and Araguaia Rivers. In Mato Grosso west to the Juruena River, including the headwaters of the Xingu River.

Vieira's Titi

Plecturocebus vieirai

Description: The forehead, sideburns and beard are white and surround the blackish-pigmented face. There are a few white whiskers around the mouth and nostrils. The ears are black. There is a very narrow white stripe between the face and the forehead. The crown and nape are pale grayish agouti and the dorsum, flanks and outer surfaces of the upper limbs are pale grayish-brown agouti. The lower limbs are paler than the upper limbs, with whitish or pale gray upper sides and black undersides to the hands and feet. The underparts and the inner surfaces of the limbs are orange. The base of the tail is darker than the dorsum, but the middle section is similar to the dorsum and the tip is whitish.

Similar species: Red-bellied Titi has an orange beard and sideburns, dark orange underparts, buff hands and feet, and buff or pale brown outer surfaces of the limbs.

HB:	30.0–34.8 cm
Tail:	41–51 cm
Wt:	955 g (type specimen)

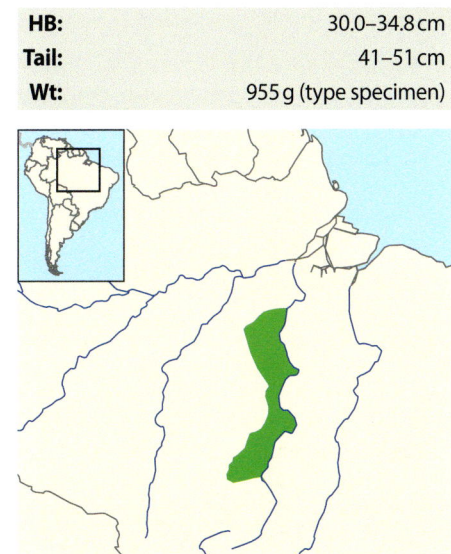

Habitat: The type locality is tropical evergreen rainforest comprising medium-sized trees with lianas and a discontinuous canopy. The understory is relatively dense and well stratified.

Distribution: Brazil, where found in the states of Mato Grosso and Pará. Occurs west of the Xingu River, south of the Iriri River, east of the Iriri and Teles Pires Rivers, and south to the headwaters of the Xingu and Teles Pires Rivers in northern Mato Grosso State and south-central Pará State.

Plecturocebus titi monkeys compared *pp. 317–318*

VU Brown Titi
Plecturocebus brunneus

Description: One of the darkest titi monkeys with brownish or reddish-brown agouti upperparts and sides of the body. The underparts are brownish or reddish. It has dark reddish-brown to blackish sideburns, forehead, forearms, legs, hands and feet, and base of the tail. The remainder of the tail is buff mixed with blackish, and the tip is paler.

HB:	♂ 30.0–34.5 cm; ♀ 30.0–32.5 cm
Tail:	♂ 37.1–42.0 cm; ♀ 38–44 cm
Wt:	♂ approx. 850 g; ♀ No info. available

Similar species: None in range and unlikely to be confused with any species with adjacent ranges.

Habitat: Lower levels of lowland Amazonian rainforest, including isolated forest fragments, at 100–400 m.

Distribution: Found only in the state of Rondônia, in Brazil, between the Mamoré, Madeira and Jiparaná Rivers. The southern limit of its range is thought to be the Serra dos Pacaás.

LC **Prince Bernhard's Titi** *Plecturocebus bernhardi*

Description: The forehead and crown are gray to grayish-black. The upper and outer surfaces of the head, trunk, and limbs are grayish-black, mixed with brownish-agouti or red-brown on the back. The ears are black with whitish tufts. The sideburns, underparts, and inner surfaces of the limbs are dark orange and the hands and feet are white, contrasting with the grayish-black lower limbs. The tail is black except for a contrasting white tip.

Similar species: Milton's Titi (*p. 339*) has a contrasting gray stripe on the forehead, dark orangey-brown sideburns and throat, a uniformly dark gray back and flanks, and a predominantly orange tail.

HB:	36–38 cm
Tail:	approx. 55 cm
Wt:	700 g–1.2 kg

Habitat: Lowland terra firme rainforest, including disturbed, secondary forest and rubber plantations.

Distribution: Restricted to the Brazilian states of Amazonas, Rondônia and Mato Grosso. Occurs between the Madeira and Aripuanã rivers, with the northern boundary at their confluence; the eastern boundary is defined by the Madeira and Jiparaná Rivers in the states of Amazonas and Rondônia. Extends south-west, south of Ji-Parana River near its headwaters, in the municipality of Alto Alegre dos Parecis, Rondônia.

LC # Madidi Titi

Plecturocebus aureipalatii

Description: Distinguished by its golden crown, which extends to just above the ears. The forehead is dark and slightly less golden, and there is no dividing line between the crown and the forehead. The facial skin is black, with a few whitish hairs around the nose. The whiskers and eyebrows are black and the ears are paler, with golden-tipped hairs on the tops of the ears. The upperparts, the sides of the body and the upper limbs are light agouti-brown. The underparts are orange, becoming more deeply colored on the neck, throat and sideburns where the hairs are thicker, while the lower limbs are deep orange-burgundy with the hands and feet deeper burgundy in color. The tail is brown or black above and paler below, with a clearly defined paler whitish tip.

Similar species: Rio Beni Titi (*p. 319*) has white ear-tufts and is more uniformly colored, with the tail lacking a white tip. **Urubamba Brown Titi** (*p. 341*) has a black forehead and brown coloration, and lacks the golden crown. **Toppin's Titi** (*p. 340*) is grizzled brown, with a buff rather than a golden crown.

Habitat: Floodplain forest and piedmont forest on the lower hills of the Andes below 500 m. Most common in dense forest with a higher density of lianas.

HB:	♂ approx. 29 cm; ♀ approx. 32 cm
Tail:	♂ approx. 52 cm; ♀ approx. 48 cm
Wt:	♂ approx. 1 kg; ♀ approx. 900 g

Distribution: The western lowlands of Bolivia, from the piedmont forest at the base of the Andes in the south into lowland humid riverine and floodplain forests. Occurs on the western side of the Beni River, but the eastern and northern limits of its range are unclear. Extends west into S Peru to at least the Tambopata River.

DD Milton's Titi

Plecturocebus miltoni

Description: The forehead has a white (or light gray) stripe contrasting with the dark agouti-gray crown and ears. The sideburns are dark orangey-brown. The ears have discrete white or light gray tufts. The upperparts, the flanks and the outer surfaces of the limbs are uniform gray-agouti. This extends onto the first 10% of the tail, while the rest of the tail is light orange. Some individuals show a blackish tone to the first two-thirds of the tail or sparse white hairs along the tail. The hands and feet are white or light gray and contrast with the gray-agouti on the outer surfaces of the limbs. The throat is dark orangey-brown, and the chest, belly and inner surfaces of the limbs are light orange and contrast with the throat and sideburns.

HB:	29.0–34.8 cm
Tail:	39.5–48.0 cm
Wt:	1.04–1.50 kg

Similar species: Prince Bernhard's Titi (*p. 337*) lacks a contrasting gray stripe on the forehead, and has dark orange underparts. **Ashy Titi** (*p. 331*) has grayish to blackish-agouti sideburns and underparts, and the 'Parecis' form has a reddish-chestnut back.

Habitat: Middle and upper strata of the ombrophilous alluvial forest in the vicinity of the Roosevelt River, where it has been recorded as using the under-canopy of the forest.

Distribution: Brazil, where restricted to the lowland interfluvial region between the Roosevelt (west) and Aripuanã (east) Rivers in the states of Mato Grosso and Amazonas. Appears to be replaced by Prince Bernhard's Titi on the west bank of the Roosevelt River.

LC Toppin's Titi

Plecturocebus toppini

Description: A grizzled-brown titi resembling Coppery Titi (*p. 324*), the crown hairs being buff-tipped and the front edge of the hairy part of the forehead being black, creating an indistinct blackish frontal band. It has a grayish to brownish back, but individuals with an orange back have also been reported. The belly and lower limbs are red, and on the hind legs the red is rather more extensive than in Coppery Titi and covers the knee. The ears are dark reddish-brown. The tail is a mixture of gray and blackish hairs, and those on the first two-thirds are tipped with black rather than white or buff. Specimens from the Atalaya region have a dark tail with a whitish tip.

Similar species: **Urubamba Brown Titi**, with which it may be sympatric, is generally darker with a noticeably darker face and brownish underparts. **Coppery Titi** (*p. 324*) has a predominantly buff to white tail. **Madidi Titi** (*p. 338*), which has an adjacent range, has a golden crown, and its tail is brown or black above and paler below with a pale whitish tip.

Habitat: Assumed to occur mainly in the lower levels of lowland Amazonian rainforest.

Distribution: Believed to occur in SE Peru, SW Brazil and NW Bolivia. The Urubamba

HB:	27–41 cm
Tail:	39–41 cm
Wt:	♂ 1.0–1.2 kg; ♀ No info. available

River is reported to be the western limit of its range, and it is replaced along the Ucayali River by Red-crowned (*p. 325*) and possibly Coppery Titis, but the exact northern limits are unknown. The eastern barrier is probably the Ituxi River and the southern barriers are the Tambopata and Madre de Dios Rivers.

LC Urubamba Brown Titi · *Plecturocebus urubambensis*

Description: The forehead has a jet-black band extending to behind the ears. The ears are black and covered with long black hairs. The cheeks are brownish-agouti with long black tips to the hairs, and appear black from a distance. The chin is brown-agouti and the facial skin black with white hairs on the nose and around the mouth. The upper and outer surfaces of the body, including the crown, upper arms and the legs, are brownish-agouti. The hands and inner surfaces of the forearms are black, while the outer surfaces of the forearms are black up to the elbow, mixed with a small amount of agouti hairs. The knees are darker than the rest of the legs, and the feet are black. The underparts including the inner surfaces of the upper arms and legs are brown-agouti and paler than on the back. The tail is almost black, mixed with some brown-agouti hairs, and the lower half of the tail is lighter with a grayish tip. Some specimens appear brown-agouti rather than black on the forearms.

Similar species: The sympatric **Toppin's Titi** is paler grizzled brown and has reddish-brown rather than black sideburns. **Madidi Titi** (*p. 338*) has a dark rather than black forehead and a distinct golden crown.

HB:	30 cm (type specimen)
Tail:	40 cm (type specimen)
Wt:	No info. available

Habitat: Thought to occur mainly in the lower levels of lowland Amazonian rainforest.

Distribution: Peru, where it has been found on the left bank of the Urubamba River and on the left bank of the upper Madre de Dios River, on the eastern border of Manu National Park, and from farther downriver. Appears to be absent from the Tambopata area and there are no records from east of the upper Madre de Dios River, where it seems to be replaced by Toppin's Titi. May occur in the lowland forest area between the right bank of the Tambo River and the left bank of the Urubamba River, and between the left bank of the Manu River and the left bank of the upper Madre de Dios River. Sympatric with Toppin's Titi on the left bank of the upper Madre de Dios River.

CR Groves's Titi

Plecturocebus grovesi

Description: The upperparts are agouti-gray with a slightly reddish tinge to the center of the back. The back is slightly darker than the lighter gray crown, body sides and outer surfaces of the limbs. The forehead is agouti-black and contrasts with the agouti-gray crown which, in turn, contrasts slightly with the back in lacking reddish tones. The sides of the face, the inner surfaces of the limbs and the underparts (throat, chest and abdomen) are bright reddish-brown. Light yellow hair on the cheeks strongly contrasts with the bright reddish-brown hair on the sides of the face. The hands and feet are dirty white and contrast with the darker gray outer sides of the limbs. The tail is almost entirely black apart from a pale gray tip. This combination of characteristics is unique among titi monkeys.

No information on measurements and weight available

Similar species: Most closely resembles **Red-bellied Titi** (*p. 334*) but in that species the centre of the back has agouti-brownish tones, the underparts are light ochre or orange, the cheeks are uniform light ochre or orange, and the distal half of the tail is buff.

Habitat: Forested habitats, many of which are under threat from encroaching agriculture.

Distribution: The six type specimens come from the vicinity of Alta Floresta, in the north of Mato Grosso, in Brazil. It is believed that the entire distribution lies between the Teles Pires River in the east, the Juruena and Arinos Rivers in the west and the transition from forest to the Cerrado in the south. This distribution is clearly separated from that of other, closely related titi monkeys.

Cheracebus titi monkeys

The six species of white-collared titi monkeys are illustrated here for comparison. They occur in the Amazon and Orinoco River Basins, in Brazil, Colombia, Ecuador, Peru, and Venezuela. Byrne *et al.* (2020) carried out an extensive review of the taxonomy and identification of this genus and concluded that there are in fact five rather than six species as previously thought. However, since then Rengifo *et al.* (2022) have described the population of collared titis in the Nanay-Tigre interfluvium in the Amazon Basin of Peru as a new species, Aquino's Collared Titi *Cheracebus aquinoi*.

Medem's Titi
Cheracebus medemi
(*p. 344*)

White-collared Titi
Cheracebus torquatus
(*p. 345*)

White-chested Titi
Cheracebus lugens
(*p. 346*)

Yellow-handed Titi
Cheracebus lucifer
(*p. 347*)

Juruá Collared Titi
Cheracebus regulus
(*p. 348*)

Aquino's Collared Titi
Cheracebus aquinoi
(*p. 349*)

Cheracebus titi monkeys compared *p. 343*

VU **Medem's** (Colombian Black-handed) **Titi** *Cheracebus medemi*

HB:	♂ approx. 33 cm; ♀ 23.2–36.0 cm
Tail:	♂ approx. 48 cm; ♀ 42.5–49.3 cm
Wt:	♂ approx. 1.1 kg; ♀ 1.15–1.46 kg

Description: A dark titi with a white collar and predominantly blackish head, sideburns, upperparts, underparts, hands, feet and tail. Individuals with yellow hands have also been recorded across its range (Byrne *et al.*, 2020).

Similar species: None in range, but see **Yellow-handed** (*p. 347*) and **White-chested** (*p. 346*) **Titis**, the ranges of which are adjacent: the upper surface of the hands of those species are yellow-orange and yellowish, respectively.

Habitat: Lowland Amazonian rainforest from 100 m to 450 m.

Distribution: The Colombian Amazon between the Caquetá and Putumayo Rivers in the Intendencia del Putumayo and the southern part of the Intendencia de Caquetá.

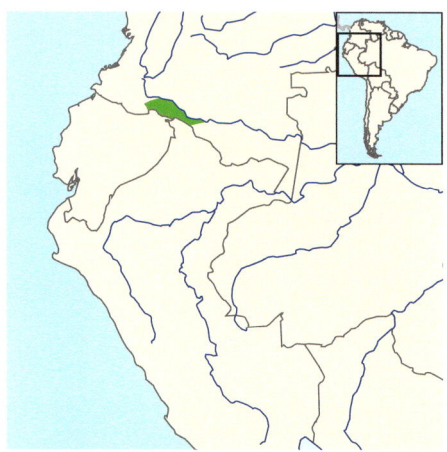

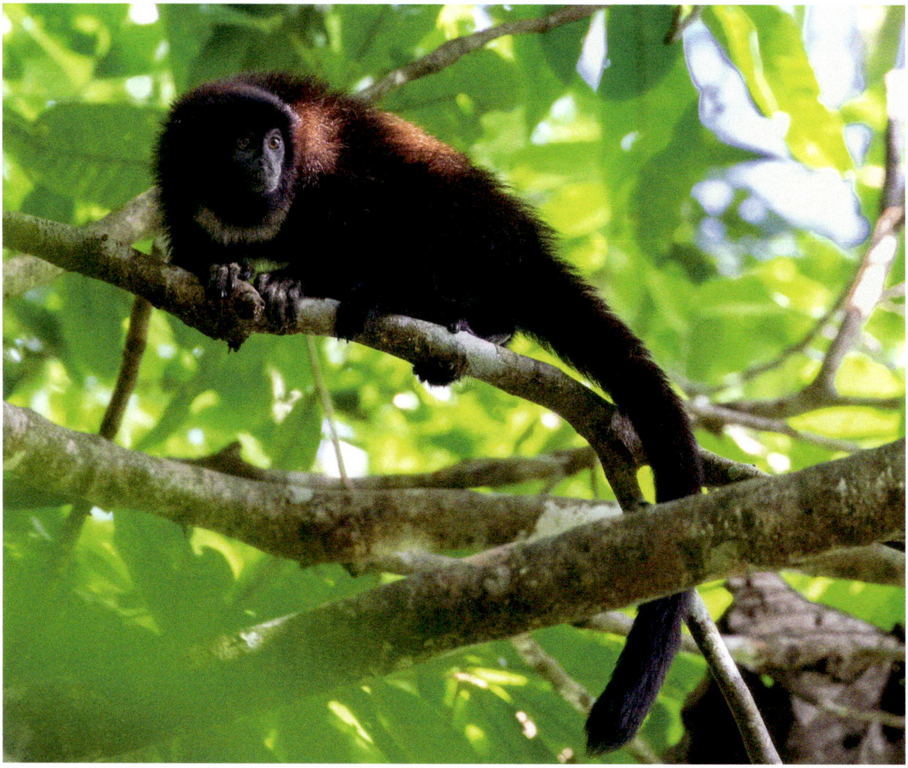

LC White-collared Titi

Cheracebus torquatus

Description: White-collared Titi is now considered to be the *Cheracebus* titi formerly known as Rio Purús Titi *C. purinus* (see Byrne *et al., 2020*). The forehead is blackish, and the crown is bright reddish and normally contrasts strongly with the dorsal colors. The dorsal color varies from dull brown through reddish-brown to reddish. The chest and belly are uniform rusty reddish, but can be deeper red or even more brownish. The collar is white and well defined. The hands are yellowish, the forearms and feet blackish and the legs the same as the chest and belly, blending into the color of the dorsal region on the outer upper hind legs. The tail is black mixed with reddish, the base of the hairs being reddish and the tips black.

Similar species: None in range, but see **White-chested** (*p. 346*), **Yellow-handed** (*p. 347*) and **Juruá Collared** (*p. 348*) **Titis**, the ranges of which are adjacent.

HB:	♂ 29.5–46.0 cm; ♀ 36.8–42.0 cm
Tail:	♂ 39.0–48.5 cm; ♀ 37–51 cm
Wt:	1.0–1.5 kg

Habitat: Terra firme rainforest at up to 200 m, but known also from blackwater seasonally flooded forest in Lago Uauaçú.

Distribution: Amazonian Brazil south of the Solimões River, east of the lower Juruá River and west of the Purús/Tapauá river system. The southern limit of its range is unclear.

LC White-chested Titi

Cheracebus lugens

Description: A highly variable titi monkey with three distinct forms, the third of which was previously considered to be White-collared Titi (*p. 345*) (see Byrne *et al.*, 2020).

Form 1: The upperparts, including the crown, are blackish to dark brown without reddish tones, although the crown may be slightly redder; the chest and belly are blackish, the collar is well defined, the hands yellowish, the forearms, feet and tail blackish, and the legs blackish to dark brown.

Form 2: Resembles Form 1, but has dark mahogany-red to claret-brown upperparts and dark brown to blackish chest and belly.

Form 3: The upperparts are dark mahogany-red to dark brown rufous, the crown is similar to the rest of the upperparts but may be deep dark reddish, and the chest and belly are dull rusty brownish-red. The white collar is poorly defined and often very thin or tipped with red. The hands are yellowish, the forearms and feet blackish, and the tail black mixed with reddish. The legs are more reddish than the chest and belly and blend into the dorsal color on the outer hindlegs.

Similar species: None in range, but see **White-collared** (*p. 345*) and **Yellow-handed Titis**, the ranges of which are adjacent.

Habitat: Terra firme forests, evergreen and montane forests, and transitional forests between tropical rainforests and Caatinga.

HB:	♂ 31.2–35.5 cm; ♀ 30–40 cm
Tail:	♂ 42.0–48.5 cm; ♀ 41–49 cm
Wt:	1.0–1.5 kg

Distribution: The range of this species has been revised following a taxonomic review of the 'collared titi' group (Byrne *et al.*, 2020). It occurs in S Venezuela south of the Orinoco River, west to the Caura River and east to the Caroni River in Bolívar State, in the lowlands of E Colombia north of the Caquetá River as far north as the Tomo River, and in Brazil north of the Japurá and Solimões rivers, west from the Negro and Blanco Rivers into Venezuela.

LC Yellow-handed Titi

Cheracebus lucifer

Description: The feet, tail, head, sideburns, and underparts except for the throat are entirely dark brown and almost blackish, the back and sides of the body being slightly browner and with distinctly or weakly banded hairs. The throat is white and the hands are yellow-orange.

HB:	♂ 36–39 cm; ♀ 34.9–41.0 cm
Tail:	♂ 46.5–48.0 cm; ♀ 45.0–45.6 cm
Wt:	♂ approx. 1.5 kg; ♀ No info. available

Similar species: None in range, but the ranges of the other five *Cheracebus* titis are adjacent (see *pp. 344–346* & *pp. 348–349* for information on these species).

Habitat: Usually 15–25 m from the ground in humid tropical forest in the Amazonian lowlands up to 300 m. Often found in terra firme primary forest on the edge of flooded forest, and in dense vegetation along the edge of rivers and lakes. Avoids swamps and palm forest.

Distribution: Brazil between the Solimões and Japurá rivers; Colombia between the Caquetá River below the mouth of Caguán River, and Putumayo and Amazon Rivers in the departments of Caquetá, Putumayo and Amazonas. Ecuador between the upper Aguarico and Putumayo Rivers in Napo Province, and in Peru in northern Loreto, north of the Amazon River between the Putumayo and Napo Rivers.

347

Cheracebus titi monkeys compared *p. 343*

LC Juruá Collared Titi

Cheracebus regulus

Description: Has a well-developed white collar across the throat. The rest of the underparts and the upperparts are brown or blackish. The hairs above and behind the ears are banded, the sideburns brownish, and the strongly contrasting crown is reddish. The inner arms are blackish, the hands orange, and the feet and the tail blackish.

Similar species: None in range, but see **White-collared** (*p. 345*) and **Yellow-handed** (*p. 347*) **Titis**, the ranges of which are adjacent.

HB:	♂ 38–44 cm; ♀ 37–45 cm
Tail:	♂ 48–49 cm; ♀ 44–49 cm
Wt:	1.0–1.5 kg

Habitat: Lowland Amazonian rainforest up to 200 m.

Distribution: Brazil in the state of Amazonas between the upper Solimões River, the lower Javarí River, and the west bank of the Juruá River from the mouth at the Solimões River to about 7°S.

NE # Aquino's Collared Titi   *Cheracebus aquinoi*

Description: Readily identified from all other *Cheracebus* titis by a combination of characteristics including its overall reddish-brown coloration and intense reddish-brown crown. The neck, forelimbs to elbow, and hindlimbs to knee are brownish or reddish-brown, the forearm to wrist dark brownish. The hands are creamy-white above with a yellow-orangish tone in some individuals. The legs have reddish-brown hairs, the thighs are whitish and the feet have black hairs. The underparts are reddish-chestnut. The collar is cream-colored and resembles a bow-tie

No information on measurements and weight available

with a band that does not extend laterally. The facial skin is blackish but has a slightly whitish appearance as a result of being sparsely covered with short white hair. There is a narrow black band between the face and the crown. The proximal third of the tail is blackish-chestnut, and the distal two-thirds of the tail are black and contrast with the reddish-brown body (Rengifo *et al.*, 2022).

Similar species: None in range, the closest congener being **Yellow-handed Titi** (*p. 347*), which occurs to the north of the Napo and Amazon Rivers.

Habitat: No information available but assumed to be similar to Yellow-handed Titi.

Distribution: Known from just 15 localities in a small area in the Peruvian department of Loreto, in the interfluvium of the Nanay and Tigre Rivers. To the north-west the species' distribution is unclear, but it may be limited by the confluence of the Quebrada Alemán and the Pucacuro River, a right bank tributary of the Tigre River.

Callicebus titi monkeys

The five species of Atlantic Forest masked titi monkeys are illustrated here for comparison. They are all endemic to Brazil and are known from NE Brazil, south of the São Francisco River, south through the states of Bahia, Espírito Santo and Rio de Janeiro as far west as the Paraná and Paranaíba Rivers, and south to the Tieté River in the state of São Paulo.

Black-fronted Titi
Callicebus nigrifrons

Masked Titi
Callicebus personatus
(buff form)
(*p. 352*)

Masked Titi
Callicebus personatus
(orange form)
(*p. 352*)

Southern Bahian Titi
Callicebus melanochir
(*p. 353*)

Blond Titi
Callicebus barbarabrownae
(*p. 354*)

Coimbra-Filho's Titi
Callicebus coimbrai
(*p. 355*)

Black-fronted Titi

Callicebus nigrifrons

Description: The forehead and crown are blackish, the back of the crown grading into coarsely banded brownish-agouti or orangey-brown on the nape. The throat and chest are pale brownish-agouti, the ears, hands and feet black, and the tail orange.

Similar species: Masked Titi (*p. 352*) has a blackish forehead, cheeks, ear-tufts and throat, and lacks contrast between the crown and the orange or buff nape.

Habitat: Atlantic Forest in both mature rainforest and disturbed fragments.

Distribution: SE Brazil, including much of the states of São Paulo, southern Minas Gerais and eastern Rio de Janeiro. North of the Tietê River and east of the Paraná and Parnaíba Rivers, and on both margins of the upper São Francisco River. In the east as far as the Mantiqueira and Espinhaço ranges.

HB:	♂ 30.0–39.5 cm; ♀ 34.5–36.0 cm
Tail:	♂ 45.5–48.0 cm; ♀ 49–50 cm
Wt:	♂ 1.05–1.65 kg; ♀ 1.0–1.6 kg

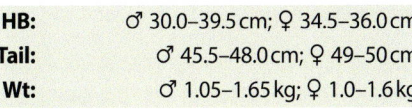

Callicebus titi monkeys compared *p. 350*

VU **Masked** (Atlantic) **Titi**

Callicebus personatus

Description: Two color forms exist (see illustrations on *page 350*). One has a largely uniform buff body and tail, the other a largely uniform orange body and tail. The forehead, crown, sideburns, throat, lower arms, hands and feet are blackish.

Similar species: Black-fronted Titi (*p. 351*) has a blackish front of the crown grading into agouti on the nape without any line of demarcation, and has a pale brownish-agouti throat and chest. **Southern Bahian Titi**, which has an adjacent range, lacks Masked Titi's distinctive sharply defined blackish forehead.

Habitat: Primary and disturbed Atlantic Forest at up to 1,000 m.

Distribution: SE Brazil from the state of Espírito Santo south to northern Rio de Janeiro State. Inland into north-west Minas Gerais as far east as Teófilo Otoni and the east bank of the Jequitinhonha River. It is unclear whether this or another species of titi occurs north-west of this river. It occurs west along the Doce River valley to the

HB:	♂ 35–42 cm; ♀ 31–40 cm
Tail:	♂ 47–55 cm; ♀ 41.8–56.0 cm
Wt:	♂ 1.05–1.65 kg; ♀ 1.0–1.6 kg

Mantiqueira Mountains in Minas Gerais, but the northern limit of its range is uncertain.

Southern Bahian Titi

Callicebus melanochir

Description: The least colorful *Callicebus* titi, being predominantly grayish-agouti, buff or pale brownish-agouti with finely banded hairs. The hands, feet and facial fringe are blackish.

Similar species: None in range, but **Masked Titi**, which has a distinctive sharply defined blackish forehead, and **Blond Titi** (*p. 354*), which is much paler with a buff crown, sideburns, throat, trunk and limbs, have adjacent ranges.

Habitat: Atlantic Forest in areas of degraded habitat.

Distribution: Brazil in the state of Bahia between the right bank of the Paraguaçu River in the north and the left bank of the Mucuri River in the south. Range extends west as far as Montes Claros in the state of Minas Gerais.

HB:	♂ No info. available; ♀ 33–37 cm
Tail:	♂ No info. available; ♀ 39.5–51.0 cm
Wt:	♂ No info. available; ♀ approx. 1.4 kg

CR Blond Titi

Callicebus barbarabrownae

Description: Pale buffy-gray, with a black supraocular stripe. The forehead and crown are predominantly buff, raised buff hairs on the crown having fine blackish tips. The sideburns, nape and shoulders are pale buff, and the ears and facial skin are blackish. The hairs on the back and sides of body, including the forearms and legs, are banded reddish-yellow and blackish-brown, with paler upper arms and thighs. The underparts are almost entirely buff and the hands and feet are blackish. The tail is mainly orange with a yellowish base.

Similar species: None in range, but **Southern Bahian Titi** (*p. 353*), which is much darker, and predominantly grayish-agouti or pale brownish-agouti, lacking the buff crown, sideburns, throat, trunk and limbs of Blond Titi, and **Coimbra-Filho's Titi**, which has a black forehead and crown, have adjacent ranges.

Habitat: Dense fragments of Caatinga from 240 m to 900 m.

HB:	♂ No info. available; ♀ 33–36 cm
Tail:	♂ No info. available; ♀ 39.5–43.0 cm
Wt:	No info. available

Distribution: Brazil in the states of Bahia and Sergipe, from the Paraguaçu River Basin to the border between Bahia and Sergipe along the margins of the Real River. Replaced by Southern Bahian Titi to the south of the Paraguaçu River in the municipality of Igrapiuna. Occurs farther west in Feira de Santana, on both banks of the Paraguaçu River.

EN Coimbra-Filho's Titi

Callicebus coimbrai

Description: The forehead, crown and ears are black, while the sideburns, cheeks, back of the head, and nape are pale buff. The trunk is buff, the tail orange, and the hands and feet blackish. The front part of the back has a saddled appearance.

Similar species: Distinguished from all other *Callicebus* titis by the black forehead, crown and ears, contrasting sharply with the buff sideburns, cheeks, back of head, nape and trunk.

Habitat: Up to 300 m in primary tropical rainforest, disturbed evergreen and semi-deciduous forests and in dry forests, including forest fragments.

Distribution: Brazil in coastal rainforests at the mouth of the São Francisco River in Sergipe and the mouth of the Itapicuru River in northern Bahia. Occurs south to Lamarão do Passé. It meets Southern Bahian Titi (*p. 353*) along the lower Paraguaçu River, with Coimbra-Filho's Titi restricted to the north bank, and Southern Bahian Titi to the south bank.

HB:	♂ No info available; ♀ 34.7–36.0 cm
Tail:	♂ No info available; ♀ 45.3–48.4 cm
Wt:	♂ No info available; ♀ 1.03–1.30 kg

Saki Monkeys

IUCN Primate Specialist Group recognizes five species of *Chiropotes* bearded sakis and 16 species of *Pithecia* sakis. Barnett *et al.* (2012) proposed the adoption of the name cuxiús as an alternative to bearded saki, but most authors continue to use bearded saki and this nomenclature is followed here.

Marsh (2014) provided an extensive review of the taxonomy of the genus *Pithecia* and described a number of new species which were recognized by the IUCN in their March 2021 revision of primate taxonomy. This taxonomic approach has been adopted here, although Serrano-Villavicencio *et al.* (2019) have subsequently challenged the validity of Mittermeier's Tapajós Saki *P. mittermeieri*, Rylands's Bald-faced Saki *P. rylandsi* and Pissinatti's Bald-faced Saki *P. pissinatti*, which they consider to be subspecies of Gray's Bald-faced Saki *P. irrorata*. To date there are, however, no genetic data to support this supposition, so the taxonomy here follows Marsh (2014). Marsh does not recognize **'Jamari Saki'** as a distinct species, but gives a detailed description of this form in an appendix to her paper. The females are distinct from other females in the genus, but there remains some doubt as to whether this represents a full species as the males can resemble Gray's Bald-faced, Rylands's Bald-faced and Mittermeier's Tapajós Sakis and/or exhibit characteristics of more than one of these species. All specimens were collected in the inundation area of the Jirau Dam on the Jamari River in Rondônia, Brazil. The *Pithecia* accounts here quote extensively from the extremely detailed descriptions and the distributions in Laura Marsh's paper, which is the essential reference for anyone interested in this group, and her assistance in this respect is gratefully acknowledged.

Sakis are large monkeys with a long, slender body and limbs with coarse, long fur and a long, non-prehensile bushy tail. They occur widely across the Amazon Basin and are diurnal and arboreal, and usually found in the middle and upper strata of the forest between 10 m and 25 m, where they are often difficult to find as they spend long periods sitting motionless in thick vegetation. *Pithecia* sakis normally occur in small groups of 2–8 individuals, although most groups consist of 3–4 individuals, whereas *Chiropotes* sakis have been reported in larger groups of up to 60 or more. Sakis feed mainly on seeds, but also eat fruit, green leaves and insects.

Gray's Bald-faced Saki (LEFT); Uta Hick's Bearded Saki (RIGHT)

White-faced Saki

Pithecia pithecia

Description: Mature males are black with little or no grizzling on the upperparts. The hands and feet are black with little or no hair. White crescents cover the circumference of the face, normally with a thin black line extending from the top of the forehead down to between the eyes. The facial skin is black. Juvenile and sub-adult males are brown to coppery above with gray-brown-cream grizzling on the hands, feet and back. They have a bright orange chest and shaggy, white, thin facial hair that becomes short and thick on adult males. Mature females are brownish to grayish with some off-white or cream grizzling on the upperparts, a light to bright orange chest, and black hands and feet. They have a white star-shaped mark in the middle of the forehead, distinct white to buff-orange malar stripes and dark grey to blackish facial hair.

HB:	♂ 33.0–41.5 cm; ♀ 32.3–41.5 cm
Tail:	♂ 39.8–45.5 cm; ♀ 37.0–43.5 cm
Wt:	♂ 1.65–2.50 kg; ♀ 1.55–1.75 kg

Similar species: Golden-faced Saki (*p. 358*) males have an orange to dark ochraceous facial disk, and distinct white to buff hairs along the lips; females have pinkish to gray facial skin and bright orange malar stripes.

Habitat: Humid tropical forest and in both primary and secondary forests, with the highest densities in mature forest.

Distribution: Venezuela, Guyana, Suriname French Guiana, and N Brazil in Roraima, Amapá and parts of Pará.

MALE

FEMALE

LC Golden-faced Saki

Pithecia chrysocephala

Description: The facial skin is black and bare around the eyes, nose and chin, and the upper lip has thick cream to light orange hairs. Younger males resemble White-faced Sakis of the same age, being buff gray-brown above including the arms, legs and hands, with a bright orange chest. Sub-adult males are more grizzled and grayer than females and have a shaggy less well-defined orange facial disk than adult males. Females are brownish to black above with some grizzling and have buff to tan limbs and tail. The hands and feet are black. The facial skin around the eyes, nose and chin is bare and pinkish to gray. The facial hair forms a black horseshoe with a white or buff-orange star at the top, and pale or whitish crown hair extends over the forehead. Orange malar stripes extend from under the eyes. The chest and belly are dull to bright orange.

Similar species: White-faced Saki (*p. 357*) males have a white to buff facial disk; females have a white star-shaped mark in the center of the forehead and dark to black facial hair and white to buff-orange malar stripes.

Habitat: Primary and secondary forests, but tends to avoid flooded forests.

HB:	♂ 28–46 cm; ♀ 30–33 cm
Tail:	♂ 34–45 cm; ♀ 33–40 cm
Wt:	approx. 1.9 kg

Distribution: Brazil, north of the Amazon River, along both sides of the Negro River east to the Nhamunda River, where populations on the east side may be a mix of White-faced and Golden-faced Sakis. The distributions of White-faced and Golden-faced Sakis farther north in the states of Amapá, Roraima and Pará are unclear.

MALE

FEMALE

LC Hairy Saki

Pithecia hirsuta

Description: Males and females are black with some grizzling, but the grizzling is shorter and sparser than that on other sakis. Females are more grizzled than males but both sexes generally have whitish hands. Males have a predominantly gray-brown to blackish gray-brown head, brownish to blackish chest hairs, and a black chest ruff below an area of bare skin. The lips and malar stripes are white to cream. There are small bare spots of pinkish to light-colored skin above the eyes, while the rest of the facial skin is largely black. Males have a bulbous sac under the neck. Females have white malar stripes but lack the males' conspicuous white hairs across the lips. The hair around the face is blackish. The skin on the face is pinkish around the eyes and above the muzzle and chin.

HB:	♂ 48–50 cm; ♀ 41–43 cm
Tail:	♂ 41–43 cm; ♀ No info. available
Wt:	No info. available

Similar species: Hairy Saki is much plainer than **Geoffroy's Monk Saki** (*p. 362*). The latter has brown hairs covering the whole face, although juvenile males can have a whitish face. The juvenile male Hairy Saki has grayish-brown facial rings contrasting with white malar stripes and obvious white lips, and almost resembles the adult. Female Geoffroy's Monk Saki has soft, shaggy brown and white facial hair with white malar stripes.

Habitat: Found in both terra firme and white-water seasonally flooded forests, as well as in palm swamp forests. Most often in the middle and upper canopies.

Distribution: Occurs in an area bordered in the east by the Negro River in Brazil, north of the Solimões River in Brazil and Peru, north of the Napo River in Peru, and south of the Caquetá River in Colombia (the Japurá River in Brazil). It is not known how far west the range extends.

MALE

VU Miller's Saki

Pithecia milleri

Description: Males have black upperparts, limbs and tail with long pale yellowish- or whitish-tipped hairs. The face is sparsely covered by short whitish hairs and the front half of the head by short reddish-brown hairs. The underparts are thinly haired. The front of the neck is naked, the chest is dark brown with yellowish-brown tips, and the throat and belly are covered by dark brown hairs with whitish tips. The hands are yellowish-white, and the feet are whitish grizzled with black. Juvenile and sub-adult males have soft reddish-brown fur. Males have white hairs along the malars and lips as on Hairy Saki, but more conspicuous white under the eyes and often above the eyes as well. They have dark sleeves on the forearms, and cuffs on the hind limbs, where there are patches of brown mixed with black. The ruff ranges in color from light tan to black. Females have a shaggier appearance, with a whiter face sometimes with a distinct white band across the forehead. They also have longer, shaggier white malar stripes, and indistinct white across the lips and whitish hands and feet.

Similar species: More grizzled than **Hairy Saki** (*p. 359*), and females are noticeably grayer. There is more variation between males and females than in Hairy Saki.

HB:	♂ 33–48 cm; ♀ 39–42 cm
Tail:	40.0–49.6 cm
Wt:	♂ 2.6–2.8 kg; ♀ approx. 2.2 kg

Habitat: Terra firme and white-water seasonally flooded forests, as well as palm swamp forests. Most often in the middle and upper canopies.

Distribution: Colombia from the foothills up to 700 m around Florencia, east to La Macarena, and south of the Caquetá River to at least Puerto Leguizamo. In Ecuador found south of the Putumayo River and north of the Napo River, but it is unclear how far east it occurs.

MALE

FEMALE

LC Burnished Saki

Pithecia inusta

Description: Both sexes have a diamond shape above the eyes in the center of their forehead, although it is more distinct and less haired in females. Males are black with extensive bold grizzling, while younger males have light grizzling across the back. The ruff is dark brown at the base with light tan tips on younger animals, but becomes brighter with buff-tan or orange tips on older males. The chest is sparsely covered in black hair. The forearms can appear to have a triangle where half or more of the forearms are not grizzled at all and the wrists are often black or lack grizzling. The hands and feet are white, some individuals with a distinct dark 'V'-shaped marking on the hind feet. The face is covered in short tan to off-white hairs, some animals having white malar stripes blending in with the rest of the facial hair. The muzzle is bare, with dark skin, and with scattered short white hairs along the lips. Transitional males and even sub-adults may appear to have a solid white face like that of older adult females. Older males have a dark line running up the forehead. Females are grizzled above, with a light tan-colored ruff and a mostly white to tan face, although juvenile females have shaggy grayish to brownish hair covering the face. The forearms are grizzled with white and have white wrist cuffs.

HB:	♂ 37.0–42.5 cm; ♀ 37.5–42.0 cm
Tail:	♂ 30.5–48.8 cm; ♀ 37.5–54.5 cm
Wt:	No info. available

Similar species: Most closely resembles **Geoffroy's Monk Saki** (*p. 362*), but has a uniformly colored face, whereas juveniles and females of that species are normally two-toned brown and white.

Continued on next page...

MALE

FEMALE

Burnished Saki (*continued*)

Habitat: Found in both terra firme and white-water seasonally flooded forests, as well as in palm swamp forests. Most often in the middle and upper canopies.

Distribution: Peru in the Ucayali River watershed, on both sides of the river south from Sarayacu, especially in the upper reaches towards the Urubamba River/ Tambor River split. Occurs also in the foothills along the Pachitea River below Oxapampa and in the upper Juruá River region in Brazil.

LC Geoffroy's Monk Saki · *Pithecia monachus*

Description: Highly variable. Males are black above with little grizzling. The hairs on the forearms and chest have whitish tips, while the ruff has dark roots and light tan to orange tips. Younger males are more grizzled, with a ruff which appears more extensive owing to the grizzling across the arms. The hair from the whorl on the nape does not extend as far over the face as on some sakis, and is darker with less grizzling. Males have black wrists, and the hands and the feet are mottled black and white, becoming whiter with age. Adult males' faces are brown, darkening with age. The facial hair is flattened, with a crease up the forehead ending in a star between and just above the eyes. They have white malar stripes and fine white hairs along the upper lip. The muzzle is black, often with fine whitish hairs. They also have brown tops to their faces, with a white band around the bottom of the face.

HB:	♂ 39.8–46.0 cm; ♀ 37–46 cm
Tail:	♂ 30.5–50.0 cm; ♀ 41.1–48.0 cm
Wt:	♂ 2.5–3.1 kg; ♀ 1.3–2.5 kg

FEMALE

Younger males have a white headband with some brown down the center. Females are black and slightly more grizzled than males, especially on the forearms. Young females can have brownish to whitish mottled faces. Older females have a soft, dark brown forehead with shaggy white below, appearing two-toned. Malar lines are shaggy and white, and can blend in with the white lower cheek hair. Females have a darker brown or black ruff than males, occasionally with lighter tips.

Similar species: Older males resemble adult female **Burnished Saki** (*p. 361*), the top of the face being brown and the bottom white, although the facial hair appears more as a band around the face rather than completely covering the face. Older males have a more pronounced throat sac than other sakis.

Adult females normally have a brown-topped forehead with white below, rather than the nearly entirely white face of adult female Burnished Saki.

Habitat: Found in both terra firme and white-water seasonally flooded forests, as well as in palm swamp forests. Most often in the middle and upper canopies.

Distribution: Peru between Solimões River, lower to middle Ucayali River, and the lower Yavarí River, south to at least Orellana/Sierra Divisor, and the lower reaches of the Javarí to Juruá Rivers in Brazil. In Peru also in the Tahuayo River region north to the Maniti and Orosa Rivers, the Yavarí-Mirim River and Yavarí River region north-east to San Fernando, Leticia and Tabatinga.

MALE

DD Cazuza's Saki

Pithecia cazuzai

Description: Males have black upperparts with hairs sparsely tipped with white, and have extensive grizzling over the shoulders and arms. The chest has a small black ruff under the bare throat patch. The wrists and ankles have brown cuffs, while the hands and feet are off-white grizzled with brown/black hairs. The facial skin of adult males is black. In sub-adults it is unpigmented except for black over the nose and muzzle and under the chin, with a black diamond in the center. There is a loose arch of white hairs over the forehead and down the sides of the face, and a distinct line through the center of the forehead. Thin white malar stripes continue up under the eyes. The hair on the lips is white. Females also have black upperparts with sparse white grizzling. Their hands and feet are white, with slightly brownish cuffs on the wrists. The underparts are sparsely covered in fine black hairs with a short dark brown ruff with tan tips. The hair on the head forms a white arch which ends below the ears, where it becomes black and

HB:	♂ approx. 60 cm; ♀ 35–48 cm
Tail:	♂ approx. 30 cm; ♀ 43–49 cm
Wt:	♂ approx. 3.2 kg; ♀ approx. 2.75 kg

white along each side of the face. There is a white diamond on the forehead with bare skin in the middle. The facial skin is black, except above the eyes where the eyelids or the skin are unpigmented or pinkish. The thick, white malar stripes are conspicuous and the lips are lined by white hairs. Younger females are more grizzled above, have a lighter ruff, a whiter facial arch covering most of the face, and pinker or less pigmented facial skin.

Similar species: Males differ from **White-faced Saki** (*p. 357*) in being coarsely black with very light tips of white rather than silky black, and in having a short, black ruff. Their faces have a diffuse white ring that easily distinguishes them from all sakis with white facial hair. **Equatorial Saki** is easily distinguished by its heavily grizzled appearance and bright orange to ochre ruff.

Habitat: Found in both terra firme and white-water seasonally flooded forests, as well as in palm swamp forests. Most often in the middle and upper canopies.

Distribution: Brazil, where found only south of the Solimões River on either side of the Juruá River at Fonte Boa and Uarini.

FEMALE

Equatorial Saki

Pithecia aequatorialis

Description: Male is black with long grizzled white tips to the hairs, and a bright orange to ochre ruff. The hands and feet are white. The face is black with a dense white horseshoe-shaped band of hair around it. White patches above each eye connect to the white band around the face. It has white malar stripes and lip hairs. Young males have a distinctly white head. Females are grizzled gray over black and appear grayer. The orange ruff is less extensive than on males. The forearms are a mixture of black, white and brown hair. The hair around the face creates a shaggy grayish-white band with distinct white malar stripes. The nose and muzzle are black and the hands and feet white.

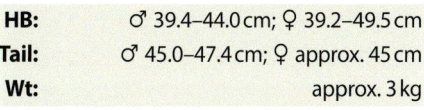

HB:	♂ 39.4–44.0 cm; ♀ 39.2–49.5 cm
Tail:	♂ 45.0–47.4 cm; ♀ approx. 45 cm
Wt:	approx. 3 kg

Similar species: Males are distinct from **Napo** (*p. 366*) and **Isabel's** (*p. 367*) **Sakis** in having a white arch around the face and an extensive bright orange ruff, together with a generally grizzled appearance. Napo Sakis have white eyespots above the eyes and a white headband between the ears over the crown. Isabel's Saki has white eyespots. Napo's and Isabel's Sakis have an orange ruff, but it is not so extensive as on Equatorial Saki. Females are distinct from Napo and Isabel's Sakis in having more extensive white around the face and being much grayer in general.

Habitat: Moist subtropical lowland forest, including seasonally flooded white-water forest.

Distribution: Peru south of the Napo and Curaray Rivers as far as the east bank of the Tigre River in the west. On both sides of the Amazon River at Iquitos, but those on the east side are largely released animals or escapes. There is also a report from Ecuador: from Enkerida, in Pastaza Province.

FEMALE

365

LC Napo Saki

Pithecia napensis

HB:	36.2–43.0 cm
Tail:	43.0–47.8 cm
Wt:	approx. 3 kg

Description: Males are large and bulky, with black dorsal hair with white grizzling. There is no grizzling on the legs. The hands and feet of both sexes are white. The partial facial disk of dense white hair extends to the ears, and is white with gray stippling towards the bottom. White eyebrow patches extend on to the striking solid white crown. Sub-adult males have white eyespots and a slight white headband between the ears. The facial skin is unpigmented except for the black nose and muzzle. It has an enlarged conspicuous bright orange to rusty ruff. The remainder of the underparts is largely naked, with sparse grayish-black hair. Females appear grayer, and the ruff is dark brown with light tan to buff tips. The facial hair forms a ring covering the face. The facial skin is darker than in males, with areas of unpigmented, lighter skin. There are distinct white malar stripes from under the eyes, following the curve of the muzzle and continuing under the chin to form a half circle.

Similar species: The white on and around the face is never so distinctly marked in a ring as it is on female **Equatorial Saki** (*p. 365*). Equatorial Saki is less bulky, has a more extensive orange ruff and has a clearly defined white horseshoe-shaped ring around the face. Napo Saki is most similar to **Isabel's Saki** (see that species for specific differences).

Habitat: Found in both terra firme and white-water seasonally flooded forests, as well as in palm swamp forests. Most often in the middle and upper canopies.

Distribution: Eastern Ecuador south of the Napo River, from Coca in the west into Yasuni National Park in the east. The distribution along the Napo farther east is unclear, but it has been found north of the Curaray River as far south as where it meets the Napo River. In the west this species occurs to the west of the Coca River and west of the Napo River in Baeza and Estribaciones at up to 1,500 m, south to the Bobanaza, Copataza, Pastaza and Maca Rivers. Found also in Peru between the Curaray and Napo Rivers to the border with Ecuador. It is unclear whether animals on the north bank of the Marañón River to the west of the Tigre River are Napo or Isabel's Saki.

IMMATURE MALE

Isabel's Saki

Pithecia isabela

Description: Males are black with short white grizzling, but can appear coppery and almost shiny. The whirl of hair at the base of the neck is black with little stippling. The hands and feet are white, with a distinct black 'V'-shaped marking extending from the wrists and ankles down the backs of the hands and feet. The forearms have light grizzling while the hind limbs have almost no grizzling. The tail is lightly grizzled. The facial ring is black in adult males, and dark brown in young males, with some light stippling mainly on the forehead. The facial disk is dark with light whitish or grayish stippling, and it lacks white bands. The eyespots can be small spots, whitish patches or long stripes which merge with a distinct line between them, extending to the facial disk. Both sexes have white malar stripes and some white hairs covering the nose. The ruff is dark rusty-orange but appears brown. Females are similar, with slightly more grizzling than males. The ruff is black with dark brown tips. The facial skin is black, and the disk is black in young females and blackish gray-brown in older females. Adults may have white in the facial disk, white eyebrows and a small white star-shaped marking just above their eyes.

HB:	36.3–43.4 cm
Tail:	40.0–48.6 cm
Wt:	No info. available

Similar species: Napo Saki is white on the crown, with white grizzling throughout the facial disk, while Isabel's Saki has smaller spots closer to the eyes, and little or no white grizzling around the rest of the face. Napo Saki is also bulkier, with a more extensive bright orange ruff, while Isabel's normally has a shorter, darker brown-orange ruff. Male Isabel's Saki is less grizzled. Distinguished from **Equatorial Saki** (*p. 365*) by fully white half-circles around its face, with an extensive orange ruff on its chest. Younger males are identified by their plain face except for small white spots above the eyes, and they have little grizzling across the back. Females have much less grizzling than Napo and Equatorial Sakis.

Habitat: Found in both terra firme and white-water seasonally flooded forests, as well as in palm swamp forests. Most often in the middle and upper canopies.

Distribution: Peru along the Samiria River at Santa Elena and at the Base Atun and Estacion Biologica Pithecia. Also along the Yanayacu River, at Quebrada Sapote near the Ucayali River, and may be present near Sarayacu, on the west side of the Ucayali River.

IMMATURE MALE

LC **Buffy Saki** *Pithecia albicans*

Description: Predominantly orange to blond. Adults are paler above and darker below on the head, arms and legs, the chest and belly being sparsely haired. The back is black from the nape to the tip of the totally black tail. The wrists are usually black, while the hands and feet are off-white. Adults have a large throat patch with a light orange ruff that is most pronounced on males. Males have largely black facial skin with small bare pinkish skin patches above the eyes, and distinct white to cream eyebrows and upper-lip hairs. The muzzle has sparse whitish hairs. Females have white malar stripes, often extending to the cheeks as short white to cream hairs. Older females have more extensive white covering the face than do younger females, sometimes including a ring of whitish hairs around the outside of the face, which is thereby made to appear almost entirely covered in white hairs.

Similar species: None in range.

HB:	♂ 39.5–57.0 cm; ♀ 38–52 cm
Tail:	♂ 40–57 cm; ♀ 40.5–53.0 cm
Wt:	♂ 2.7–3.7 kg; ♀ 2.2–2.8 kg

Habitat: Lowland terra firme tropical rainforest, but also in flooded forests. Unlike other members of the genus, this species is reported to use exclusively the upper-canopy level.

Distribution: Brazil between the lower Purús River, along both banks of the Tefe River, and the south-bank tributaries of the upper Solimões–Amazon Rivers.

MALE

DD Vanzolini's Bald-faced Saki

Pithecia vanzolinii

Description: Both sexes have black upperparts with white to cream grizzling, and contrasting pale yellowish-buff arms and legs. Both have a buff-orange ruff, but the male's is more distinct. The facial skin of males is bare and black, with white to cream malar stripes. The facial skin of females is black, fringed with soft black hair, and normally has a star of white on the forehead. Juvenile females have a shaggier facial appearance. The body hair of females is not so thick or wavy as that of males.

Similar species: Gray's Bald-faced Saki (*p. 370*), males of which have a pronounced bright orange to dull orange/light tan ruff, and brownish to white forearms.

Habitat: Lowland terra firma and seasonally flooded white-water forests.

Distribution: Brazil between the south bank of the Juruá River and the south bank of the Tarauacá River, in the south-west of the states of Amazonas and Acre. Most records are from the south-east side of the Juruá

HB:	♂ 36.0–41.5 cm; ♀ 27–42 cm
Tail:	♂ 40–52 cm; ♀ 46–52 cm
Wt:	No info. available

River, but discovered also in the Riozinho da Liberdade Extractive Reserve in the upper Juruá Basin in Acre, 100 km to the south-east of its previous known range.

MALE (BOTH)

DD Gray's Bald-faced Saki

Pithecia irrorata

Description: Males are black with long bands of variable white grizzling, and have a distinct white headband (but not so distinct as that of Rylands's Bald-faced Saki). The forearms can be brownish to white, animals from Peru generally appearing browner than other populations. The hands and feet are white. The face is largely pink or unpigmented, with a dark muzzle outlined by white malar stripes. Western animals from Urubamba River have a darker face as adults, while eastern populations from Acre, in Brazil, have a pink face with a black muzzle, but are more heavily grizzled. Males have a pronounced bright orange to dull orange/light tan ruff. Young males can be browner, with grizzling throughout the upperparts. They have some white hair on the face above the eyes, and a less distinct light headband than adults, and pink-black facial skin. Females are browner across the front of the shoulders and the chest, and do not have a defined ruff. Young females have black hair around the face, with some white in the center of the forehead and a small amount above the eyes. Older females in Peru can have a grayer face, while females in Acre can have a whiter body. Young females have pinkish-black facial skin, becoming blacker with age. They are smaller and lack the black face and more prominent whitish grizzling of Rylands's Bald-faced Saki, and they have an obvious orange ruff.

Similar species: Males have darker facial skin as they age, but never so black as **Rylands's Bald-faced Saki** (*p. 373*). Young male Gray's Bald-faced Sakis have pink-black rather than black facial skin, with lighter pink highlighting around the eyes, but it is not so contrasting as in juvenile Rylands's Bald-faced or **Mittermeier's Tapajós** (*p. 372*) **Sakis**. Females differ from other bald-faced sakis in having black hair surrounding their face. This becomes grizzled with white as they reach the adult stage. Females in Acre can have whiter body grizzling resembling Rylands's Bald-faced Saki. Female Rylands's Bald-faced Saki has an almost naked black

HB:	♂ 37.5–49.0 cm; ♀ 36–53 cm
Tail:	♂ 42.0–51.4 cm; ♀ 33–50 cm
Wt:	♂ approx. 2.9 kg; ♀ 2.2–2.4 kg

face with sparse black hairs along the sides of the face, and can have distinct white eyebrows and pinkish to gray eyespots. **Vanzolini's Bald-faced Saki** (*p. 369*) has a buff-orange ruff and distinctly buff forearms and hind limbs.

Habitat: Lowland terra firme and seasonally flooded white-water forests.

Distribution: Brazil along the west side of the Purús River, south to Acre. It is uncertain whether it occurs east of the upper Madeira River in southernmost Rondônia. Peru in the Manu region, west of the Manu River and south of the upper Madre de Dios River, west of its confluence with the Manu River. Found also along the lower Urubamba River and likely to occur north-east of the Urubamba/alto Madre de Dios/Manu Rivers to the Brazil border.

MALE (BOTH)   Additional photo *p. 356*

VU Mittermeier's Tapajós Saki

Pithecia mittermeieri

HB:	♂ 45–46 cm; ♀ 35–37 cm
Tail:	♂ approx. 45 cm; ♀ 41–49 cm
Wt:	No info. available

Description: Adult males are black with long white bands of grizzling, and the forearms are densely covered in shorter white hairs. Wrist cuffs are grizzled with white. The fronts of the hind legs are black and the hands and feet are white. The ruff is bright orange. White hair extends towards the face and can form a wide white band on adults. Long white hairs extend over the side of the face from the nape and shoulders, creating a blackish hole over the ears. Adult males have scattered white hairs along the sides of the face, the eyebrows, the malar stripes and the lips; these diminish with age. The facial skin is black, although young males have prominent pink eye-spots that can also extend below the eyes. Sub-adult males can be gray-faced with less intensely pink eye-spots, while juveniles can be grayish, coppery-brownish or black with a white fringe, or colorful with both white hairs and bright pink eye-spots. Females are usually less grizzled, but become more grizzled with age. Adult females normally have a darker ruff than males, but it can be light tan or orange. Some sub-adult females have black fur around the face, and very little grizzling. The facial skin is black, with scattered hairs over the muzzle. Distinguished from other sakis by the males' bright orange ruff, the grayish face of a sub-adult, and contrasting white and black coloration with dense buff to white forearms.

Similar species: Females never appear so white as **Rylands's Bald-faced Saki**. Older adults can be quite white but retain a bright orange ruff, unlike Rylands's Bald-faced Saki. See also **Gray's** (*p. 370*) and **Pissinatti's** (*p. 374*) **Bald-faced Sakis**.

Habitat: Lowland terra firme and seasonally flooded white-water forests. Known also from a small fragment of savanna woodland surrounded by cattle-ranching and soybean plantations.

Distribution: Brazil south of the Amazon River between the Madeira and Tapajós Rivers, including the Aripuanã River. Mainly north of Aripuanã in Mato Grosso, although may have occurred along the Madeira River as far as the Mamore and Guaporé Rivers, and also north of the Jamari River and in southern Rondônia. Recently discovered in Lambari D'Oeste 76 km south of the previously known range.

FEMALE

VU Rylands's Bald-faced Saki

Pithecia rylandsi

Description: One of the largest sakis with a bare face with black pigmentation. Young males are black and moderately to heavily grizzled with white. Older males become almost entirely white. The forelimbs can be buff to white, and the hind limbs may have black on the inner legs. The hands and feet are white. The hair extending onto the forehead from the crown is white and thicker than in other sakis. Long white hairs extend across the side of the head from the nape and shoulders, creating a blackish hole over the ears, particularly on the sides of older males' heads. Males have white hair on the lips. The ruff is black or dark brownish. Males have pale or pinkish spots over the eyes. Juvenile males have reddish-black facial skin that appears lighter than that of adults, although it can be black. Adult females can be intensely white and the hair on the forehead can appear to create a white shield. Their forearms may be more buff than those of males, and this can extend across the chest. The ruff can be light buff or grayish. The facial skin is black and hairless, with reddish

No information on measurements and weight available

undertones. Young females have a grayish face, sometimes with a ring of light black hairs around it, with light gray eyebrow spots and light skin around the eye. They can have obvious white eyebrows. All females have distinct, white, shaggy malar stripes. Adult females have light eyebrow spots and bright malar stripes.

Similar species: Distinguished from other bald-faced sakis by the large size, very black face, extreme white grizzling and lack of a bright orange ruff. See **Gray's Bald-faced** (*p. 370*) and **Mittermeier's Tapajós Saki** for specific differences from those species.

Habitat: Lowland terra firme and seasonally flooded white-water forests.

Distribution: NW Bolivia, SE Peru, and possibly in the south of Rondônia and the west of Mato Grosso States, in Brazil (also reported from semi-deciduous forest in Vila Bela da Santíssima Trindade, in Mato Grosso State). Bolivia from the west of Pando Department, but may also be in north-east Beni and northern La Paz. In Peru found mainly north of the Madre de Dios River, but may occur also south of the river.

FEMALE

DD Pissinatti's Bald-faced Saki

Pithecia pissinattii

Description: Both sexes have distinct grizzling on the upperparts, this being most prominent on younger males and females. The grizzling varies from whitish through buff-cream to tan, particularly on the forearms and shoulders. There are often white cuffs on the wrists. The hind legs are brownish to blackish, sometimes with dark bands without white grizzling. The hands and feet are white. The underparts are black, with a distinct ruff ranging from bright orange in males to dull orange/light tan in females. Males have a white band of hair draping down over the forehead from the crown. Both sexes have a pinkish to red bare face, becoming black on older females, with white malar stripes extending up under the eyes. The lips and muzzle may have sparse indistinct white hairs. Females are more tan-colored than males on the back, arms and legs. Their back legs can be partly black

HB:	♂ 45–55 cm; ♀ 41–53 cm
Tail:	♂ 42–52 cm; ♀ 43–49 cm
Wt:	No info. available

FEMALE

along the front edge above the knee, and they often have dark wrist cuff bands. The faces of adult females have white hairs covering the forehead, cheeks and eyebrows. Older females often have a grayish-white band around the face, with a diamond-shaped piece of skin in the center of the forehead between the eyes. White malar stripes are more obvious on females. The skin above and below the eyes is light pinkish, with the rest of the bare parts of the face black.

Similar species: Mittermeier's Tapajós Saki (*p. 372*) that occurs on the eastern side of the Madeira River, adult males of which have a black face and a brighter orange ruff. Juvenile males are very different, as they, unlike young Mittermeier's Tapajós Saki, do not have bright pink eye spots that give way to white facial hair.

Habitat: Lowland terra firme and seasonally flooded white-water forests.

Distribution: Brazil south of the Solimões River in the northern area between the Purús and Madeira Rivers. It is not known how far south it occurs and whether its range meets that of Gray's Bald-faced Saki.

VU Red-nosed Bearded Saki

Chiropotes albinasus

Description: Distinctive: the only *Chiropotes* saki with bright red lips and a red triangular nose patch. The coat is uniformly black and thick, and the tail is long and bushy. It has black coronal tufts and a beard, which is most prominent in males.

Similar species: None in range.

Habitat: Tall terra firme forest, occasionally in fragmented areas and inundated forests, and in the transition zone between forest and savanna.

Distribution: Brazil, where restricted to Amazonas, the north of Mato Grosso, Rondônia and western Pará, although with a patchy distribution, and absent from two-thirds of Rondônia despite being present in other areas close to the border with Bolivia.

HB:	♂ 39–46 cm; ♀ 36–51 cm
Tail:	♂ 36–45 cm; ♀ 36–48 cm
Wt:	♂ 2.7–3.7 kg; ♀ 2.2–2.8 kg

375

EN **Black Bearded Saki** *Chiropotes satanas*

Description: Resembles Red-nosed Bearded Saki (*p. 375*), being largely black, but does not overlap in range and has blackish facial skin, which may be mottled with pink or white, and a dark brown to blackish back and upper limbs. Both sexes have a long bushy tail, and have black coronal tufts and beards which are most prominent in males.

Similar species: None in range.

Habitat: Primarily terra firme forests. Found also in coastal mangrove forests and forested areas in the transitional zone between Amazon Forest and the Cerrado at the southern and eastern borders of its distribution. Most abundant in larger tracts of forest, but small groups can also be encountered in smaller forest fragments.

Distribution: Eastern Amazonia in Brazil between the right bank of the Tocantins River and the eastern limits of the Amazon rainforest in the Brazilian states of Pará and Maranhão.

HB:	♂ 38–42 cm; ♀ 34–39 cm	
Tail:	♂ 36–42 cm; ♀ 36–41 cm	
Wt:	♂ 2.5–4.0 kg; ♀ 2.0–3.5 kg	

LC Rio Negro Bearded Saki
Chiropotes chiropotes

Description: The back and upper limbs are olivaceous, and the extremities of the limbs are brown. The face is blackish and sparsely haired. Like other *Chiropotes* sakis it has prominent coronal tufts and beard and a long bushy tail. The form *israelita*, previously split as a separate species, Brown-backed Bearded Saki, is now treated as a synonym.

Similar species: None in range.

Habitat: Prefers terra firme forests, including tall forest, savanna forest, and mixed semi-deciduous forest.

HB:	♂ 36–46 cm; ♀ 35–46 cm
Tail:	♂ 30–46 cm; ♀ 34–45 cm
Wt:	♂ 2.2–4.0 kg; ♀ 2.0–3.5 kg

Distribution: North of the Amazon River in S Venezuela (Bolivar) and N Brazil (Amazonas and Roraima) west of the Branco River. South to the Negro River in Brazil and north to the Orinoco River in Venezuela.

VU Uta Hick's Bearded Saki *Chiropotes utahickae*

Description: Reddish-brown to buff-colored, with the head, tail and lower limbs darker than the rest of the body. The face is blackish. Both sexes have a long bushy tail, and black coronal tufts and beard which are most prominent in males.

Similar species: None in range.

Habitat: The fluvial plain of Amazonia in tall terra firme humid forests, including disturbed forests.

Distribution: Brazil in the Amazon lowlands, between the Xingu, Amazon and Tocantins–Araguaia Rivers, although the exact limits of its range are unclear.

HB:	36–42 cm
Tail:	37–58 cm
Wt:	2–4 kg

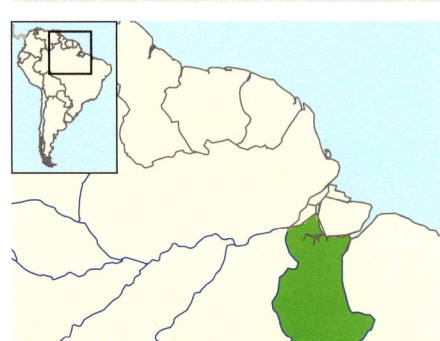

Additional photo *p. 356*

LC Guianan Bearded Saki

Chiropotes sagulatus

Description: Orange to reddish-brown back and upper limbs, with a black head, nape and lower limbs. The face is blackish and sparsely haired. It has prominent coronal tufts, a beard, and a long completely hairy tail.

Similar species: None in range.

Habitat: Lowland rainforest, terra firme forests including montane savanna forests, and occasionally seasonally flooded forests.

Distribution: North of the Amazon River and east of the Branco River in N Brazil and the Guianas. Largely absent from W Guyana and the coastal plain in Suriname, and French Guiana. In NE Brazil found throughout the state of Amapá and possibly absent only from wetter regions along the coast.

HB:	33–46 cm
Tail:	39–46 cm
Wt:	approx. 3 kg

Uakaris

The IUCN Primate Specialist Group recognizes eight species of uakari and follow Silva *et al.* (2022) who proposed, based on phylogenetics, that the four existing subspecies of Bald Uakari *Cacajao calvus*, three that are predominantly reddish/orange (*rubicundus, uyacali and novaesi*) and one that is 'whitish' (*calvus*), should be treated as separate species, and also described a fifth, white, form as a separate species, Kanamari Bald Uakari *C. amuna*. This treatment is followed here.

Uakaris are unusual among Neotropical monkeys in that the tail length is substantially less than their head-and-body length; their tails are non-prehensile. Uakaris are active throughout the day and prefer the upper strata of forests, but they will forage on the ground when floods recede and when food is in short supply. They feed mainly on seeds and fruits, but will also eat buds, nectar, flowers and leaves, along with small numbers of grasshoppers, ants, spiders and cockroaches. They can occur in large groups, but these often split into much smaller groups when feeding. They have large home ranges and have been recorded moving over 6 km in a single day, although the average is around 2.7 km.

Ucayali Bald Uakari

Red Bald Uakari

Cacajao rubicundus

Description: Unlikely to be confused with any other species apart from other bald uakaris, due to its bald pink face. The top of the head is covered by short whitish hairs. The rest of the body is covered by long, shaggy uniformly reddish/reddish-chestnut fur, apart from the nape and shoulders which are whitish and form a cape. The tail is short and bushy. There are three populations. Animals from the north bank of the Solimões River and the flooded forests of the Jacurapá channel have a nape with whitish hairs contrasting with the reddish-orange of the mid-back, rump, and sides of the trunk and limbs; in some individuals, the whitish hairs of the nape are gradually replaced by light orange hairs in the mantle. Specimens from the left bank of the Jutaí River have whitish hairs on the nape extending to the mantle, these are gradually replaced by the reddish-chestnut hairs of the saddle, while the pattern of the mid-back, rump, and sides of the trunk and limbs is

HB:	36–57 cm
Tail:	14–19 cm
Wt:	2.3–3.5 kg

entirely reddish-chestnut. The overall pattern of these individuals is reddish-chestnut and differs from the reddish-orange of specimens from the Jacurapá Channel. Individuals from the Auatí-Paraná region have a similar reddish-chestnut coloring to those from the Jutaí River on the sides of the trunk and limbs, but with an evident whitish or pale buffy color throughout the dorsum, from the nape to the tail in the male and to the saddle in the one female examined by Silva *et al.* (2022). The arms, legs, and sides of the trunk in the adult males are reddish-chestnut interspersed with largely yellowish hairs.

Similar species: None in range.

Habitat: Restricted primarily to seasonally flooded white- and black-water forests.

Distribution: Restricted to three disjunct populations along the middle Solimões River in Brazil; the flooded forests between the Jacurapá Channel and the Solimões River; the left bank of the Jutaí River, a south bank tributary of the Solimões River; and along the Auatí-Paraná Channel which connects the Solimões and Japurá Rivers.

NE Ucayali Bald Uakari

Cacajao ucayalii

Description: Unlikely to be confused with any other species apart from other bald uakaris, due to its bald pink face. Overall reddish-chestnut or reddish-orange without the contrasting whitish or yellowish coloration of the dorsum found in the other red uakaris. Specimens collected along the Javarí-Mirim River (in Loreto District in Peru) appears reddish-chestnut with some dark-reddish or blackish hairs on the limbs and dorsal surface of the tail. Specimens collected along the right bank of the Amazon River opposite the mouth of the Napo River are paler toned reddish-orange with some yellowish hairs on the dorsum.

Similar species: None in range.

Habitat: Terra firme forest and palm swamps.

Distribution: Occurs mainly in the Ucayali-Javarí interfluve of Peru, with isolated populations reported beyond this area. In Brazil it is found only along the Môa River in the Serra do Divisor National Park.

HB:	36–57 cm
Tail:	14–19 cm
Wt:	2.3–3.5 kg

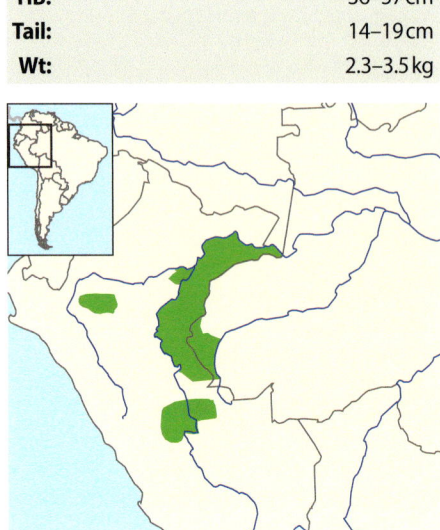

Additional photo *p. 380*

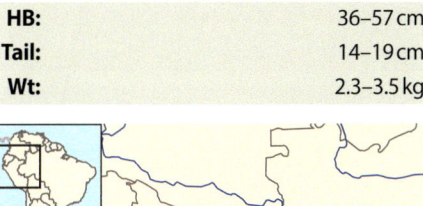 Novaes's Bald Uakari

Cacajao novaesi

Description: Unlikely to be confused with any other species apart from other bald uakaris, due to its bald pink face. The overall coloration is reddish-orange or reddish-chestnut, contrasting with the whitish and yellowish coloration of the dorsum, which in turn is gradually replaced by light orange hairs on the saddle, limbs, and flanks. The arms and legs are reddish-orange, interspersed with yellowish hairs.

Similar species: None in range.

Habitat: Restricted primarily to seasonally flooded white- and black-water forests.

Distribution: Known only from the Gregório and Tarauacá interfluve of Brazil.

HB:	36–57 cm
Tail:	14–19 cm
Wt:	2.3–3.5 kg

LC White Bald Uakari

Cacajao calvus

Description: Unlikely to be confused with any other species apart from other bald uakaris, due to its bald pink face. The top of the head is covered by short whitish hairs. The rest of the body is covered by long, shaggy fur, which forms a cape over the animal's shoulders. The upperparts are off-white or pale gray. The underparts are reddish or yellowish, but variation in the overall pelage coloration is geographically consistent in two disjunct populations. The tail is short and bushy. Males from Mamirauá have a yellowish and greyish-white pattern on the nape, dorsum, and sides of the trunk. The upperparts contrast with the orange or golden-orange pattern of the underparts, especially on the chest and limbs. Some adult females from the Auatí-Paraná Channel have the same contrast, while others have a more uniform buffy-yellowish or whitish pattern on the upperparts. The beard is dark reddish-brown in both males and females. Animals from the right bank of the Jutaí River show a grayish pattern in the nape, dorsum and sides of the trunk, rather than the whitish (or yellowish-white) of other individuals. Specimens from the Jutaí River have whitish or yellowish-orange underparts, including the inner side of the limbs.

Similar species: None in range.

HB:	36–57 cm
Tail:	14–19 cm
Wt:	2.3–3.5 kg

Habitat: Restricted primarily to seasonally flooded white- and black-water forests.

Distribution: Occurs in two areas in Brazil: the Mamirauá Reserve between the Solimões and Japurá Rivers east of the Auati-Paraná Channel; and on the left bank of the middle Juruá River, and along the right banks of the Riozinho River and the lower Jutaí River.

Kanamari Bald Uakari
Cacajao amuna

Description: Unlikely to be confused with any other species apart from other bald uakaris, due to its bald pink face. The top of the head is covered by short whitish hairs. The rest of the body is covered by long, shaggy fur, forming a cape over the animal's shoulders. Largely white above with a few of the hairs having a greyish terminal portion or being entirely greyish. This creates a light greyish-white appearance throughout the dorsum. This contrasts slightly with the nape and the proximal portion of the tail where the hairs are entirely white. The flanks have hairs that are entirely white or whitish with the terminal portion greyish. The sparse hairs on the chest and belly are whitish. The arms are yellowish-white, paler on the forearms. The legs are whitish on the outer surface, but cream-colored on the inner legs. The tail is short and bushy, whitish above and cream-colored below. The hands and feet have short orange hairs interspersed with greyish-white hairs. The beard is reddish-orange.

Similar species: None in range.

Habitat: Restricted to seasonally flooded white-water forest.

HB:	36–57 cm
Tail:	14–19 cm
Wt:	2.3–3.5 kg

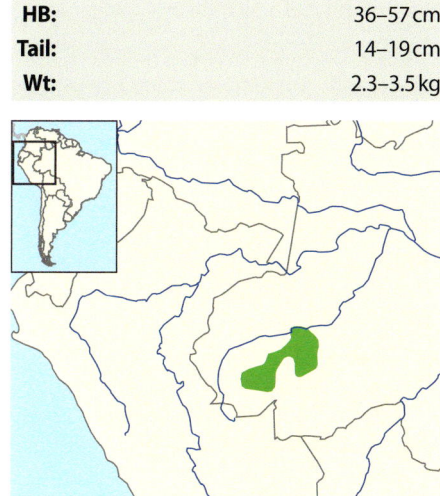

Distribution: Occurs along the right bank of the Tarauacá River, extending to the Envira and Jurupari Rivers. Found also in the headwaters of the Pauini and Moaco Rivers. The southern limit of the range is unknown but the species has yet to be found along the upper Tarauacá River.

VU **Neblina Black-headed Uakari** *Cacajao hosomi*

Description: Has a hairless black face, and a black-haired head with the hair overlapping the face to give a bearded appearance. The face is broad and flat. Has long coarse black body hair, with reddish-brown middle and lower back, flanks, thighs and tail. The tail is short and bushy.

Similar species: None in range.

Habitat: Seasonally flooded black-water forest, terra firme forest, white-sand forests, scrub forest and montane forests.

Distribution: Found in the north-west Amazonian regions of S Venezuela and NW Brazil. Occurs between the Negro River in Brazil and Venezuela in the south and west, and the Marauiá River in Brazil in the east. The northern limits of its range are believed to be the Canal Cassiquiare and Orinoco River in Venezuela, although it is unclear whether it occurs also to the north of the Canal Cassiquiari.

HB:	30–50 cm
Tail:	13–21 cm
Wt:	2.0–4.5 kg

Black-faced (Golden-brown) Uakari   *Cacajao melanocephalus*

Description: The head, beard, nape and arms are blackish, resembling those of Neblina Uakari. The middle of the back, however, is buff, pale orange or golden. The lower back, thighs and short bushy tail are reddish-orange. Animals from the border of Brazil and Colombia are darker, with a blacker nape, upper mantle and arms.

HB:	35–56 cm
Tail:	14–16 cm
Wt:	1.9–2.8 kg

Similar species: None in range.

Habitat: Seasonally flooded lowland forest, palm swamp forest and terra firme forest, including dry Caatinga forest, at elevations of 100–150 m.

Distribution: Found in S Venezuela (south of the Casiquiare River), in SE Colombia (from the Serranía de La Macarena, in the west, east to the Guayabero–Guaviare interfluvium and the lower Apaporis River, and south to the Caquetá River), and in NW Brazil (south of the Negro River east to its confluence with the Solimões River, and south to the Japurá River).

LC Aracá Black-headed Uakari

Cacajao ayresi

Description: Much darker than Neblina Black-headed Uakari (*p. 386*) and with less contrasting blackish middle and lower back, tail and thighs. Like that species it has a hairless black face and black-haired head, the hair overlapping the face to give a bearded appearance. It has short, coarse black body hair and the non-prehensile tail is short and bushy.

Similar species: None in range.

Habitat: Seasonally flooded black-water forest, terra firme forest, white-sand forests, scrub forest and montane forests.

Distribution: NW Brazil, where restricted to a small area that encompasses the Curuduri River Basin and the middle to lower Aracá River north of the Negro River.

HB:	36.3–38.0 cm
Tail:	15.5–18.0 cm
Wt:	2.00–2.45 kg

Howlers

The IUCN Primate Specialist Group recognizes 14 species of howler, 13 of which occur in South America.

Howlers are stocky, large-headed and rather short-limbed monkeys with a conspicuous beard and with a prehensile tail frequently carried in a coil. They are very vocal particularly around dawn, during the late afternoon and during rainstorms. Males give a series of deep grunts which develop into long deep roars. Females give higher-pitched roars.

Howlers live in troops of between 5 and 44 animals, troop size varying among species. They feed on fruit, flowers, both young and mature leaves, and other plant matter.

Bolivian Red Howler (FEMALE AND INFANT)

LC **Colombian Red Howler** *Alouatta seniculus*

Description: The back and sides are usually orange to bright reddish-brown, while the head, shoulders, limbs, tail and underparts are darker coppery-red to purplish-red. Adult males often have a blackish beard, limbs and tail. Has a forward-facing 'V'-shaped crest of long hairs on the crown. There can be considerable variation in coloration, and the tip of the tail may be paler than the rest of the tail.

HB:	♂ 51–63 cm; ♀ 48–57 cm
Tail:	♂ 57–80 cm; ♀ 52–69 cm
Wt:	♂ 5.4–9.0 kg; ♀ 4.1–7.0 kg

Similar species: None in range.

Habitat: Found at up to 3,200 m in primary lowland rainforest, dry deciduous forest, Andean cloud forest (including oak forest in Colombia), gallery forest and mangrove swamps, as well as in várzea forest.
In Ecuador it is found in tropical and subtropical evergreen rainforest from 20 m to 2,000 m, but it is most common below 700 m.

Distribution: The Colombian Andes, NW Venezuela, E Ecuador, E Peru and the Amazonian Basin of Brazil north of the Solimões River and south of the Negro River.

LC Juruá Red Howler

Alouatta juara

Description: Generally darker than Colombian Red Howler, with which it was previously lumped. Reddish-brown or dark red, the mantle being lighter yellowish-red or golden. The arms and thighs are dark red and become darker towards the hands and feet. The base of the tail is reddish-brown and the distal half of the tail is paler and more golden in coloration.

Similar species: None in range.

Habitat: Evergreen rainforest, and more common in seasonally flooded forest (várzea) than in terra firme forest.

Distribution: Throughout the upper Amazon in Brazil in the states of Acre and Amazonas, to the south of the Solimões River. Some taxonomic authorities still consider Juruá Red Howler to be conspecific with Colombian Red Howler.

No information on measurements and weight available

VU **Purús Red Howler**

Alouatta puruensis

Description: Generally darker than Colombian Red Howler (*p. 390*), with which it was previously lumped. Adult males are dark reddish except for the back, which has lighter golden hues, and the tail, which is dark reddish with golden tips towards its base and lighter with a golden hue towards the tip. Females are light red with a pale golden shiny dorsum. The beard is dark reddish-brown. The hind limbs are light red, the upper part of the limbs being lighter than the lower parts, which are almost brown with a reddish tint to the feet. The tail is reddish-yellow. Young males have a short black beard. The limbs and tail have tawny hues and are lighter than the remainder of the dark reddish body.

Similar species: None in range.

Habitat: Evergreen rainforest; more common in seasonally flooded forest (várzea) than in terra firme forest.

No information on measurements and weight available

Distribution: The upper Amazon Basin in the Brazilian states of Acre, Amazonas, Mato Grosso and Rondônia, and also in E Peru.

Ursine Red Howler

Alouatta arctoidea

Description: Large and coppery-red, with contrasting maroon head, shoulders, limbs and upper section of tail. There is a front-facing pale 'V'-shaped marking on the top of the head. Males are larger than females and have a blacker beard. Formerly treated as a subspecies of Colombian Red Howler (*p. 390*).

Similar species: None in range.

Habitat: At 10–1,160 m in the Llanos in seasonally inundated forest, in deciduous forest patches, in lowland tropical or sub-montane humid rainforest and gallery forest. May also occur at up to 2,000 m in cloud forests of coastal Venezuela.

Distribution: N Venezuela along the coast, east of Lake Maracaibo, from Falcón to the state of Miranda. Also widely distributed throughout the Venezuelan Llanos north of the Orinoco River and west through the Apure Basin north of the Meta River, although it has

HB:	45–65 cm
Tail:	55–68 cm
Wt:	♂ 6–8 kg; ♀ 4.5–7.0 kg

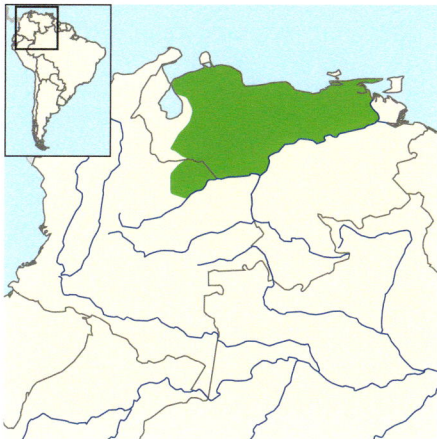

been suggested that this population may prove to be a distinct and as yet undescribed species.

LC Guianan Red Howler

Alouatta macconnelli

Description: Bright golden-brown to dark reddish-brown, with light golden fur extending from the back on to the flanks and a dark reddish-brown dorsal band. The limbs and the base of the tail are also bright golden-brown to dark reddish-brown, the forelimbs being slightly darker than the hind limbs. The distal portion of the tail is paler than the rest of the tail. Adults have a well-developed beard, and the nape is dark red. Young animals are darker than the adults and generally dark red in color, with a lighter golden-red dorsal region, although the dorsum is never so bright or pale as that of adults.

Similar species: None in range.

Habitat: Primary lowland rainforest, secondary forest, gallery forest, swamp forest, savanna forest, dry deciduous forest and mangroves. Normally in the middle to upper canopy, although it can be encountered at lower levels and even on the ground.

Distribution: North of the Amazon River, east of the Negro River in Brazil and east and south of the Orinoco River in Venezuela. Found

HB:	♂ 51–63 cm; ♀ 48–57 cm
Tail:	♂ 57–80 cm; ♀ 52–69 cm
Wt:	♂ 5.4–9.0 kg; ♀ 4.1–7.0 kg

also in N Brazil (Amapá, Amazonas, Pará, Roraima), French Guiana, Suriname, Guyana and Venezuela, and on Trinidad.

Bolivian Red Howler

Alouatta sara

Description: The only red howler in its range. Reddish-orange, with slightly darker rufous arms, legs and upper section of the tail.

Similar species: Overlaps in range with Black-and-gold Howler (p. 401), but that species is sexually dimorphic and looks distinctly different, in having black males and yellowish-brown or grayish-yellow females.

Habitat: Up to 1,000 m in tropical forest, including riverine forest and seasonally flooded forest, alongside Black-and-gold Howler east of the Beni River. Appears to prefer humid forest areas and seasonally flooded forest along major rivers, while Black-and-gold Howler is normally found in drier, semi-deciduous forest and gallery forest in areas of savanna and Chaco.

Distribution: Bolivia from the department of Pando south along the Andean Cordillera and east into central Bolivia as far east as the Mamoré-Guaporé River. Also in E Peru along the Madre de Dios, Tambopata and Urubamba Rivers.

HB:	54.0–71.2 cm
Tail:	51.6–60.0 cm
Wt:	6–9 kg

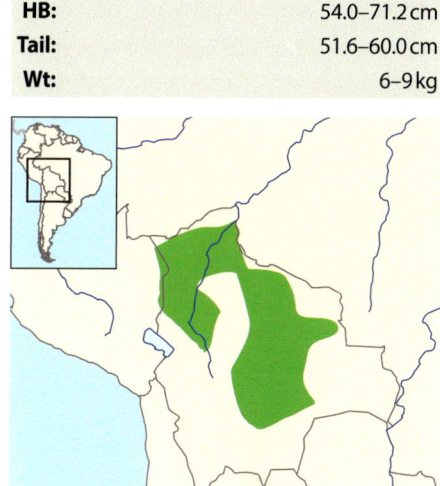

Additional photo *p. 389*

LC Amazon Black Howler

Alouatta nigerrima

Description: All-black, with a short crest running from the middle of the crown to the back of the crown, and a long dense beard. There is little difference between the sexes.

Similar species: Spix's Red-handed Howler (*p. 398*) has dark red limbs, tail and back.

Habitat: Tall evergreen terra firme forest, seasonally inundated forest and forest patches in savanna areas, as well as secondary forest in abandoned rubber plantations.

Distribution: Brazil in the states of Amazonas and Pará south of the Amazon River, west from the Tapajós and Juruena rivers, to the Madeira and Aripuanã rivers. Isolated populations occur north of the Amazon River (not mapped) on the lower Trombetas River and just south of the Amazon–Solimões Rivers, midway between the Madeira and Purús Rivers.

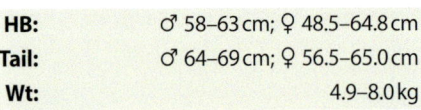

HB:	♂ 58–63 cm; ♀ 48.5–64.8 cm
Tail:	♂ 64–69 cm; ♀ 56.5–65.0 cm
Wt:	4.9–8.0 kg

Eastern Red-handed Howler

Alouatta belzebul

Description: Largely black, with dull rusty-red hands, feet and tip of the tail. The thick beard is black or in some individuals reddish-brown. Adult males have a rusty-red scrotum. There is regional variation: in some areas both males and females may be uniformly black, while in others the males are completely black and the females have red extremities. All-red and all-brown individuals have been recorded but are rare.

Similar species: None in range, but see **Maranhão Red-handed Howler** (*p. 399*), which has brownish flanks.

HB:	♂ 58–65 cm; ♀ 37–50 cm
Tail:	♂ 56–70 cm; ♀ 45–57 cm
Wt:	♂ 6.5–8.0 kg; ♀ 4.8–6.2 kg

Habitat: Occurs in a mix of habitats, including lowland Amazon rainforest, semi-deciduous terra firme and flooded forest. Feeds in the upper levels of the forest canopy.

Distribution: NE Brazil to the east of the Xingu and Iriri Rivers. In the lower Amazonian states of Amapá, Pará and Maranhão, and in Atlantic Forest in the states of Rio Grande do Norte, Piauí, Pernambuco, Paraíba and Alagoas. The limits of this species' range in comparison with Spix's Red-handed Howler (*p. 398*) along the lower Amazon River are poorly understood.

397

VU Spix's Red-handed Howler

Alouatta discolor

Description: Dark brown to black, with contrasting dark reddish back, limbs and tail.

Similar species: Amazon Black Howler (*p. 396*) is uniformly black.

Habitat: Lowland terra firme forest, seasonally inundated rainforest, palm swamps and secondary forests.

Distribution: Brazil, along the south bank of the Amazon River and south through the Curuá River Basin between the Tapajós and Juruena Rivers in the west and the Xingu and Iriri Rivers in the east. The limits of this species' range in comparison with Eastern Red-handed Howler (*p. 397*) along the lower Amazon River are poorly understood.

HB:	♂ 57–65 cm; ♀ 46.5–57.0 cm
Tail:	♂ 65.0–67.5 cm; ♀ 60.5–65.0 cm
Wt:	♂ 6.5–8.0 kg; ♀ 4.8–6.2 kg

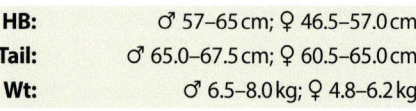

Maranhão Red-handed Howler

Alouatta ululata

Description: Sexually dimorphic, the male being black with reddish-brown flanks, hands, feet and tip of the tail. The female is yellowish-brown with sparse grayish hairs, and appears olivaceous in the field.

Similar species: None in range, but see **Eastern Red-handed Howler** (*p. 397*), which has black flanks.

Habitat: In Maranhão in open, transitional *babaçu* palm forest and in the Serra da Ibiapaba in humid forest on the steep eastern slopes. In Ceará in areas of semi-deciduous and dry forest. Found also in mangroves.

Distribution: Maranhão, Piauí and Ceará in coastal NE Brazil. The western limit of its range appears to be Humberto de Campos, on the coast of Maranhão, and Serra da Ibiapaba is the eastern limit. The range extends as far south as the Poti River, in Piauí.

HB:	56.5 cm (type specimen)
Tail:	56 cm (type specimen)
Wt:	No info. available

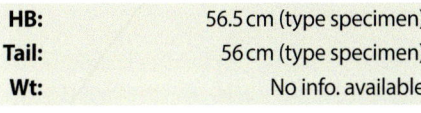

ADULT (BELOW); JUVENILE MALE (INSET)

VU Brown Howler

Alouatta guariba

Description: A howler with subspecific and regional variations in coloration. Adult males have dark rufous to yellowish-rufous upperparts, with darker arms, legs, and tail. The beard is reddish-brown to rufous, and brightest at the back of the neck and behind the ears. Adult females are fully dark brown or reddish-brown, with a dark brown to red head. Adult males from southern populations are redder than the ones from northern populations, while adult females show an opposite trend in which darker individuals occur farther to the south. As a consequence, there is a greater difference between males and females in southern populations. Fortes & Bicca-Marques (2008) reported an unusually colored population in Rio Grande do Sul consisting of reddish-cream adult males with redder beard, back of neck and tail, yellowish-cream females with golden head, back of neck and limbs, and yellowish-cream as opposed to dark brown juvenile males.

Similar species: None in range, but overlaps and may hybridize with **Black-and-gold Howler** in Paraná; although the two species should not be confusable in the field, hybrids may cause difficulties.

Habitat: Up to 700 m in lowland, sub-montane and montane forest of the Atlantic coast of Brazil, and inland in semi-deciduous

HB:	♂ 50–60 cm; ♀ 44–54 cm
Tail:	♂ 52–67 cm; ♀ 48–57 cm
Wt:	♂ 5.3–7.2 kg; ♀ 4.1–5.0 kg

forest in the states of São Paulo, Minas Gerais and Paraná. Has been found also in subtropical and temperate, mixed broadleaf and monkey-puzzle (*Araucaria*) forests in Paraná and Rio Grande do Sul.

Distribution: The Atlantic Forest of Brazil, from southern Bahia south to Rio Grande do Sul in S Brazil and Misiones in extreme NE Argentina.

MALE

⊤ Black-and-gold (Paraguayan) Howler · *Alouatta caraya*

Description: Males in the north of the range in Goiás and Bahia are entirely black, while those from Mato Grosso and Paraná have a brown back and hindquarters and a black head and forequarters. Males from São Paulo and Minas Gerais are brownish-black with yellowish belly, hands, feet and tip of tail. Young animals and adult females range in color from grayish-white to golden-brown, and some young animals can appear blond. Even largely black males may have reddish-brown hairs on the beard, back and tail and some brown coloration on the back, limbs and tail. The females may have a dark brown to grayish line extending from the head to the base of the tail. The face is dark and the fur long and stiff, creating a prominent beard.

HB:	♂ 60–65 cm; ♀ approx. 50 cm
Tail:	♂ 60–65 cm; ♀ 54.5–60.0 cm
Wt:	♂ 5.3–9.6 kg; ♀ 3.6–6.5 kg

Similar species: None in range, but overlaps and may hybridize with **Brown Howler** in Paraná; although the two species should not be confusable in the field, hybrids may cause difficulties. Also overlaps with **Bolivian Red Howler** (*p. 395*) east of the Blanco River in Bolivia, but that species is reddish-orange.

Habitat: A wide range of habitats, including Cerrado and semi-arid Caatinga scrub, gallery and riparian forests, semi-deciduous and deciduous forests, highland savannas, and inundated and terra firme lowland and sub-montane Chiquitania forests.

Distribution: Widely distributed in the Cerrado and dry forests of central Brazil and also in lowland Bolivia east of the Beni River, in E Paraguay and parts of the southern and central regions of the Paraguayan Chaco, in NE Argentina and possibly extreme NW Uruguay.

MALES (LEFT); FEMALE (RIGHT)

VU Mantled Howler

Alouatta palliata

Description: Silky black (or sometimes dark brown) and generally short-haired, with a mantle of longer gold or yellowish-brown hairs along the flanks and sometimes appearing as a saddle across the lower back. The mantle hairs of the South American subspecies, *aequatorialis*, are shorter than those of the four Central American subspecies. The face is naked, black and bearded, and there is a slight, straight transverse crest along the crown. Adult males have a white scrotum.

Similar species: **Brown-headed** (*opposite*) and **Colombian Black** (*p. 404*) **Spider Monkeys** are smaller headed, lack a beard and are much longer limbed.

Habitat: Upper canopy of primary and secondary rainforest at up to 2,500 m. Also in dry lowland forest, riparian forest, coffee plantations and coastal mangroves.

Distribution: The foothills, lowlands and lower montane areas west of the Andes from the Caribbean coast of Colombia, through Colombia and Ecuador, and into the Tumbes and Piura regions of N Peru. Range extends through Central America to S Guatemala, with an isolated population in E Mexico.

HB:	♂ 47–63 cm; ♀ 46–60 cm
Tail:	♂ 60–70 cm; ♀ 55–66 cm
Wt:	♂ 4.5–9.0 kg; ♀ 3.1–7.6 kg

Spider Monkeys

The IUCN Primate Specialist Group recognizes eight species of spider monkey, seven of which occur in South America. All are slim, long-limbed monkeys with a long prehensile tail, and lack a beard. The Colombian subspecies of **Central American Spider Monkey** *Ateles geoffroyi grisescens* is no longer considered valid, and the species is therefore excluded from this book.

Group size varies by species, some species being found in groups of up to 55 animals, although in many species group size is 20–30 animals, and larger groups tend to split into sub-groups often containing as few as 5–8 animals. They feed primarily on fruit, but will also eat young leaves and flowers, young seeds, floral buds, bark and decaying wood and very occasionally small insects. Some species have also been recorded as visiting salt licks, and when doing so some of the group remain in the trees, presumably as lookouts, while others descend to the ground to lick the soil and drink the salty water. They spend most of the time traveling and feeding in the upper canopy, but are also often seen in the middle canopy.

VU Brown-headed Spider Monkey | *Ateles fusciceps*

Description: A dark, coarse-haired spider monkey with varying amounts of brown on the cheeks and crown, and a few white hairs on the chin and lips. Black to dark brown with a yellow-brown forehead, becoming brown to black on the nape. The belly can be paler in some individuals. The facial skin is black but may have a mask of pale unpigmented skin around the eyes and nose.

HB:	30.5–63.0 cm
Tail:	63.5–85.5 cm
Wt:	6.6–10.5 kg

The head is small and black to reddish-brown with tufts of hair protruding sideways from in front of the eye and over the eye. The very long prehensile tail is densely furred towards the proximal two-thirds with a clear narrowing towards the tip.

Similar species: Mantled Howler (*opposite*) is robust with shorter limbs and tail and has a large head with a beard.

Continued on next page...

Brown-headed Spider Monkey (*continued*)

Habitat: Foothills in the northern and central lowlands inhabiting humid, tropical and subtropical forests from 1,000 m to 2,000 m but generally below 1,200 m. Prefers the upper stratum of forest but has been recorded at mid-levels and even in the understory.

Distribution: NW Ecuador from Esmeraldas Province to the NW of Pichincha and Santo Domingo Provinces, extending to the western borders of Imbabura and Carchi Provinces. There are past records from the Colon Colonche Mountain Range, and farther west in Chimborazo Province, and a recent report from the NW of Manabí Province.

EN Colombian Black Spider Monkey   *Ateles rufiventris*

Description: Very similar to Brown-headed Spider Monkey (*p. 403*) with which it was previously lumped. A dark, coarse-haired spider monkey with varying amounts of brown on the cheeks and crown, and a few white hairs on the chin and lips. Glossy black, with a few white or gold hairs on the face and a slight brownish tinge to the forehead. The facial skin is black.

Similar species: Mantled Howler (*p. 402*) is robust with shorter limbs and tail and has a large head with a beard.

Habitat: Recorded in dry forest, humid forest and cloud forest to 2,500 m in Colombia.

Distribution: Found in the Pacific lowlands from SW Colombia, north along the western side of the Cauca River and into E Panama.

HB:	30.5–63.0 cm
Tail:	63.5–85.5 cm
Wt:	6.6–10.5 kg

EN Black Spider Monkey

Ateles chamek

Description: A short-haired glossy-black spider monkey with black facial skin. Often shows a few white hairs on the forehead, cheeks and muzzle. Juveniles appear grayish, with pink feet, and whitish skin around the eyes. A form occurring in Mato Grosso in Brazil, *longimembris*, has a triangular pink muzzle including chin, a black face, and a triangular patch of backward-pointing black hairs on the forehead. Another form in an area east of the Purús River and south of the Ipixuna River reportedly differs in having a triangular forehead patch or blaze, with black, backward-directed hairs and triangular flesh-colored muzzle.

Similar species: Reported to hybridize with **White-whiskered Spider Monkey** (*p. 407*) on the east bank of the upper Tapajós River in Brazil, with hybrids having the distinctive white whiskers and frontal blaze of that species. May overlap also in range with **White-bellied Spider Monkey** (*p. 408*) in Peru, where hybridization has also been reported, but that species has pale underparts that contrast strongly with the dark upperparts.

Habitat: Widespread and common, where not hunted, in tall primary rainforest, including terra firme, subtropical semi-

HB:	♂ 45–60 cm; ♀ 40–52 cm	
Tail:	♂ 80–88 cm; ♀ 70–80 cm	
Wt:	♂ approx. 7 kg; ♀ approx. 5 kg	

deciduous and riverine and flooded forests. Has been found also in transitional forest–savanna border areas.

Distribution: NE Peru into N & central Bolivia and W Brazil. Replaced by White-bellied Spider Monkey on the west bank of the lower Ucayali River in Peru.

VU Red-faced Black Spider Monkey

Ateles paniscus

Description: Glossy black and long-furred, with sparser fur on the underside. The face is naked, pink or even reddish, with a few pale hairs on the muzzle. Young animals have a darker face. The tail is distinct from those of all other spider monkeys, the proximal two-thirds being thickly furred, and the outer third tapering abruptly to the tip.

Similar species: None in range.

Habitat: Primary forest with tall mature trees and a closed canopy, and only occasionally encountered in degraded forest or around the forest edge. Appears to be absent from coastal forests in Guyana and Suriname. In Brazil found in lowland, sub-montane and montane forest.

Distribution: North of the Amazon River, east of the Negro and Branco Rivers in E & S Guyana, east into Suriname and French Guiana, except for the lowland coastal plains, and south into Amapá, Pará (north of the Amazon River) and NE Amazonas in Brazil.

HB:	♂ 51.5–58.0 cm; ♀ 42–66 cm
Tail:	♂ 72.0–85.2 cm; ♀ 64–93 cm
Wt:	(means): ♂ 9.1 kg; ♀ 8.4 kg

White-whiskered Spider Monkey

Ateles marginatus

Description: Small and all-black, with whitish whiskers which create a white line that gives it a white-cheeked appearance. Also has a distinctive white triangular patch on the forehead. The face is reddish in adults and pale fleshy-pink in juveniles. An undescribed form of spider monkey, with dark chestnut-brown-black fur and larger white whiskers and forehead patch, has been reported from the range of White-whiskered Spider Monkey in the dry savanna forests and Cerrado of northern Mato Grosso.

Similar species: Reported to hybridize with **Black Spider Monkey** (*p. 405*) on the east bank of the upper Tapajós River, with hybrids having the distinctive white whiskers and frontal blaze of White-whiskered Spider Monkey rather than an all-black face.

Habitat: Lowland rainforest, riverine, semi-deciduous and dry savanna forests

Distribution: Brazilian Amazon between the east bank of the Tapajós River and the west bank of the Xingu River, south of the Amazon River.

HB:	♂ 50–70 cm; ♀ 35–58 cm
Tail:	♂ 75–90 cm; ♀ 62–77 cm
Wt:	♂ approx. 10.4 kg; ♀ approx. 5.8 kg

EN White-bellied Spider Monkey

Ateles belzebuth

Description: Blackish or dark brown above, with yellowish-white underparts including the inner surfaces of the limbs, the backs of the thighs and the underside of the tail. The dark face is hairless, and many individuals have a yellowish-brown or white triangular patch on the forehead. Juveniles have dark fur and a reddish face.

HB:	46–50 cm
Tail:	74–81 cm
Wt:	(means): ♂ 8.3 kg; ♀ 7.9 kg

Similar species: May overlap overlap in range with **Black Spider Monkey** (*p. 405*) in Peru, where hybridization has also been reported, but that species is all-black and lacks pale underparts that strongly contrast with dark upperparts.

Habitat: Terra firme primary forest, including montane tropical, subtropical, semi-deciduous and lowland riverine forests.

Distribution: NE Peru, E Ecuador, up to 1,300 m in lowland Colombia to the east of the Cordillera Oriental, southern Venezuela, and NW Brazil as far east as the Branco River. The distribution is fragmented, and the Brazilian and Venezuelan populations appear to be separated from those in SW Colombia, Ecuador and NE Peru.

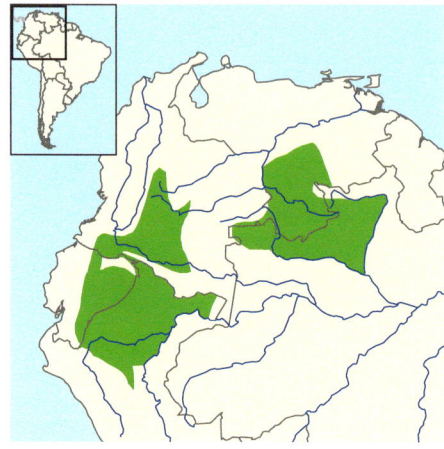

CR Variegated Spider Monkey

Ateles hybridus

Description: Ranges in color from light brown to dark brown above, with a darker head. The abdomen, inner sides of the limbs and underside of the tail are white or yellowish-buff, and there is a conspicuous triangular white patch on the forehead. There are two subspecies, which differ in color, SSP. *hybridus* having white to light yellow underparts, and hind limbs that are similar in color to the upperparts, while SSP. *brunneus* is darker brown above, contrasting more strongly with the white to yellow belly.

Similar species: None in range.

Habitat: In Venezuela, found in the upper canopy of old, tall trees in lowland primary evergreen rainforests at 20–700 m. Has been recorded also in montane tropical, riverine and semi-deciduous forest.

Distribution: The nominate subspecies, SSP. *hybridus*, occurs in the south-eastern areas of the Sierra Nevada de Santa Marta, in the Perijá Mountains in Colombia and into W Venezuela. There is also an isolated

HB:	♂ 47–50 cm; ♀ 45–48 cm
Tail:	♂ 76–81 cm; ♀ 74–76 cm
Wt:	♂ 7.9–8.6 kg; ♀ 7.5–10.5 kg

population in the Parque Nacional Guatopo in NE Venezuela.
SSP. *brunneus* occurs between the lower Cauca and Magdalena Rivers in Colombia.

SSP. *brunneus*

Woolly Monkeys

FAMILY | **Atelidae**
(SUBFAMILY | Atelinae)

The IUCN Primate Specialist Group recognizes two species of woolly monkey with four previously recognized species now being lumped with Humboldt's Woolly Monkey. Descriptions of these subspecies and their distributions are included under Humboldt's Woolly Monkey.

Woolly monkeys are rangy and long-legged, with a long prehensile tail. They are diurnal and occur in mixed-sex groups. Group size varies with species but can be up to 70 individuals. They feed on a variety of ripe fruits, flowers, leaves, lichens, bromeliads, epiphyte roots and bulbs, and also insects. They generally feed in the middle to upper canopy.

Humboldt's Woolly Monkey ssp. *lugens*

VU Humboldt's (Common) **Woolly Monkey** *Lagothrix lagothricha*

Description: Five subspecies are recognized with a wide variation in color even within subspecies.

ssp. *lagothricha*: From light blond in the Amazon Basin to uniformly dark brown and sometimes almost black in the highlands. Most are predominantly brown above, often with a paler head with a pale crown stripe, and dark gray hands and feet. Adults frequently have long black chest hair, and sometimes have pale tufts at the base of the tail. The black face often contrasts with the paler cheeks and crown.

ssp. *lugen*s: The body, including the limbs are dark gray to brown or blackish-brown. The

HB:	♂ 40–65 cm; ♀ 45–58 cm
Tail:	♂ 53–80 cm; ♀ 53–72 cm
Wt:	♂ 7–10 kg; ♀ 5–7 kg

darker head has a blackish cap and a dark gray crown stripe. The tail is blackish-gray to black.

ssp. *poeppigii*: The upperparts range from pale yellow-gray through various shades of brown to almost black, although most have a silvery sheen. The head, chest and limbs are usually black and the remainder of the belly is deep reddish-brown. Animals on the west bank of the Juruá River in Brazil are ochre-blond, with the head significantly paler than the back. The underparts, including the limbs and the tip of the tail, are ochre-white. The facial skin is black with white whiskers.

ssp. *cana*: Generally grayish-brown with a contrasting dark gray to black head. The underparts are blackish-gray with a dark reddish tinge. The hands, feet and tail are dark brown. Lowland populations are normally pale gray with more contrasting head, limbs and tail.

ssp. *tschudii*: Dark blackish-gray with a reddish tinge, and black head, limbs and tail. Individuals from Bolivia are mainly dark smoky gray with noticeably darker underparts and a black head and face.

Continued on next page...

ssp. *cana*

411

Humboldt's Woolly Monkey (*continued*)

Similar species: Yellow-tailed Woolly Monkey is mahogany-brown with a distinct yellow band on underside of the tail.

Habitat: Found in primary forest, including tropical and subtropical terra firme and seasonally inundated forests, and occasionally in secondary lowland or high-elevation rainforests at up to 3,000 m. Occasionally found in disturbed/fragmented forests. ssp. *lugens* is found also in mauritia palms. In Bolivia, ssp. *tschudii* occurs in low-elevation Andean cloud forest, on steep slopes with abundant mosses, tree ferns and bromeliads.

Distribution: Colombia within a narrow strip from the border with Ecuador to the border with NW Venezuela, mainly to the east of the Cordillera Oriental. In SE Colombia as far north as the northern bank of the Guaviare River. W Brazil west of the Tapajós River. NE Ecuador and the highlands of E Ecuador. From N Peru south through E Peru to the Madre de Dios River and the Tambopata River Basin. An isolated population of ssp. *tschudii* or possibly an undescribed taxon, occurs between 700 m and 2,500 m in Madidi National Park, in Bolivia.

ssp. *lagothricha*

CR (Peruvian) **Yellow-tailed Woolly Monkey** *Lagothrix flavicauda*

Description: A distinctive largely deep mahogany-brown woolly monkey, although slightly darker on the nape and lower back, with a yellow band on the underside of the final 40–50% of the tail. The face is brown, with a triangular buff patch on the nose. Long-furred particularly on the relatively short legs, giving it a robust muscular appearance. Males have a tuft of yellowish or pale brown fur on the scrotum.

Similar species: Humboldt's Woolly Monkey (*p. 411*) lacks a yellow band on the underside of the tail.

Habitat: Exclusively primary premontane, montane and cloud forest from 1,400 m to 2,700 m.

HB:	40–54 cm
Tail:	56–63 cm
Wt:	♂ 8.3–10.0 kg; ♀ 5–7 kg

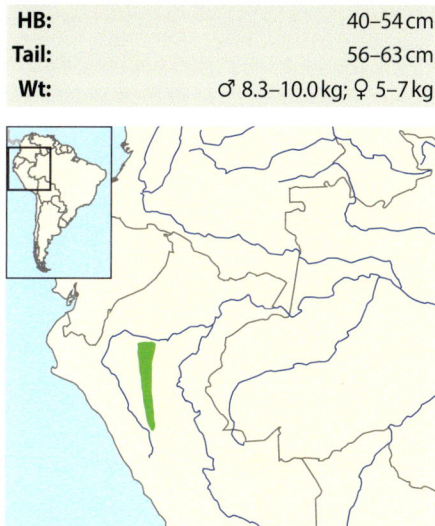

Distribution: Eastern slopes of the Andes of Peru, where it is largely restricted to scattered localities in the departments of San Martín and Amazonas, south and east of the Marañón River. Discovered also in Huánuco Department and farther to the south-east in montane forests on the eastern side of the Huallaga River.

413

Muriquis

Two species occur, both in Brazil, but do not overlap in range. They are long-limbed and long-tailed, and look very similar, but are unlikely to be confused with any other primate in their range. Northern Muriquis have been extensively studied in the forests around the Caratinga Biological Station in Minas Gerais (Strier & Boubli, 2006), where they occur in mixed-sex groups of up to approximately 60 individuals. They generally spend early morning and late afternoon feeding, in the late morning traveling between feeding areas, although they will feed throughout the day at one site when fruit is plentiful. They feed mainly on leaves and fruits, the remainder of the diet comprising flowers, bamboo, bark, buds and ferns.

CR Southern Muriqui

Brachyteles arachnoides

Description: Beige-colored with some brown or gray-brown coloration. The face and palms and the soles of the feet are black, and the black face has a grayish border. The arms have a metacarpal hook with no external thumb.

Similar species: None in range.

Habitat: Dense evergreen montane forests at 400–1,800 m, and in one isolated population in seasonal semi-deciduous forest in Fazenda Barreiro Rico.

Distribution: Brazilian Atlantic Forest in the states of Rio de Janeiro, São Paulo and northern Paraná.

HB:	47.9–63.0 cm
Tail:	65.0–75.3 cm
Wt:	8.5–10.2 kg

CR Northern Muriqui

Brachyteles hypoxanthus

Description: Beige-colored with some brown or gray-brown coloration, but often appears darker than Southern Muriqui. The face and palms and the soles of the feet are black, and the black face has a whitish border. The face loses some of its pigmentation at maturity, when it becomes spotty pink, although the black is generally retained around the eyes. Unlike Southern Muriqui, it does have a vestigial thumb.

Similar species: None in range.

Habitat: Evergreen, semi-deciduous and deciduous forest from 250 m to 1,350 m. Mature forests as well as logged and degraded forest, and secondary growth.

Distribution: Occurs north of the range of Southern Muriqui in the Atlantic Forest in the Brazilian states of Rio de Janeiro, Minas Gerais, Espírito Santo and Bahia. Absent from the lowland forests in southern Bahia and northern Espírito Santo.

HB:	46.1–51.4 cm
Tail:	73.4–81.0 cm
Wt:	♂ 9.2–9.6 kg; ♀ 6.9–8.8 kg

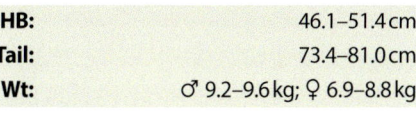

Cetaceans (dolphins)   FAMILIES | **Pontoporiidae** (Franciscana), **Delphinidae** (Guiana Dolphin & Tucuxi) **& Iniidae** (Amazon River Dolphin)

Four species of river dolphin are generally considered to occur in South America, although Wilson & Mittermeier (2014) treat three forms of Amazon River Dolphin as separate species. Cañizales (2020) and Emin-Lima *et al.* (2022) recognize a fourth form, *Inia humboldtiana*. These splits are not recognized by IUCN and there are no known external differences between the four forms.

Dolphins feed mainly on fish but Amazon River Dolphin has heterodont dentition, unique in cetaceans, which enables it also to take armored prey, including crabs and even turtles.

Another 48 species of cetaceans have been recorded in the waters around South America. These marine species fall outside the scope of this book but are listed on *pages 420–421*.

VU Franciscana   *Pontoporia blainvillei*

Description: Small and very pale, with a very long and slender beak, and a steep rounded forehead. The dorsal fin is low and rounded, with a convex tip and trailing edge. Pale brown to gray above, with a contrasting slightly darker cape. Paler yellowish-gray below, including lower flanks. Large broad spatulate flippers with undulating trailing edges. Individuals north of Rio de Janeiro tend to be smaller than those occurring farther south. Normally in small groups of 2–3 individuals, but has been recorded in groups of up to 30. Shy, and avoids boats.

Similar species: Guiana Dolphin has a much shorter beak and a more triangular dorsal fin.

Habitat: Largely restricted to turbid, shallow coastal waters less than 30–40 m deep.

HB:	♂ 117–136 cm; ♀ 148–162 cm
Wt:	20–40 kg

Favors tropical and temperate waters, generally avoiding deep, clear and cold waters.

Distribution: Atlantic coast from Espírito Santo in Brazil south to Golfo San Matias (northern Patagonia) in Argentina.

NT Guiana Dolphin

Sotalia guianensis

Description: Previously lumped with Tucuxi, but is much larger but predominantly coastal. The dorsal fin is broad-based, low, triangular and often slightly hooked. The moderately long, rather narrow beak resembles that of Common Bottlenose Dolphin. The head has a distinct melon shape with a gently sloping forehead. Dorsal coloration, including the flippers, is usually dark gray, with a slight blue or brown tinge that fades into pale gray on the lower jaw and belly. May exhibit a broad but ill-defined dark gray band from behind the eye to the base of the flippers. Like Tucuxi, it is a fast acrobatic swimmer which makes short dives. Often occurs singly or in small pods, but groups of up to 50 individuals have been recorded.

Similar species: Tucuxi (*p. 418*) is much smaller, with darker cape and paler gray flanks. **Franciscana** has a longer beak, square as opposed to pointed flippers, and more uniform sides to the body. **Common Bottlenose Dolphin** [not illustrated] looks similar but is larger, with a tall, falcate dorsal fin.

Habitat: Shallow coastal waters, including bays and estuaries, where it feeds on a wide variety of fish. It has been recorded beach hunting by driving fish on to sloping beaches.

HB:	♂ up to 210 cm; ♀ up to 220 cm
Wt:	up to 80 kg

Distribution: Occurs mainly in shallow coastal waters along the Atlantic coast of South America as far south as Santa Catarina, in southern Brazil, possibly extending inland into the lower Orinoco River in Venezuela and the lower stretches of some other rivers. Its distribution is patchy and discontinuous.

EN Tucuxi

Sotalia fluviatilis

Description: Small and compact, with a broad-based, low, triangular and often slightly hooked dorsal fin. The beak is moderately long and rather narrow. The head has a distinct melon shape, with a gently sloping forehead. Dorsal coloration, including the flippers, is usually dark gray with a slight blue or brown tinge that fades into pale gray, or white with a slight pinkish tinge on the lower jaw and belly. May exhibit a vague dark gray band from behind the eye to the base of the flippers. A fast acrobatic swimmer with short dives. Occurs in schools of up to 20 individuals, although pods are normally much smaller. Relatively shy, tending to stay away from boats. The young are grayish.

Similar species: **Guiana Dolphin** (*p. 417*) is much smaller, with paler cape and darker gray flanks. Young **Amazon River Dolphin** can look similar in coloration, but should be readily identified by its longer beak and inconspicuous dorsal fin.

Habitat: Found exclusively in freshwater and riverine habitats, including all three of the main river types (clear water, white water and black water) in the Amazon Basin. Does not enter flood zones, and generally occurs in water depths of 3 m or more in rivers and at least 1.8 m in lakes. Less common in rapids and other turbulent waters.

HB:	♂ up to 149 cm; ♀ up to 152 cm
Wt:	up to 52 kg

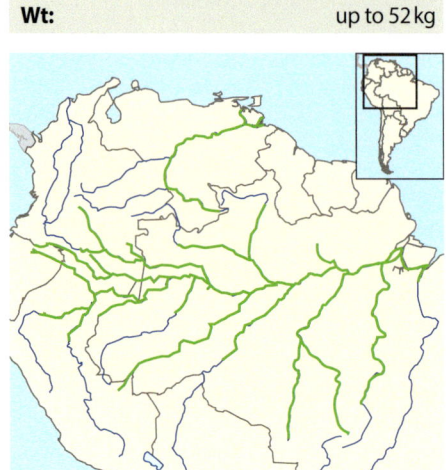

Distribution: Amazon River and most of its major tributaries in S Peru, Colombia, Ecuador and Brazil. It is unclear whether past records from the lower Orinoco River in Venezuela, and claims from farther upriver, relate to Tucuxi or to Guiana Dolphin, which was previously considered to be a coastal form of Tucuxi.

EN Amazon (Pink) River Dolphin

Inia geoffrensis

Description: A large-bodied, robust river dolphin with a steep bulbous forehead and long beak. Color is highly variable, from off-white to blue-gray or pink above and paler below. Some are totally pink, hence the alternative name of Pink River Dolphin. A slow swimmer which rarely breaches, but may race across the surface. Normally found singly or in pairs, sometimes in company with Tucuxi. The young are usually uniform dark gray.

Similar species: Young individuals can be confused with **Tucuxi**, but that species has a much more conspicuous dorsal fin and tends to stay in deeper water.

Habitat: Freshwater waterways from the upper reaches of rivers to brackish waters in estuarine areas, and often congregates around confluences of rivers and in bays. It will enter flooded forest, unlike Tucuxi, but often abandons smaller channels and lakes during the dry season.

Distribution: Four geographically isolated forms are recognized:

Amazon River Dolphin SSP. *geoffrensis*: Throughout most of the Amazon Basin in Brazil, Peru, Ecuador and Colombia

Araguaian River Dolphin SSP. *araguaiaensis*: In the Araguaia–Tocantins Basin in the states of Goiás, Tocantins, Mato Grosso and Pará, in Brazil.

Bolivian River Dolphin SSP. *boliviensis*: In the Bolivian lowlands, and in the SW Brazilian Amazon River, in the upper

HB:	♂ 219–255 cm; ♀ 182–255 cm
Wt:	♂ 113.5–207.0 kg; ♀ 72–154 kg

Madeira River and a further eight rivers and their tributaries that drain into the Madeira River. Has been recorded also between Porto Velho and Borba, in Brazil.

Orinoco River Dolphin SSP. *humboldtiana*: In the Orinoco Basin of Colombia and Venezuela, including the Apure and Meta Rivers as far upstream as Puerto Ayacucho.

SSP. *geoffrensis* (LEFT AND TOP RIGHT); SSP. *boliviensis* (BOTTOM RIGHT)

Marine cetaceans

A further 48 species of cetacean have been recorded in the inshore waters of South America. These are essentially marine species and although it is beyond the scope of this book to cover them in detail, as most are unlikely to be encountered, they are, however, listed here for completeness. For further information on their identification and distribution see Shirihai & Jarrett (2006) and Burgin *et al.* (2021).

LC	**Southern Right Whale**	*Eubalaena australis*
LC	**Pygmy Right Whale**	*Caperea marginata*
LC	**Humpback Whale**	*Megaptera novaeangliae*
VU	**Fin Whale**	*Balaenoptera physalus*
EN	**Blue Whale**	*Balaenoptera musculus*
EN	**Sei Whale**	*Balaenoptera borealis*
LC	**Bryde's Whale**	*Balaenoptera brydei*
LC	**Common Minke Whale**	*Balaenoptera acutorostrata*
NT	**Antarctic Minke Whale**	*Balaenoptera bonaerensis*
VU	**Sperm Whale**	*Physeter macrocephalus*
LC	**Pygmy Sperm Whale**	*Kogia breviceps*
LC	**Arnoud's Beaked Whale**	*Berardius arnuxii*
DD	**Shepherd's Beaked Whale**	*Tasmacetus shepherdi*
LC	**Cuvier's Beaked Whale**	*Ziphius cavirostris*
LC	**Southern Bottlenose Whale**	*Hyperoodon planifrons*
LC	**Longman's Beaked Whale**	*Indopacetus pacificus*
LC	**Gervais's Beaked Whale**	*Mesoplodon europaeus*
LC	**Strap-toothed Whale**	*Mesoplodon layardii*
DD	**Andrew's Beaked Whale**	*Mesoplodon bowdoini*
DD	**Spade-toothed Whale**	*Mesoplodon traversi*
DD	**Hector's Beaked Whale**	*Mesoplodon hectori*
LC	**Gray's Beaked Whale**	*Mesoplodon grayi*

LC	**Blainville's Beaked Whale**	*Mesoplodon densirostris*
LC	**Pygmy Beaked Whale**	*Mesoplodon peruvianus*
DD	**Killer Whale**	*Orcinus orca*
LC	**Dusky Dolphin**	*Sagmatias obscurus*
LC	**Hourglass Dolphin**	*Sagmatias crucifer*
LC	**Peale's Dolphin**	*Sagmatias australis*
LC	**Commerson's Dolphin**	*Cephalorhynchus commersonii*
NT	**Chilean Dolphin**	*Cephalorhynchus eutropia*
LC	**Southern Right-whale Dolphin**	*Lissodelphis peronii*
LC	**Rough-toothed Whale**	*Steno bredanensis*
LC	**Risso's Dolphin**	*Grampus griseus*
NT	**False Pygmy Whale**	*Pseudodorca crassidens*
LC	**Pygmy Killer Whale**	*Feresa attenuate*
LC	**Melon-headed Whale**	*Peponocephala electra*
LC	**Long-finned Pilot Whale**	*Globicephala melas*
LC	**Short-finned Pilot Whale**	*Globicephala macrorhynchus*
LC	**Common Bottlenose Dolphin**	*Tursiops truncatus*
LC	**Atlantic Spotted Dolphin**	*Stenella frontalis*
LC	**Pantropical Spotted Dolphin**	*Stenella attenuata*
LC	**Striped Dolphin**	*Stenella coeruleoalba*
LC	**Clymene Dolphin**	*Stenella clymene*
LC	**Spinner Dolphin**	*Stenella longirostris*
LC	**Common Dolphin**	*Delphinus delphis*
LC	**Fraser's Dolphin**	*Lagenodelphis hosei*
LC	**Spectacled Porpoise**	*Phocoena dioptrica*
NT	**Burmeister's Porpoise**	*Phocoena spinipinnis*

Pigs

FAMILY | **Tayassuidae** (peccaries)

Three species of peccary occur in South America. A fourth species, the **Giant Peccary** *Pecari maximus*, was described from the Brazilian Amazon and N Bolivia (van Roosmalen *et al.*, 2007), but it is not recognized as a separate species as there are insufficient data to support its status as distinct from Collared Peccary *Dicotyles tajacu*. All three species are largely diurnal and crepuscular, and generally rest up during the heat of the day. They may also, however, be active at night, particularly where hunted and/or in extremely hot areas. Peccaries feed largely on fruit, cacti, roots, seeds and leaves, but will take invertebrates, birds' and turtles' eggs, snakes, small mammals and carrion. They regularly visit salt licks and mounds of leaf-cutter ants to eat the mineral-rich soil. Peccaries are extremely vocal, and their vocalizations include low-pitched barks, clicks, snorts and growls. If threatened, White-lipped Peccaries move their canine teeth together, creating a loud clicking sound. White-lipped Peccary occurs in much larger groups (40–400 animals) than Collared Peccary, which is generally found in groups of 2–5 animals although groups of up to 30, and exceptionally up to 50, have been recorded. Chacoan Peccary generally occurs in small groups of 2–10 individuals, but solitary individuals do occur.

Introduced Wild Boar *Sus scrofa* (*p. 455*) has become established in parts of South America.

Collared Peccary

W White-lipped Peccary

Tayassu pecari

Description: A large peccary with a triangular head, stocky body and thin legs. Typically blackish-brown, but in some areas individuals can be dark reddish-brown, and has a conspicuous white patch along the lower jaw, cheeks and throat. The fur is long and coarse, and there is a mane of long stiff hairs along the dorsal ridge from the crown to the rump. Young animals are paler, with a grizzled reddish-brown coloration and indistinct pale throat patch, not unlike Collared Peccary.

Similar species: Collared Peccary (*p. 425*) is smaller, with a conspicuous white collar, and has plain cheeks and lower jaw. **Chacoan Peccary** (*p. 424*) has a grizzled appearance and white collar and lacks the conspicuous white cheek and lower jaw. Feral **Wild Boar** (*p. 455*) is considerably larger, lacks a pale patch along the lower jaw, and has a conspicuous tail and ears.

Habitat: Up to 2,000 m on the eastern slopes of the Andes, much of its range being within moist tropical and subtropical forests. Occurs also in dry forests, including the Gran Chaco, savannas, dry grasslands, seasonally flooded forest-savanna systems, and in coastal mangroves.

Distribution: Northern and central South America, as far south as Entre Ríos Province

HB:	90–139 cm
Tail:	1.0–6.5 cm
Sh:	40–60 cm
Wt:	25–40 kg

in N Argentina (although now extirpated from most of N Argentina) and Rio Grande do Sul in S Brazil. Absent from NE Brazil. Range extends discontinuously through Central America and into Mexico.

EN Chacoan Peccary

Catagonus wagneri

Description: A large peccary with long bristles, giving it a shaggy appearance. Usually grizzled brownish-gray with dark, stiffer hairs running down the back, and a whitish collar across the chest and shoulders. It is larger-headed and longer-legged than the other peccaries.

Similar species: Collared Peccary is smaller with a less shaggy appearance and a smaller head. **White-lipped Peccary** (*p. 423*) lacks the grizzled appearance and white collar and has a conspicuous white cheek and lower jaw.

Habitat: Restricted to the driest, hottest parts of the Chaco. Found in xerophytic thorn forest with emergent trees, a dense shrub layer and a ground cover of bromeliads and cacti. Occurs also, at lower densities, in open woodland.

Distribution: Dry Chaco of W Paraguay, SE Bolivia and N Argentina.

HB:	90–117 cm
Tail:	2.4–10.2 cm
Sh:	52–69 cm
Wt:	29.5–40.0 kg

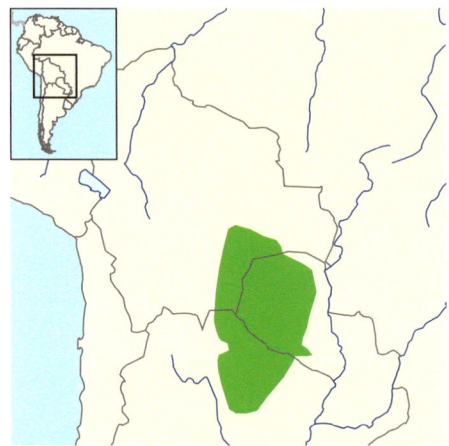

LC Collared Peccary

Dicotyles tajacu

HB:	84–106 cm
Tail:	up to 10 cm
Sh:	30–50 cm
Wt:	15–28 kg (exceptionally to 42 kg)

Description: A relatively small peccary with a triangular head, stocky body and thin legs. Typically grizzled dark blackish-gray, darkest on the dorsal crest and legs, with a whitish collar across the chest and shoulders. The collar is highly variable, being barely discernible on some individuals. Some populations appear reddish or dark brownish. Young are paler reddish-tan with a distinct dark brown dorsal stripe and pale collar.

Similar species: Chacoan Peccary is similar in appearance but is larger, noticeably large-headed, and has long bristles giving it a much shaggier appearance. **White-lipped Peccary** (*p. 423*) is larger, with ungrizzled blackish coloration and a conspicuous white cheek and lower jaw. Feral **Wild Boar** (*p. 455*) is considerably larger, lacks the pale collar, and has a conspicuous tail and ears.

Habitat: A wide range of habitats including rainforest, montane cloud forests, dry forests, scrublands, savannas, wetlands and deserts.

Distribution: Much of northern South America, including the Pacific coastal forests of Colombia, Ecuador and Peru, the Llanos and lowland forest of Venezuela, Trinidad, the Guianas and Suriname, the entire Amazon Basin, all of Brazil (although the range is fragmented in the south and east), and the Gran Chaco of Paraguay and Bolivia, and also in the Gran Chaco and upper Paraná and Paraguay River Basins in N Argentina. A population remains in Misiones, Argentina. Range extends through Central America and E & W Mexico into south-central & SW USA.

Camelids

Two species of camelid, Guanaco and Vicuña, occur in a wild state in South America, with a further two, Llama and Alpaca, occurring as domesticated forms (see *page 460*).

Guanacos graze and browse on a wide range of plants. Typical family groups comprise 10–15 individuals, and herds can be made up of a number of family groups. Groups of non-breeding males average 25 animals, but range from 3 to 60. Mixed groups form in winter and average 60 animals, but groups of as many as 500 individuals have been recorded.

Vicuñas are predominantly grazers and over much of their range eat mainly grasses. They usually occur in family groups of 6–7 animals. Groups of non-territorial males tend to consist of fewer than 30 animals, but groups of up to 155 have been recorded.

Guanaco

LC Vicuña

Lama vicugna

Description: A slender, long-necked camelid with pale cinnamon to reddish-brown upperparts and white underparts, including the insides of the legs. The northern ssp. *mensalis* has a white bib of long, coarse hairs at the base of the neck. The southern ssp. *vicugna* is larger and lacks the long chest hairs. Both subspecies occur in Chile and Bolivia.

Similar species: Guanaco (*p. 428*) is larger and more bulky, and has a darker face. The domesticated form, **Alpaca** (*p. 460*), can appear Vicuña-like when shorn.

Habitat: The puna between the treeline and snowline at 3,200–4,800 m. Frequently in dry xerophytic shrub steppes, but equally at home around high-altitude wetlands.

Distribution: High-altitude areas of Peru, Bolivia, Argentina and Chile, and has been introduced into Ecuador (Reserva de Produccion de Fauna Chimborazo). Found from 9°30'S in N Peru to 29°30'S in northern-central Chile.

HB:	125–190 cm
Tail:	15–25 cm
Sh:	85–90 cm
Wt:	38–45 kg

LC Guanaco

Lama guanicoe

Description: A large, elegant, long-necked camelid, generally larger than the similar Vicuña, although there is considerable variation in size across its range. Northern animals are light brown with ochre-yellow tones, while southern individuals are darker reddish-brown. The face is gray or black and the underparts are white.

Similar species: Vicuña (*p. 427*) is smaller and more slender, and has a paler face. The domesticated form, **Llama** (*p. 460*), can have Guanaco-like coloration.

Habitat: A wide range of habitats from sea level to 5,000 m. In desert and xeric shrublands, montane grasslands (including puna and Andean steppe), semi-arid grasslands and savannas (including Chacoan grasslands), and temperate forests.

Distribution: Occurs widely but discontinuously from Calipuy National Reserve in N Peru (8°30' S) to Navarino Island (55°S) in S Chile. In Bolivia, found in the Chaco region and in the southern highlands between Potosi and Chuquisaca. Has been reported in the Chaco region of NW Paraguay. In Argentina, found in much of Patagonia, with populations more

HB:	190–250 cm
Tail:	23–27 cm
Sh:	90–130 cm
Wt:	90–140 kg

fragmented in the northern provinces. In central & N Argentina restricted to the Andean mountains and foothills up to the Bolivian border, and has been reported in the arid Chaco in north-west Córdoba.

Additional photo *p. 426*

Deer

Twenty-one species currently occur in South America, including the introduced Common Fallow Deer, Western Red Deer, Chital, Sambar and Pere David's Deer (see *page 456*). Gutiérrez *et al*. (2017) and Tirira (2017) both recognize a number of additional native species, but in the absence of more widespread acceptance of these splits IUCN taxonomy is followed here. Tirira (2017) splits **Peruvian White-tailed Deer** *Odocoileus peruvianus*, **Andean White-tailed Deer** *O. ustus*, **Gualea Red Brocket** *Mazama gualea*, **Zamora Red Brocket** *M. zamora* and **La Murelia Brown Brocket** *M. murelia*.

Gonzalez & Duarte (2020) suggest that two additional species of brocket deer should be recognized, but this treatment has yet to be widely adopted and is therefore not followed here. Some authorities consider Mérida Brocket *M. bricenii* to be a synonym of Little Red Brocket *M. rufina*, but it is treated here as a separate species in line with the taxonomy of the IUCN Deer Specialist Group.

Many of the larger South American deer are diurnal and most active in the early morning and late afternoon/evening, when they often emerge from cover to feed out in the open. Where levels of disturbance and hunting are high, they may become more nocturnal. They are generally solitary or found in small groups. Most pudus and brockets tend to be solitary and are more crepuscular or nocturnal, but can also be active by day.

Deer are herbivores, grazing on grass, herbaceous plants, low shrubs and fallen fruits and even taking fungi and nuts, and browsing on leaves and twigs.

White-tailed Deer (MALE)

LC **White-tailed Deer** *Odocoileus virginianus*

Description: Five subspecies of this widespread deer occur in South America. Usually gray-brown to orangey-brown, being grayer in winter and redder in summer. There is a white band around the nose and a conspicuous white ring around each eye. The chin, throat, inner thighs, belly and under-tail are white. The ears are long and narrow, and males have curved antlers with three branches. Fawns are reddish-brown with white spots and stripes, which are lost after 3–4 months. Adults have a flat back and run with the head held high, unlike brockets which run with the head held at the same level as their back. Normally encountered singly or in small groups. When disturbed, gives a whistling call and bounds away with the white tail raised and fanned (see *page 429*).

Similar species: North Andean Huemul (*p. 444*) is more compact and darker, with a dark rump: males have only two-pronged antlers.

Habitat: A wide range of habitats, from temperate forests and semi-arid environments to subtropical and tropical forests, including montane forests, rainforests, deciduous forests and savannas.

HB:	♂ 145–155 cm; ♀ 130–145 cm
Tail:	16–19 cm
Sh:	♂ 85–95 cm; ♀ 70–80 cm
Wt:	♂ 50–80 kg; ♀ 35–50 kg

Distribution: North and north-west South America from N Brazil and the Guianas, through Venezuela, Colombia and W Ecuador into W Peru. Range extends through Central America, Mexico, most of the USA and into Canada.

MALE (LEFT); FEMALE (RIGHT)

Additional photo *p. 429*

NT Pampas Deer

Ozotoceros bezoarticus

Description: A slim medium-sized deer with reddish-brown or yellowish-gray upperparts and legs, the coloration being richest on the back. The face, crown and tail are darker, with whitish areas around the eyes. The chest, throat, underparts, and underside of the tail are also whitish. There is no marked difference between summer and winter coats. Fawns up to two months old are chestnut, with rows of white spots on each side of the back and from the shoulders to the thighs. Males have three-pronged antlers. It is paler and taller than any other deer in its range, and unlikely therefore to be confused.

Similar species: None in range.

Habitat: Found singly or in groups of up to six (but exceptionally up to 50) animals in a wide range of grassland habitats, including Pampas, savannas and seasonally flooded grasslands, along with Cerrado scrub, and also in coastal salt marshes in Argentina.

Distribution: Now restricted to highly fragmented populations in W, N & central Argentina, E Bolivia, central & S Brazil, Paraguay and Uruguay.

HB:	♂ 90–120 cm; ♀ 85–90 cm
Tail:	10–14 cm
Sh:	♂ 65–70 cm; ♀ 60–65 cm
Wt:	♂ 24–34 kg; ♀ 22–29 kg

MALE (LEFT); FEMALE (RIGHT)

DD Common Red Brocket

Mazama americana

Description: A large brocket with a characteristic rounded body and thin neck and legs. The upperparts are largely bright reddish-brown, with a dark grayish-brown face and neck and whitish throat and chest. The belly is orange. The lower legs can appear dark, almost black at times. The ears are broad and the male's antlers are short, straight and unbranched and similar in length to the ears. The short tail is white on the underside.

Similar species: Amazonian (*p. 438*) and **Common** (*p. 437*) **Brown Brockets** overlap with Common Red Brocket across large parts of its range, but are smaller and lighter, with grayish-brown rather than rich reddish-brown coloration. The area of overlap between this species and **Mexican Red Brocket** is unclear, and the distinguishing features of the two species still require further clarification. Common Red Brocket, however, is generally larger and heavier, and appears darker overall. Overlaps in range with several smaller red brockets, most of which should be identifiable by their smaller size. **Small Red Brocket** (*p. 434*), however, may not be safely identifiable in the field.

Habitat: A wide range of forested areas at up to 5,000 m, including tropical and subtropical

HB:	90–145 cm
Tail:	12–16 cm
Sh:	60–80 cm
Wt:	30–35 kg (exceptionally to 65 kg)

forests, and found also in gallery forests and savannas close to the forest edge.

Distribution: Colombia, Venezuela, Trinidad and the Guianas south to N Argentina, although absent to the west of the Andes and from NE Brazil.

MALE (LEFT); FEMALE (RIGHT)

DD Mexican Red Brocket

Mazama temama

Description: A medium-sized brocket. The upperparts are reddish-brown, with a grayish neck and a brown or reddish head and throat. The rest of the underparts are white. The ears are broad, and the male's antlers are short, straight and unbranched and of a similar length to the ears. The short tail is white on the underside.

Similar species: Very similar to **Common Red Brocket**, with which it was previously lumped, and the distinguishing features of the two species still require further clarification.

Habitat: In central America occurs in primary and secondary forest, including perennial forest, cloud forest, flooded forests and low-dry forest, and has been recorded from sea level up to the páramo at 2,800 m.

Distribution: In South America known only from W Colombia. Range extends through Central America and into S & E Mexico.

HB:	80–110 cm
Tail:	10–14 cm
Sh:	60–70 cm
Wt:	12–32 kg

FEMALE

VU Small Red Brocket

Mazama bororo

Description: Similar in body shape and structure to Lesser Brocket (*p. 439*), but with coloration that strongly resembles that of Common Red Brocket and is morphologically intermediate between the two. Mainly reddish, with a gray neck. The chin, throat, chest and belly are whitish and the lower hind legs are blackish.

Similar species: Very similar in appearance to **Common Red Brocket** (*p. 432*) and may not be safely identifiable in the field.

Habitat: Up to 1,200 m in lowland, sub-montane and montane Atlantic Forest in the Serra do Mar.

Distribution: Known from nine sites in a small area between 23°S and 26°S, in the south of São Paulo State, east of Paraná and extreme north-east of Santa Catarina State, in Brazil.

HB:	approx. 85 cm
Tail:	11–14 cm
Sh:	50–60 cm
Wt:	approx. 25 kg

MALE

 Mérida Brocket

Mazama bricenii

Description: Small, long-haired and reddish, with an orange throat, and a dark head and legs. The male's antlers are short and relatively straight and are up to 6 cm long.

Similar species: Common Red Brocket (*p. 432*) is much larger and has a grayish throat and neck.

Habitat: From 1,000 m to 3,500 m. Main habitats are páramo and tropical montane cloud forests above 1,500 m.

Distribution: Patchily distributed in the Andes of N Colombia and W Venezuela.

HB:	80–95 cm
Tail:	8–9 cm
Sh:	45–50 cm
Wt:	8–13 kg

MALE (BOTH)

VU **Common** (Peruvian) **Dwarf Brocket** *Mazama chunyi*

Description: A small brocket appearing quite dark, with more rounded ears than other brockets. The upperparts are dark reddish-brown, with the center of the back and flanks appearing slightly brighter. The head, neck and legs appear blackish, while the throat, chest and belly appear orange. The ears have more conspicuous white edges than other brockets. The male's antlers are short and relatively straight.

Similar species: None in range.

Habitat: From 1,000 m to 4,000 m in elfin forest and grasslands, montane and sub-montane forests.

Distribution: Restricted to S Peru (Junín, Cusco and Puno) and N Bolivia (La Paz and Cochabamba).

HB:	70–75 cm
Tail:	No info. available
Sh:	approx. 38 cm
Wt:	approx. 11 kg

MALE (ABOVE); FEMALE (BELOW)

LC **Common Brown Brocket** (Gray Brocket) *Mazama gouazoubira*

Description: Small to medium-sized and relatively uniform light grayish-brown, although some populations are darker and grayer, particularly in forest. The throat and chest are paler gray. The rump and upper tail are orange, and the ears are large and rounded. The male's antlers are short spikes up to 12 cm long.

Similar species: Common Red Brocket (*p. 432*) is larger and bright reddish-brown, with a dark face and neck. **Lesser Brocket** (*p. 439*) is smaller and reddish-chestnut. **Amazonian Brown Brocket** (*p. 438*) is smaller and has a brown rump and upper tail and smaller ears.

Habitat: Found mainly in woodland, brushy areas and forest edge, but avoids dense forest and absent from the Amazon region. Found in both dry and moist Chaco regions. Although active throughout the day, it emerges from cover to feed in more open areas, including grassland and plantations, at night.

Distribution: South of the Amazon region and east of the Andes in Brazil, Bolivia, Paraguay,

HB:	85–105 cm
Tail:	8–9 cm
Sh:	50–65 cm
Wt:	11–25 kg

Uruguay, and N Argentina as far south as Entre Ríos Province.

MALE

FEMALE

LC Amazonian Brown Brocket

Mazama nemorivaga

Description: A small to medium-sized deer with a long muzzle, small ears, protruding eyes, small and spike-like antlers (on males) and a dark brown coat. The flanks are paler brown. The male's antlers are up to 11 cm long.

HB:	75–100 cm
Tail:	6–11 cm
Sh:	approx. 50 cm
Wt:	14–16 kg

Similar species: Common Red Brocket (*p. 432*) is larger and bright reddish-brown, with a dark face and neck. Amazonian and **Common Brown** (*p. 437*) **Brockets** should be identifiable based on range but the latter is larger and has larger ears and an orange rump and upper tail.

Habitat: Mainly in tropical and subtropical broadleaf moist forests in the Amazon region, although also in desert and xeric scrubland in coastal states in N Venezuela. It appears to avoid seasonally flooded forests.

Distribution: Amazonian Forest and transitional border areas in N Brazil, French Guiana, Suriname, Guyana, Venezuela, Colombia, Ecuador, Peru and possibly Bolivia.

FEMALE

VU Lesser Brocket

Mazama nanus

Description: Small and reddish-chestnut, with a slightly darker face and legs. Males have short straight antlers.

Similar species: Common Red Brocket (*p. 432*) is much larger. **Common Brown Brocket** (*p. 437*) is larger and grayish-brown.

Habitat: Rarely seen but closely associated with mixed wet forests, including monkey-puzzle (*Araucaria*) forest. Mainly in mountainous regions, but apparently occurs also outside mountainous habitats throughout its distribution. Prefers habitats with dense lower strata, abundant bamboo understory and secondary-growth forests.

Distribution: SE Paraguay, Misiones Province in Argentina, and in Brazil in the extreme south of Minas Gerais, São Paulo State (except Serra do Mar), south and south-east Mato Grosso do Sul, Paraná, Santa Catarina and northern Rio Grande do Sul.

HB:	approx. 70 cm
Tail:	9–12 cm
Sh:	45–50 cm
Wt:	14–16 kg

MALE

VU Little Red Brocket

Mazama rufina

Description: One of the smallest brockets, with a small rounded body, and very similar in coloration to the much larger Common Red Brocket (*p. 432*), normally being largely reddish with an almost black head and legs. There are white spots around the nostrils, on the chin and on the underside of the tail. Robust, with short legs and small ears. Male's antlers are short spikes up to 8 cm in length.

Similar species: Northern Pudu is much darker, smaller (half the weight), shorter-legged and shorter-necked.

Habitat: Remnant forest patches and páramo from 1,500 m to 3,600 m.

Distribution: From the Central Andes in Colombia to Huancabamba Valley in northern Peru. Restricted to the western slopes of the Andes in Colombia, but on both slopes in Ecuador. In Peru found on the eastern side of the Andes.

HB:	80–95 cm
Tail:	8–9 cm
Sh:	36–48 cm
Wt:	10–15 kg

MALE

DD Northern Pudu

Description: The world's smallest deer, with short legs, a short tail and rounded ears. Rufous, with a dark brown back, dark brown legs and a black face. Fawns are unspotted. Adult males have short spike-like antlers up to 9 cm long.

Similar species: Little Red Brocket is reddish, larger (double the weight), longer-legged and longer-necked.

Habitat: Elfin, cloud and montane forests, and in the northern part of its range also in humid grasslands above the treeline. At 2,000–4,000 m and exceptionally to 4,500 m. Remains in cover when in forest and as a consequence is rarely seen. Occasionally enters cultivated areas to feed.

Distribution: Patchily distributed through the montane forests and humid grasslands of the Andes in Colombia, Ecuador and Peru. Absent from the Marañón dry forest lying between the Ecuadorian and Peruvian populations south of the Marañón River.

HB:	approx. 75 cm
Tail:	approx. 3 cm
Sh:	25–38 cm
Wt:	5–6 kg

FEMALE

NT Southern Pudu

Pudu pudu

Description: The only small deer in its range. Rufous in summer and dark brown in winter, with short, thick, paler legs. It has small rounded ears, small eyes and a short inconspicuous tail. Adult males have short spike-like antlers up to 10 cm long. Fawns are rufous with lines of small white spots on their back for the first three months, at which point they attain adult coloration.

Similar species: None in range.

Habitat: Up to 1,700 m in both mature and disturbed forests, particularly those with dense understory, such as bamboo. These include primary and secondary evergreen forest, Alerce conifer forests and eucalyptus plantations. Feeds along forest edges and tracks.

Distribution: From 35°10'S to 46°45'S, in S Chile from Maule to Aisen, and in SW Argentina. Common on Isla Chiloé in Chile.

HB:	75–90 cm
Tail:	approx. 4 cm
Sh:	30–40 cm
Wt:	9–10 kg (exceptionally to 14.8 kg)

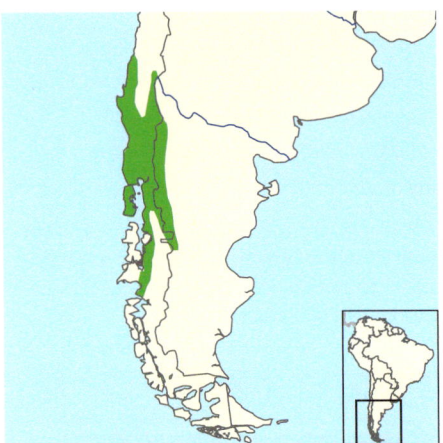

MALE

Marsh Deer   *Blastocerus dichotomus*

Description: South America's largest native deer. Long-legged, with a shaggy, reddish-chestnut coat, with a paler throat, chest and belly, a distinctive black muzzle, an indistinct white ring around each eye, black socks below the knees and a short tail with a black tip. Adult males have branched antlers with 4–5 points. Has long, broad hooves that are joined by a membrane which spreads out to prevent it from sinking into swampy ground.

Similar species: Unlikely to be confused with any other deer but, perhaps surprisingly, can resemble **Maned Wolf** (*p. 170*) if seen poorly.

Habitat: Generally found singly or in groups of 2–3 animals in marshy habitats, including seasonally flooded areas, but tends to avoid forests.

HB:	♂ approx. 180 cm; ♀ approx. 165 cm
Tail:	13–15 cm
Sh:	♂ 115–130 cm; ♀ 100–115 cm
Wt:	♂ 110–130 kg; ♀ 70–100 kg

Distribution: Now restricted to fragmented populations in east-central & NE Argentina, Paraguay, west-central & S Brazil, E Bolivia and SE Peru.

MALE (ABOVE); FEMALE (BELOW)

VU **North Andean Huemul**

Hippocamelus antisensis

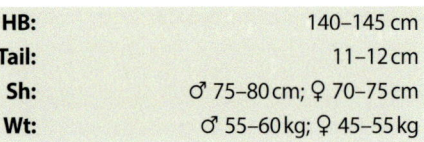

Description: Generally light brown to gray, the top of the tail being gray while the under-tail and the throat are white. There is a dark 'Y'-shaped mark on the forehead and a whitish band around the muzzle. The insides of the ears are white. Males have short 'V'-shaped antlers, the front prong being shorter than the rear one.

Similar species: White-tailed Deer (*p. 430*) is noticeably larger and males have larger curved antlers with three branches.

Habitat: Mountain slopes with rocks and cliff-like outcrops, and alpine grasslands and fertile valleys, often with a water source nearby, *e.g.* a small ravine, lagoon or marsh. From 2,000 m to 3,500 m in Argentina and up to 5,000 m in Peru and Bolivia. Occurs in family groups of 4–9 animals.

Distribution: The High Andes from the north of Peru to NE Chile, although its distribution is extremely fragmented throughout the high Andes of Bolivia, N Chile and NW Argentina. Absent from SW Bolivia.

HB:	140–145 cm
Tail:	11–12 cm
Sh:	♂ 75–80 cm; ♀ 70–75 cm
Wt:	♂ 55–60 kg; ♀ 45–55 kg

MALE (NOTE: THIS INDIVIDUAL HAS SHED ITS RIGHT ANTLER)

Hippocamelus bisulcus

Description: Rusty-brown in summer and grayish-yellow in winter, occasionally with pale spotting. Males have a dark chevron on the face and muzzle. It has large ears, and the male's antlers have two branches and are up to 25 cm long. The under-tail is white.

HB:	155–165 cm
Tail:	13–15 cm
Sh:	♂ 85–90 cm; ♀ 80–85 cm
Wt:	♂ 70–75 kg; ♀ 60–70 kg

Similar species: The only medium-sized deer in range and likely to be confused only with the introduced **Western Red Deer** (*p. 456*), which is much larger, longer-legged, and males of which have much bigger, highly branched antlers. Female Western Red Deer has a buff-white tail.

Habitat: Found in a variety of habitats from valley bottoms to steep mountain slopes up to 3,000 m, and from open grasslands to closed shrubby or forested areas, particularly forest edge and southern beech (*Nothofagus*) forests. Occurs singly or in groups of 2 or 3 (occasionally up to 17) animals. In winter, favors warmer north-facing slopes.

Distribution: The Andes of S Chile (mainly Aisen and Magallanes, with small numbers in Los Nevados de Chillan) and S Argentina.

Primarily on the eastward-facing slopes of the Patagonian Andes.

MALE

Tapirs

Three species of tapir are found in South America, a fourth occurring in South-east Asia. A fifth species, **Kabomani Tapir** *Tapirus kabomani*, was described from the Brazilian Amazon by Cozzuol *et al.* (2013) based on its smaller size, darker fur and mane, and cranial and dental differences, but the validity of the species has been challenged and this form is not currently recognized as a distinct species by IUCN or the wider scientific community. Voss *et al.* (2014) have also suggested that the taxonomy of Mountain Tapir should be revisited, but at present it is still considered a valid species.

All three species are active by day and night, spending the hottest parts of the day resting up in standing water, mud wallows or shaded areas. Mountain Tapirs are most active in the early morning and late afternoon/evening. They are largely solitary, although young animals may stay with their mothers well after they have weaned, and in areas where they are common groups of 2–3 Lowland Tapirs can be encountered around salt licks, at permanent water and in adjacent agricultural areas. They are browsers, feeding on a wide range of fruits, leaves, shoots, herbs, grasses and flowers.

Lowland Tapir

Central American (Baird's) Tapir

Tapirus bairdii

Description: The largest terrestrial mammal in South America. Dark brown or grayish-brown with cream-colored markings on the cheeks and throat. Very young animals are reddish-brown with bold white spots and stripes.

HB:	190–230 cm
Tail:	<10 cm
Sh:	approx. 120 cm
Wt:	180–350 kg

Similar species: None in range.

Habitat: Generally, in humid forests from sea level to 3,600 m, including forested areas with ponds and streams, páramo, riparian forest, palm swamps and mangroves. It will also feed on beaches at night and frequently enter the sea.

Distribution: The Chocó and Darién regions of Colombia. Has been reported alongside Lowland Tapir in dry forests and savannas of the upper Sinu River (Caribbean region), but surveys have found only the latter species here. The southernmost record in Colombia is from Departamento de Nariño on the Pacific slopes of the Andes, but there are no recent records from S Colombia. Range extends through Central America and into S Mexico but is very fragmented.

FEMALE WITH JUVENILE

EN Mountain Tapir · *Tapirus pinchaque*

Description: The smallest tapir, with long dark brown to black fur, a distinctive white-lipped appearance, and rounded ears with white tips.

Similar species: If seen well, unlikely to be confused with any other species in its range, but glimpses in thick vegetation could lead to confusion with **Andean Bear** (*p. 174*) by the unwary.

Habitat: Chaparral, tropical montane forests, páramo and riverine meadows between 1,400 m and 4,800 m. The highest densities have been found in secondary forests, but in some areas it is regularly encountered in open habitats.

Distribution: The Andes of Colombia, Ecuador and N Peru, although absent from large areas of seemingly suitable habitat. Formerly occurred in W Venezuela.

HB:	180–200 cm
Tail:	<10 cm
Sh:	80–90 cm
Wt:	150–200 kg

448

VU **Lowland** (Brazilian) **Tapir** *Tapirus terrestris*

Description: Dark blackish-brown above, with white-edged ears. The chest, vent and legs are dark brown, with the cheeks grizzled brown or gray. Has a unique mane of long black hairs which runs from the forehead to the middle of the back. Young are dark, with white or yellowish stripes and spots.

Similar species: None in range.

Habitat: Found in a wide range of habitats, including tropical lowland forests, dry Chaco and Cerrado forests, palm forests, savanna wetlands, grasslands and lower montane grasslands and forests at up to 2,000 m.

Distribution: Lowland areas in Argentina, Bolivia, Brazil, Colombia, Ecuador, French Guiana, Guyana, Paraguay, Peru, Suriname and Venezuela. The species has disappeared from the dry inter-Andean valleys of the northern Andes and is becoming increasingly rare as agriculture expands through parts of the western and southern Amazon Basin. In the Cerrado, now restricted to a few small populations in protected areas; populations in the Pantanal are also declining.

HB:	191–242 cm
Tail:	<10 cm
Sh:	83–118 cm
Wt:	180–300 kg

ADULT (BELOW); JUVENILE (INSET)

Additional photo *p. 446*

Two species of manatee occur in rivers and/or coastal areas in the northern half of South America. They are slow-moving and generally lethargic, making short shallow dives. They feed during the day and at night on a wide variety of submerged, floating and emergent vegetation.

VU Amazonian Manatee

Trichechus inungris

Description: Dark to blackish-gray with variable white or pink patches below, although these can be absent. Broad-headed with a bulbous muzzle.

Similar species: May overlap with **West Indian Manatee** in the mouth of the Amazon, but is slenderer, 25% shorter and weighs less than a third of that species' weight. It is also smoother-skinned and generally darker. Unlikely to be confused with river dolphins (*pp. 418–419*) given good views, the head shape and coloration both being distinctive.

Habitat: Usually solitary or in small groups. Generally, away from human settlements in both white- and black-water rivers within the Amazon Basin, but also in brackish waters in the lower Amazon River.

Distribution: Patchily distributed in nutrient-rich flooded forest throughout 300,000 km² of the Amazon River drainage, from Colombia, Ecuador and Peru to the mouth of the Amazon River in Brazil, the range being

HB:	♂ up to 300 cm; ♀ up to 270 cm
Wt:	up to 450 kg

strongly influenced by seasonal flooding; some individuals make long-distance movements from flooded areas during the wet season to deeper lakes during the dry season.

U **West Indian** (American) **Manatee** *Trichechus manatus*

Description: Much longer and bulkier than Amazonian Manatee. Also much paler than its smaller relative, often showing scars and damaged skin. Its broad neck narrows into a thick slightly downcurved muzzle.

Similar species: Amazonian Manatee, with which it may occasionally overlap near the mouth of the Amazon River. Overlaps in range with a number of cetaceans (*p. 416*) but, given good views, the head shape and lack of a dorsal fin should be diagnostic.

Habitat: Shallow marine and estuarine areas within a few hundred meters of the coast, and the lower and middle reaches of some larger rivers.

Distribution: The Atlantic coast from Colombia as far south as the state of Sergipe, in Brazil, and inland along the lower and middle reaches of the Orinoco River in Venezuela, and of the larger rivers in N Colombia. Found also along the Caribbean coasts of Central America and S & E USA

Lth:	250–390 cm
Wt:	up to 1,620 kg

and many of the Caribbean islands, although absent from the coasts of NE Mexico and Texas.

INTRODUCED MAMMALS

Many non-native mammals have been introduced into South America, either deliberately for food, fur or for hunting, some of which have subsequently become naturalized and established self-sustaining populations. Since these may be encountered and potentially be confused with native species, they are included here for completeness. The IUCN codes relate to the species' native ranges.

Lagomorphs (Rabbits & Hares) (see *p. 79*) FAMILY | **Leporidae**

EN European Rabbit *Oryctolagus cuniculus*

Description: Small and grayish-brown with paler underparts, including a white throat and chin. The ears are short and dark without black tips. The tail is short, of the same color as the rest of the upperparts above, and white below.

HB:	36–38 cm
Tail:	6.5–7.0 cm
Wt:	1.5–3.0 kg

Similar species: European Hare is much larger, longer-limbed, tan to brown above, and has long black-tipped ears.

Habitat: Found mainly around woodland edges and areas with sandy or soft soils.

Distribution: Native to SW Europe. Established in extreme S Argentina, and in Chile near Arica, from Coquimbo to Valdivia, and in Magallanes.

LC European Hare *Lepus europaeus*

Description: The only hare in range with tan to brown upperparts, slightly paler chest and flanks, and a white belly. Long-limbed, with long pale gray ears each with a conspicuous black tip. The tail is black above, white below.

HB:	55–68 cm
Tail:	7.5–14.0 cm
Wt:	3.5–5.0 kg

Similar species: European Rabbit is smaller, with shorter limbs, and plain ears. Superficially similar to **Patagonian Mara** (*p. 131*), which overlaps in range but should not be confused given a good view.

Habitat: Open woodland, pastures, steppes, sub-deserts, Pampas, sand dunes, marshes and alpine slopes.

Distribution: Native to Eurasia. Established in Chile from Valparaíso to Magallanes, through much of Argentina except the far

north, in Uruguay, and in the extreme south of Brazil.

Squirrels (see *p. 83*) (see *p. 83*)

FAMILY | **Sciuridae**

LC # Pallas's Squirrel

Callosciurus erythraeus

Description: A medium-sized squirrel. Grizzled olive-brown above with reddish, maroon, orange-brown or cream underparts. The long olive-brown tail may be tipped black or reddish, or in some cases has a pale tip.

Similar species: None in range.

Habitat: Forested areas, including commercial forests, temperate and subtropical riparian forests, woodland patches and wooded corridors in otherwise unsuitable habitat.

Distribution: Native to Asia. Established in the Pampas region west of Buenos Aires in Argentina, north to within 30 km of the Paraná River Delta.

HB:	21.7–22.7 cm
Tail:	20.5–21.6 cm
Wt:	359–375 g

Beavers

FAMILY | **Castoridae**

LC # North American Beaver

Castor canadensis

Description: A large, robust, aquatic rodent with a broad, flat, paddle-like, scaly tail and fully webbed hind feet. The fur color ranges from light brown through rich chestnut-brown to almost black. The underparts are generally paler. It has small eyes and ears and prominent orange incisors. Normally swims with just the forehead, eyes and nose exposed, although will occasionally swim with the nape and back also visible above the surface, creating a single-humped appearance.

Similar species: **Coypu** (*p. 138*) generally swims showing a double-humped appearance. **Common Muskrat** (*p. 454*) is much smaller and normally swims with the top of the head, the back and sometimes the tail exposed, appearing double- or triple-humped.

Habitat: Freshwater wetlands, including lakes, ponds, and streams and rivers. Lives in family groups. Active from early evening until early morning, feeding on a variety of woody and herbaceous plants

HB:	80–90 cm
Tail:	20–30 cm
Wt:	15–30 kg

Distribution: Native to North America. Introduced to S Argentina (Tierra del Fuego) in 1946 and has now spread north of the Strait of Magellan into the Andean forests of S Chile and has colonized many of the Fuegian islands.

Muskrats

FAMILY | **Cricetidae**

LC # Common Muskrat

Ondatra zibethicus

Description: A smallish aquatic rodent with a long, laterally flattened tail covered with scales rather than fur. The fur is thick, with long glossy guard hairs. Upperparts brown, belly grayish-white; cheeks often white. The ears are largely obscured by the thick fur. Swims well and can stay submerged for up to 20 minutes.

Similar species: Coypu (*p. 138*) is larger, and generally swims showing a double-humped appearance. **North American Beaver** (*p. 453*) is much larger.

Habitat: Brackish and freshwater lakes, ponds, streams, rivers and marshes. Mainly crepuscular but can be seen throughout the day. Largely herbivorous, feeding on aquatic vegetation, but when food is scarce will feed also on mussels, turtles, mice, birds, frogs, and fish.

Distribution: Native to North America. Introduced from Canada to S Chile (Tierra

HB:	26–37 cm
Tail:	19–27 cm
Wt:	700 g–1.5 kg

del Fuego and Navarino Island) in 1948 and has spread into neighboring S Argentina.

Weasels (see *p. 186*)

FAMILY | **Mustelidae**

LC # American Mink

Neovison vison

Description: Medium-sized and dark chocolate-brown, with small white patches on the chin and on some individuals on the throat and chest. Long-bodied and short-limbed, with a dark brown bushy tail.

Similar species: Most likely to be mistaken for an **otter** (*pp. 182–185*), but otters are generally much larger, with a longer tapered tail.

Habitat: Rivers, streams and lakes, as well as swamps and marshes, often in forested areas. Carnivorous, feeding on small mammals, birds, amphibians, fish and crustaceans.

Distribution: Native to North America. Feral populations exist in S Chile and S Argentina as a consequence of escapes during 1930–1950.

HB:	♂ 33–43 cm; ♀ 30–40 cm
Tail:	15.2–20.0 cm
Wt:	♂ 850 g–1.81 kg; ♀ 450–840 g

Mongooses

LC Small Indian Mongoose
Urva auropunctata

Description: A small weasel-like carnivore, ranging in color from grizzled pale yellow-gray, to buff and rufous-brown, with pale buff around the mouth, chin and throat.

Similar species: Tayra (*p. 190*) is much larger and generally darker.

Habitat: Typically diurnal, found in forests, grasslands, agricultural and coastal areas.

Distribution: Native to Asia. Introduced to Trinidad.

HB:	♂ 22–44 cm; ♀ 21.0–38.5 cm
Tail:	15–33 cm
Wt:	500 g–1 kg

Pigs (see *p. 422*)

LC Wild Boar
Sus scrofa

Description: South American feral populations are believed to originate from European animals and the measurements and description given here relate to that form. A distinctive, dark gray to blackish-gray bristly pig; in temperate climates the coat is thicker in winter than in summer. The body is stocky with a large head, short thick neck and proportionately short thin legs. The male has upward-pointing tusks formed from its canine teeth. In some areas, such as the Pantanal in Brazil, may hybridize with feral domestic pigs.

Similar species: Collared Peccary (*p. 425*) is usually noticeably smaller. May be confused with **White-lipped Peccary** (*p. 423*) on a poor view but that species has a distinctive white lower jaw and tends to occur in large groups.

Habitat: In South America occurs in forest and forested scrubland, including areas of the Pantanal in Brazil. Usually found individually or in small family groups.

Distribution: Native to Europe, Asia and North Africa. Introduced to parts of Colombia, Ecuador, Brazil, Argentina and Chile but range poorly known.

HB:	Up to 185 cm
Tail:	15–20 cm
Sh:	70–96 cm
Wt:	♂ 140–175 kg; ♀ 35–150 kg

Wild Boar (MALE, TOP); feral domestic pig (FEMALE, BOTTOM)

Deer (see p. 429)

(see p. 429)

FAMILY | **Cervidae**

LC Common Fallow Deer

Dama dama

Description: A large but slim tawny to reddish-brown deer with white spots on the back and flanks in summer, becoming gray-brown with indistinct spots in winter. The lower legs and the underparts from the throat to the belly are white, and it has a white rump patch outlined with a characteristic black horseshoe. Several color variations do occur, including both white and black individuals. The palmate antlers are unique among deer in South America.

Similar species: None in range.

Habitat: Little known in South America, but this adaptable species occurs in a wide range of habitats, including forest, scrubland, grassland including pastures, and plantations.

Distribution: Native to Europe. Introduced into Argentina, Chile, Peru and Uruguay.

HB:	♂ 145–155 cm; ♀ 130–145 cm
Tail:	16–19 cm
Sh:	♂ 85–95 cm; ♀ 70–80 cm
Wt:	♂ 50–80 kg; ♀ 35–50 kg

MALE (LEFT); FEMALE WITH JUVENILE (INSET)

LC Western Red Deer

Cervus elaphus

Description: Large and reddish-brown to brown in summer and brown to gray in winter. There are no spots on the adult coat. Females have a buff-white tail. Stags have large, highly branched antlers and the number of branches increases with age: those on mature animals can have up to 16 points.

Similar species: May overlap in range with **South Andean Huemul** (*p. 445*), which is much smaller, shorter-legged, and males of which have two-pronged antlers.

Habitat: Montane woodland and forests.

Distribution: Native to Europe. Introduced into Argentina, from where it has spread into the montane forests from Bío Bío to Aysén Regions, in Chile.

HB:	♂ 180–205 cm; ♀ 165–180 cm
Tail:	14–16 cm
Sh:	♂ 105–130 cm; ♀ 95–115 cm
Wt:	♂ 110–220 kg; ♀ 75–120 kg

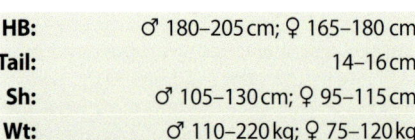

MALE (ABOVE); FEMALE (INSET)

LC Chital

Axis axis

Description: A large deer, both sexes of which are predominantly reddish brown with distinctive white spotting. The throat is white and a line of white spots runs along the body between the flank and the belly. The antlers are long and three-pronged.

Similar species: None in range.

Habitat: Open grasslands and scrubby habitats.

Distribution: Native to Asia. Well established and common in the provinces of La Pampa, Buenos Aires and Entre Ríos in Argentina, where more than 15,000 animals are believed to occur. Also found in Rio Grande do Sul State in Brazil.

HB:	♂ 150–155 cm; ♀ 140–145 cm
Tail:	25–30 cm
Sh:	♂ 90–100 cm; ♀ 65–75 cm
Wt:	♂ 70–85 kg; ♀ 45–60 kg

MALE (LEFT); FEMALE (INSET)

VU Sambar

Rusa unicolor

Description: A large dark brown deer with a shaggy coat. The tail is short, bushy and black, the belly is dark, and the inner sides of the legs are pale. The ears lack any white. The antlers are large, broadly spread and three-pronged.

Similar species: None in range.

Habitat: No information available for South America (but in native region inhabits a variety of ecosystems including savanna, forests, shrubland, grasslands and wetlands).

Distribution: Native to Asia. Two isolated feral populations occur in São Paulo State, Brazil.

HB:	160–210 cm
Tail:	25–33 cm
Sh:	(means): ♂ 140 cm; ♀ 110 cm
Wt:	♂ 180–270 kg; ♀ 130–230 kg

MALE (ABOVE); FEMALE (INSET)

EW Pere David's Deer

Elaphurus davidianus

Description: A large rufous-brown deer in summer, but woollier and duller gray-brown with a dark dorsal stripe and light cream underparts in winter. The head is long and slender with small pointed ears and large eyes. The tail is long with an obvious black tuft. The antlers are distinctive with all prongs sweeping backwards and a long rear branch running almost parallel to the back.

Similar species: None in range.

Habitat: Patagonian mountains.

Distribution: Native to China, where now Extinct in the Wild. Introduced into Argentina, where the naturalized population in Neuquén Province in the south-west is now the third largest in the world.

HB:	150–200 cm
Tail:	50–66 cm
Sh:	approx. 114 cm
Wt:	150–200 kg

MALE (ABOVE); FEMALE (INSET) (BOTH WINTER PELAGE)

Bovids

FAMILY | **Bovidae**

LC Blackbuck

Antilope cervicapra

Description: A fairly large antelope. Mature males are dark brown to velvety black above with the throat and upper chest being the same color as the upperparts. The rump, remainder of the underparts, inner legs, nose and lower muzzle are white. Females and immatures are reddish-brown above. Mature males have divergent and spiralled horns, whereas in young males they are 'V'-shaped.

Similar species: None in range.

Habitat: Wooded habitats, where active mainly during the day, browsing and grazing on low vegetation.

Distribution: Native to India. Introduced into Argentina, where more than 10,000 animals now occur in the east of the country, their stronghold being in La Pampa Province.

HB:	120–150 cm
Tail:	8.2–13.5 cm
Sh:	approx. 80 cm
Wt:	♂ 34–45 kg; ♀ 31–39 kg

MALE (ABOVE); FEMALE (INSET)

LC Alpine Ibex

Capra ibex

Description: A large bovid. Brownish-grey in summer, but thicker-coated and grayer in winter. Mature males have a dark belly. The head is uniformly colored and the ears are relatively short. The horns are curved and have distinctive rings; they form a single curve and are up to 100 cm long in males, 20 cm long in females.

Similar species: None in range.

Habitat: High mountains.

Distribution: Native to the European Alps. Introduced into the Patagonian mountains in Neuquén Province in SW Argentina.

HB:	♂ 130–150 cm; ♀ 105–125 cm
Tail:	12–15 cm
Sh:	68–85 cm
Wt:	♂ 65–125 kg; ♀ 40–70 kg

MALE (ABOVE); FEMALE (INSET)

LC Mouflon

Ovis gmelini

Description: A large, variably colored bovid with russet-brown to blonde upperparts. Animals originating from the Mediterranean region often have a white saddle patch that is most evident in winter. The tail and rump are dark and the underparts are whitish. The ears are small. Males have long, thick, spiralled horns up to 90 cm long; females may have horns up to 20 cm long but can be hornless.

Similar species: None in range.

Habitat: No information available for South America (but in native region inhabits grassland, forest, shrubland and rocky areas).

Distribution: Native to the Middle East, widely introduced to Europe. Large numbers have been introduced to hunting preserves in Neuquén Province in SW Argentina, in the Sierra de Ventania in Buenos Aires Province, and La Pampa Province.

HB:	♂ 110–150 cm; ♀ 110–130 cm
Tail:	6–11 cm
Sh:	65–75 cm
Wt:	♂ 35–70 kg; ♀ 30–50 kg

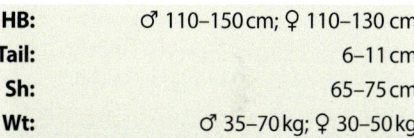

MALE (ABOVE); FEMALE (INSET)

DOMESTICATED MAMMALS

A range of domesticated species also occur in a semi-wild state in South America. These include Alpaca, Llama and Domestic Water Buffalo, which are are covered briefly here. In addition, **Horse** *Equus ferus*, **Goat** *Capra hircus*, **Dog** *Canis familiaris* and **Cat** *Felis catus* have established naturalized or feral populations in certain areas and may well be encountered. However, since these last four species are very familiar to most people – despite being highly variable in size, form and coloration – they are not covered in this book.

Camelids (see *p. 426*)   FAMILY | **Camelidae**

Llama   *Lama glama*

Description: A long-legged domesticated camelid with a rounded muzzle and cleft upper lip. The tips of its long ears curve inwards. Black, brown or white, blotched with contrasting colors.

Similar species: Unlikely to be confused with any other species, although individuals with **Guanaco**-like (*p. 428*) coloration can occur.

HB:	190–250 cm
Tail:	23–27 cm
Sh:	90–130 cm
Wt:	90–140 kg

Habitat & Distribution: Open puna habitats close to human habitations in the Andes from central Peru, through W Bolivia into N Argentina and N Chile.

Alpaca   *Lama pacos*

Description: Domesticated form of the Vicuña. Long-necked with short straight ears, slender legs and a fluffy body. Uniformly colored and covered in soft, long hair.

Similar species: Unlikely to be confused with any other species, although can appear slender and **Vicuña**-like (*p. 427*) when shorn.

HB:	125–190 cm
Tail:	15–25 cm
Sh:	85–90 cm
Wt:	38–45 kg

Habitat & Distribution: Open puna habitats from central Peru into Bolivia and N Chile.

Bovids   FAMILY | **Bovidae**

Domestic Water Buffalo   *Bubalus bubalus*

Description: A smaller domesticated form of the Asian Wild Water Buffalo *B. arnee*. Largely black to dark brown with moderately long, coarse hair. The tail has a bushy tip. Both sexes have horns curving backwards from a moderately convex forehead.

Similar species: Unlikely to be confused with any other species.

HB:	240–300 cm
Tail:	60–100 cm
Sh:	150–190 cm
Wt:	250–1,200 kg

Habitat & Distribution: Wetlands. Large herds in Colombia, Venezuela, Brazil (with records from ten states), and N Argentina.

Llama

Alpaca

Domestic Water Buffalo

Glossary

This glossary defines the technical terms used in the species accounts. For a more detailed description of the habitat types within South America see Campbell *et al.* (2021).

agouti (color) — Fur coloration in which each hair displays alternating bands of dark and light pigmentation. It is also the wild-type pigmentation for many domesticated mammals.

allopatric — Species that are geographical replacements of each other and occur in adjacent geographical areas without overlapping in range.

alluvial — Soils composed of sand and earth left behind when an area has been flooded.

anthropogenic soils — Soils constructed or deeply modified by humans.

Atlantic Forest — Highly fragmented tropical and subtropical forest stretching from NE Brazil to NE Argentina and Paraguay (see *page 13*).

Caatinga — Areas of thorny shrubs and stunted trees in semi-arid country with annual rainfall of 30–80 cm in the interior of north-east Brazil (see *page 13*).

cabruca — Traditional agro-forests in Brazil in which cacao plants are planted below thinned-out native trees.

Cerrado — The tropical savanna ecoregion of central Brazil, adjacent eastern Bolivia and north-east Paraguay, comprising savanna grasslands and wetlands, forest savanna and gallery forest (see *page 13*).

Chaco — Extensive lowland plains covering areas of Argentina, Bolivia and Paraguay (see *page 13*).

Chaparral — An ecological community composed of shrubby plants adapted to dry summers and moist winters. Often a dense impenetrable thicket of shrubs or dwarf trees.

Chiquitania — An area of tropical savanna comprising grasslands, wetlands and forest in lowland eastern Bolivia.

Chocó — Wet forests along the coast of Colombia and NW Ecuador (see *page 13*).

cline — A series of often gradual changes in a species' characteristics across its range.

cloud forest — Tropical wet montane forest normally at an altitude of 1,000–2,500 m that is characterized by a profusion of epiphytes and the presence of clouds even in the dry season.

conspecific — Belonging to the same species.

deciduous — Trees or shrubs that shed their leaves in the autumn and grow new ones in the spring.

dimorphic — Where the males and females of a species exhibit differences in their appearance.

distal — Farthest from the point of the attachment to the body, *e.g.* the area towards the tip of the tail.

dorsum — The upper surface of an animal.

elfin forest	The upper stretches of cloud forest close to the treeline where the trees are often twisted, gnarled and stunted in size.
endemic	Restricted in range to a specific area or country.
gallery forest	Forest found on the banks of a river, stream or other watercourse.
gular	On the throat.
holotype	The type (original) specimen upon which the description and name of a new species is based.
immature	Used here to reflect individuals that are younger than sub-adult.
Igapó forest	Seasonally flooded tropical evergreen forest. Mainly tall forest with more open undergrowth than várzea forests and normally found on sandy soils along black-water and clearwater rivers in Amazonia.
interfluvium	An area of higher ground between two rivers in the same drainage system.
left bank	The bank of a river on the left as one faces downstream.
Llanos	The vast grassland plain comprising flooded grasslands and savannas to the east of the Andes in Colombia and Venezuela.
malar	A marking, often a stripe, on the cheek.
mangroves	Trees or shrubs with tangled roots growing above ground and creating tangled thickets in intertidal, saline or brackish, rivers and swamps, normally in tropical regions.
matorral	Dry, small-leaved woodlands and scrublands.
mesic	A moderately moist environment.
mesophytic	Vegetation adapted to moderately moist environments.
monotypic	The only member of a grouping, *e.g.* species, family, order *etc*.
montane	Growing in or inhabiting mountainous areas.
Neotropical	Refers to Central and South America including southern Mexico and the West Indies and other Caribbean islands.
nominate subspecies	The first-named subspecies of a given species and identified by the scientific species and subspecies name being the same.
ombrophilous	Plants tolerant of very wet conditions.
Pampas	Lowland grassland plains (see *page 13*).
Pantanal	A vast seasonally flooded wetland (see *page 13*).
páramo	Humid grassland with shrubby vegetation occurring above montane or elfin forest in northern South America.
parapatric	Two closely related species or subspecies with ranges which are largely separated but which may have narrow contact zones, *e.g.* two species may have largely overlapping geographical ranges but be separated by their altitudinal ranges.
piaçava	A palm tree native to the Brazilian states of Alagoas, Bahia, Espírito Santo and Sergipe.
piedmont forest	Predominantly montane forest in an area interspersed with hills and plains.
polytypic	A species comprising of two or more subspecies.
prehensile	The ability to grasp or hold, often used in relation to the tail.
preorbital	In front of or near the eye.

primary forest	Forests of native tree species where there are no clearly visible indications of human activities, *e.g.* logging.
proximal	Closest to the point of attachment to the body.
rainforest	Tropical forest, usually consisting of tall, densely growing, broad-leaved evergreen trees in an area of high annual rainfall.
restinga	Coastal forests and scrub, often comprised of fairly low trees and rich in terrestrial bromeliads, found growing in white-sand soils in eastern Brazil.
right bank	The bank of a river on the right as one faces downstream.
riparian	Found on the banks of a river, stream or wetland.
savanna	A grassy plain in tropical or subtropical regions with a mixture of drought-resistant grassland and scattered trees.
scute	A bony external plate or scale overlaid with horn.
secondary forest	Regenerated native forest which has been cleared by natural causes *e.g.* fire, or by human intervention *e.g.* agriculture or ranching.
semi-deciduous	Plants which lose some but not all of their leaves in the cooler months, but keep their leaves in milder or warmer weather.
sub-montane	The foothills or lower slopes of a mountain range.
subocular	Below the eye.
superspecies	Two or more closely related taxa which replace each other geographically but which may be regarded as subspecies rather than full species by some authorities.
supraocular	Above the eye.
sympatric	Of similar species occurring within the same or overlapping geographical areas.
synonym	An alternative scientific name for a species, for example where the same form has been given alternative names by different authors.
taxonomy	The study of classification and the naming of species and subspecies.
temporal	The sides of the head.
terra firme forest	Forest that is never subject to inundation (flooding). The most abundant habitat in the Amazon Basin.
trunk	The central part (or core) of the body to which the head, limbs and tail are attached.
type	The original specimen upon which the description and name of a new species is based, or its/their location (type location).
várzea forest	Seasonally flooded tropical evergreen forest. Normally tall forest with rich undergrowth and normally found along white-water rivers in Amazonia.
xeric	A very dry environment.
xerophytic	Vegetation adapted to extremely dry environments.
Yungas	A narrow band of forest along the eastern slope of the Andes Mountains in Peru, Bolivia and northern Argentina. Also a genus of plants.

References

The following are the key references that were used frequently when preparing this book, or are cited specifically. A complete bibliography that includes all the references used is available for download from the "Resources" tab at https://press.princeton.edu/books/paperback/9780691174099/a-field-guide-to-the-larger-mammals-of-south-america or via the QR (quick response) code shown here.

Aulagnier, S., Haffner, P., Mitchell-Jones, A. J., Moutou, F., & Zima, J. 2009. *Mammals of Europe, North Africa and the Middle East*. A&C Black Publishers Ltd. London.

Benitez, V. V., Chavez, S. A., Gozzi, A. C., Messetta, M. L., & Guichón, M. L. 2013. Invasion status of Asiatic red-bellied squirrels in Argentina. *Mammalian Biology* **78(3)**: 164–170.

Boubli, J. P., Byrne, H., da Silva, M. N. F., Silva-Júnior, J., Araújo, R. C., Bertuol, F., Gonçalves, J., de Melo, F. R., Rylands, A. B., Mittermeier, R. A., Silva, F. E., Nash, S. D., Canale, G., de M. Alencar, R., Rossi, R. V., Carneiro, J., Sampaio, I., Farias, I. P., & Hrbek, T. 2018. On a new species of titi monkey (Primates: Plecturocebus Byrne *et al.,* 2016), from Alta Floresta, southern Amazon, Brazil. https://doi.org/10.1016/j.ympev.2018.11.012

Boubli, J. P., da Silva, M. N. F., Amado, M. V., Hrbek, T., Pontual, F. B., & Farias, I. P. 2008. A Taxonomic Reassessment of *Cacajao melanocephalus* Humboldt (1811), with the Description of Two New Species. *International Journal of Primatology* **29(3)**: 723–741.

Boubli, J. P., da Silva, M. N. F., Rylands, A. B., Nash, S. D., Bertuol, F., Nunes, M., Mittermeier, R. A., Byrne, H., da Silva, F. E., Röhe, F., Sampaio, I., Schneider, H., Farias, I. P., & Hrbek, T. P. 2017. How many pygmy marmoset (*Cebuella* Gray, 1870) species are there? A taxonomic re-appraisal based on new molecular evidence. *Molecular Phylogenetics and Evolution* **120**: 170–182.

Boubli, J. P., Janiak, C., Porter, L. M., de la Torre, S., Cortés-Ortiz, L., da Silva, M. N. F., Rylands, A. B., Nash, S., Bertuol, F., Byrne, H., Silva, F. E., Rohe, F., de Vries, D., Beck, R. M. D., Ruiz-Gartzia, I., Kuderna, L. F. K., Marques-Bonet, T., Hrbek, T., Farias, I. P., van Heteren, A. H., & Roos, C. 2021. Ancient DNA of the pygmy marmoset type specimen *Cebuella pygmaea* (Spix, 1823) resolves a taxonomic conundrum. *Zool. Res.* 2021, **42(6)**: 761–771. https://doi.org/10.24272/j.issn.2095-8137.2021.143

Buckner, J. C., Lynch Alfaro, J. W., Rylands, A. B., & Alfaro, M. E. 2015. Biogeography of the marmosets and tamarins (Callitrichidae). *Molecular Phylogenetics and Evolution* **82**: 413–425.

Burgin, C. J., Wilson, D. E., Mittermeier, R. A., Rylands, A. B., Lacher, T. E., & Sechrest, W. 2020. *Illustrated Checklist of the Mammals of the World*. Lynx Edicions, Barcelona.

Byrne, H., Costa-Araújo, R., Farias, I. P., da Silva, M. N. F., Messias, M., Hrbek, T., & Boubli, J. P. 2021. Uncertainty Regarding Species Delimitation, Geographic Distribution, and the Evolutionary History of South-Central Amazonian Titi Monkey Species (*Plecturocebus*, Pitheciidae). *Int. J. Primatol.* (2021). https://doi.org/10.1007/s10764-021-00249-9

Byrne, H., Rylands, A. B., Carneiro, J. C., Lynch Alfaro, J. W., Bertuol, F., da Silva, M. N. F., Messias, M., Groves, C. P., Mittermeier, R. A., Farias, I., Hrbek, T., Schneider, H., Sampaio, I., & Boubli, J. P. 2016. Phylogenetic relationships of the New World titi monkeys (*Callicebus*): first appraisal of taxonomy based on molecular evidence. *Frontiers in Zoology* **13**, 10 (2016). https://doi.org/10.1186/s12983-016-0142-4

Byrne, H., Rylands, A. B., Nash, S. D., & Boubli, J. P. 2020. On the Taxonomic History and True Identity of the Collared Titi, Cheracebus torquatus (Hoffmannsegg, 1807) (Platyrrhini, Callicebinae). *Primate Conservation* 2020 (**34**).

Caldara Junior, V., & Leite, Y. L. R. 2012. Geographic variation in hairy dwarf porcupines of *Coendou* from eastern Brazil (Mammalia: Erethizontidae). *Zoologia* **29(4)**: 318–336. http://doi.org/10.1590/S1984-46702012000400005

Campbell, I., Behrens, K., Hesse, C., & Chaon, P. 2021. *Habitats of the World. A Field Guide for Birders, Naturalists and Ecologists*. Princeton University Press, Princeton and Oxford.

Cañizales, I. 2020. Morphology of the Skull of *Inia geoffrensis humboldtiana* Pilleri & Gihr, 1977 (Cetacea: Iniidae): a morphometric and taxonomic analysis. *Graellsia* **76(2)**: e115. https://doi.org/10.3989/graellsia.2020.v76.253

Castelló, J. R. 2018. *Canids of the World: Wolves, Wild Dogs, Foxes, Jackals, Coyotes, and Their Relatives.* Princeton University Press, Princeton and Oxford.

Castelló, J. R. 2020. *Felids and Hyenas of the World: Wildcats, Panthers, Lynx, Pumas, Ocelots, Caracals, and Relatives.* Princeton University Press, Princeton and Oxford.

Chemisquy, A., Prevosti, F. J., Martínez, P., Raimondi, V., Cabello Stom, J. E., Acosta-Jamett, G., & Montoya-Burgos, J. I. 2019. How many species of gray foxes (Canidae, Carnivora) are there in South America? *Mastozoología Neotropical* **26(1)**: 81–97. https://doi.org/10.31687/saremMN.19.26.1.0.16

Chester, S. 2008. *A Wildlife Guide to Chile. Continental Chile, Chilean Antarctica, Easter Island Juan Fernandez Archipelago.* A&C Black Publishers Ltd, London.

Costa-Araújo, R., de Melo, F. R., Canale, G. R., Hernández-Rangel, S. M., Messias, M. R., Rossi, R. V., Silva, F. E., da Silva, M. N. F., Nash, S. D., Boubli, J. P., Farias, I. P., & Hrbek, T. 2019. The Munduruku marmoset: a new monkey species from southern Amazonia. *PeerJ.* 2019; 7: e7019. http://doi.org/10.7717/peerj.7019

Costa-Araújo, R., Silva Jr, J. S., Boubli, J. P., Rossi, R. V., Canale, G. R, Melo, F. R., Bertuol, F., Silva, F. E., Silva, D. A., Nash, S. D., Sampaio, I., Farias, I. P., & Hrbek, T. 2021. An integrative analysis uncovers a new, pseudo-cryptic species of Amazonian marmoset (Primates: Callitrichidae: Mico) from the arc of deforestation. *Scientific reports:* https://www.nature.com/articles/s41598-021-93943-w

Cozzuol, M. A., Clozato, C. L., Holanda, E. C., Rodrigues, F. H. G., Nienow, S., de Thoisy, B., Redondo, R. A. F., & Santos, F. R. 2013. A new species of tapir from the Amazon. *Journal of Mammalogy* **94(6)**: 1331–1345.

Dalponte, J. C., Silva, F. E., & de Sousa e Silva Júnior, J. 2014. New species of titi monkey, genus Callicebus Thomas, 1903 (Primates, Pitheciidae), from Southern Amazonia, Brazil. *Pap. Avulsos Zool. (São Paulo)* **54** (32). http://doi.org/10.1590/0031-1049.2014.54.32

da Rosa, C. A., de Almeida Curi, N. H., Puertas, F., & Passamani, M. 2017. Alien terrestrial mammals in Brazil: current status and management. *Biol Invasions.* https://link.springer.com/article/10.1007/s10530-017-1423-3

de Abreu Jr, E. F., Pavan, S. E., Tsuchiya, M. T. N., Wilson, D. E., Percequillo, A. R., & Maldonado, J. E. 2020. Museomics of tree squirrels: a dense taxon sampling of mitogenomes reveals hidden diversity, phenotypic convergence, and the need of a taxonomic overhaul. *BMC Evol Biol* **20**, 77 (2020). https://doi.org/10.1186/s12862-020-01639-y

D'Elía, G., Hurtando, N., & D'Anatro, A. 2016. Alpha taxonomy of *Dromiciops* (Microbiotheriidae) with the description of 2 new species of monito del monte. *Journal of Mammalogy* **97(4)**: 1136–1152. https://doi.org/10.1093/jmammal/gyw068

Defler, T. R., & Bueno, M. L. 2007. Aotus Diversity and the Species Problem. *Primate Conservation* **(22)**: 55–70. https://doi.org/10.1896/052.022.0104

Diersing, V. E., & Wilson, D. E. 2017. Systematic status of the rabbits *Sylvilagus brasiliensis* and *S. sanctamartae* from northwestern South America with comparisons to Central American populations. *Journal of Mammalogy* **98(6)**: 1641–1656. https://doi.org/10.1093/jmammal/gyx133

do Nascimento, F. O., & Feijó, A. 2017. Taxonomic revision of the tigrina *Leopardus tigrinus* (Schreber, 1775) species group (Carnivora, Felidae). *Papéis Avulsos de Zoologia.* Museu de Zoologia da Universidade de São Paulo. **57(19)**: 231–264.

do Nascimento, F. O., Cheng, J., & Feijó, A. 2020. Taxonomic revision of the pampas cat *Leopardus colocola* complex (Carnivora: Felidae): an integrative approach. *Zoological Journal of the Linnean Society* **191(2)**: 575–611. https://doi.org/10.1093/zoolinnean/zlaa043

Emin-Lima, R., Machado, F. A., Siciliano, S., Gravena, W., Aliaga-Rossel, E., de Sousa e Silva, J., & de Oliveira, L. R. 2022. Morphological disparity in the skull of Amazon River dolphins of the genus *Inia* (Cetacea, Iniidae) is inconsistent with a single taxon. *Journal of Mammalogy* **103(6)**: 1278-1289. https://doi.org/10.1093/jmammal/gyac039

Feijó, A. & Anacleto, T. C. 2021. Taxonomic revision of the genus *Cabassous* McMurtrie, 1831 (Cingulata: Chlamyphoridae), with revalidation of *Cabassous squamicaudis* (Lund, 1845). *Zootaxa* **4974(1)**: 47–78.

Feijó, A., & Cordeiro-Estrela, P. 2014. The correct name of the endemic *Dasypus* (Cingulata: Dasypodidae) from northwestern Argentina. *Zootaxa* **3887(1)**: 88–94.

Feijó, A., & Cordeiro-Estrela, P. 2016. Taxonomic revision of the *Dasypus kappleri* complex, with revalidations of *Dasypus pastasae* (Thomas, 1901) and Dasypus beniensis Lönnberg, 1942 (Cingulata, Dasypodidae). *Zootaxa* **4170(2)**: 271–297.

Feijó, A., Patterson, B. D., & Cordeiro-Estrela, P. 2018. Taxonomic revision of the long-nosed armadillos, Genus *Dasypus* Linnaeus, 1758 (Mammalia, Cingulata). *PLoS ONE* **13(4)**: e0195084. https://doi.org/10.1371/journal.pone.0195084

Ferrari, S. F., Guedes, P. G., Figueiredo-Ready, W. M. B., & Barnett, A. B. 2010. Reconsidering the taxonomy of the Black-Faced Uacaris, Cacajao melanocephalus group (Mammalia: Pitheciidae), from the northern Amazon Basin. http://dx.doi.org/10.11646/zootaxa.3866.3.3

Garbino, G. S. T. 2014. The Taxonomic Status of *Mico marcai* (Alperin 1993) and *Mico manicorensis* (van Roosmalen *et al.*, 2000) (Cebidae, Callitrichinae) from Southwestern Brazilian Amazonia. *International Journal of Primatology* **35(2)**: 529–546.

Garcia-Perea, R. 1994. The pampas cat group (genus *Lynchailurus* Severtzov, 1858) (Carnivora: Felidae), a systematic and biogeographic review. *American Museum Novitates* 3096: 1–35.

Gongora, J., Taber, A., Keuroghlian, A., Altrichter, M., Bodmer, R. E., Mayor, P., Moran, C., Damayanti, C. S., & González, S. 2007. Re-examining the evidence for a 'new' peccary species, '*Pecari maximus*', from the Brazilian Amazon. *Newsletter of the Pigs, Peccaries, and Hippos Specialist Group of the IUCN/SSC* **7(2)**: 19–26.

Gonzalez, S., & Duarte, J. 2020. Speciation, Evolutionary History and Conservation Trends of Neotropical Deer. *Mastozoología Neotropical* **27(SI)**. 35–46. https://doi.org/10.31687/saremMN_SI.20.27.1.05

Gregorin, R., Athaydes, D., Santos Júnior, J. E., & Ayoub, T. B. 2023. Taxonomic status of Tamarinus imperator subgrisescens (Lönnberg, 1940) (Cebidae, Callitrichinae). *Papéis Avulsos De Zoologia* **63**, e202363005. https://doi.org/10.11606/1807-0205/2023.63.005

Groves, C. P. 2001. *Primate Taxonomy*. Smithsonian Institution Press, Washington DC.

Gusmão, A. C., Messais, M. R., *et al.* 2016. A New Species of Titi Monkey, Plecturocebus Byrne *et al.*, 2016 (Primates, Pitheciidae), from Southwestern Amazonia, Brazil. *Primate Conservation* **33**: 21–35.

Gutiérrez, E. E., Helgen, K. M., McDonough, M. M., Bauer, F., Hawkins, M. T. R., Escobedo-Morales, L. A., Patterson, B. D., & Maldonado, J. E. 2017. A gene-tree test of the traditional taxonomy of American deer: the importance of voucher specimens, geographic data, and dense sampling. *ZooKeys* **697**: 87–131.

Helgen, K. M., Kays, R., Helgen, L. E., Tsuchiya-Jerep, M. T. N., Pinto, C. M., Koepfli, K.-P., Eizirik, E., and Maldonado, J. E. 2009. Taxonomic boundaries and geographic distributions revealed by an integrative systematic overview of the mountain coatis, *Nasuella* (Carnivora: Procyonidae). *Small Carnivore Conservation* **41**: 65–74.

Helgen, K. M., Pinto, C. M., Kays, R., Helgen, L. E., Tsuchiya, M. T. N., Quinn, A., Wilson, D. E., & Maldonado, J. E. 2013. Taxonomic revision of the olingos (*Bassaricyon*), with description of a new species, the Olinguito. *ZooKeys* **324**: 1–83.

Hyde, M., Bardales, R., Lizarazo, J., Sánchez, F. Payán, E., & Ortiz, R. 2021. First camera-trap evidence of the Western Mountain Coati *Nasuella olivacea* in San Martin, Peru. *Small Carnivore Conservation* **59**.

Kasper, C. B., da Fontoura-Rodrigues, M. L., Cavalcanti, G. N., de Freitas, T. R. O., Rodrigues, F. H. G., de Oliveira, T. G., & Eizirik, E. 2009. Recent advances in the knowledge of Molina's Hog-nosed Skunk *Conepatus chinga* and Striped Hog-nosed Skunk *C. semistriatus* in South America. *Small Carnivore Conservation* **41**: 25–28.

Kitchener, A. C., Breitenmoser-Würsten, C., Eizirik, E., Gentry, A., Werdelin, L., Wilting, A., Yamaguchi, N., Abramov, A. V., Christiansen, P., Driscoll, C. A., Duckworth, J. W., Johnson, W. E., Luo, S. J., Meijaard, E., O'Donoghue, P., Sanderson, J., Seymour, K., Bruford, M. W., Groves, C., Hoffmann, M., Nowell, K., Timmons, Z., & Tobe, S. 2017. *A revised taxonomy of the Felidae. The final report of the Cat Classification Task Force of the IUCN/SSC Cat Specialist Group. Cat News Special Issue* (11). ISSN 1027–2992

La Sala, L. F., Burgos, J. M., Caruso, N. C., Bagnato,C. E., Ballari, S. A., Guadagnin, D. L., Kindel, A., Etges, M., Merino, M. L., Marcos, A., Skewes, O., Schettino, D., Perez, A. M., Condori, E., Tammone, A., Carpinetti, B., & Zalba, S. M. 2023. Wild pigs and their widespread threat to biodiversity conservation in South America. *Journal for Nature Conservation* **73(3)**. https://www.researchgate.net/publication/369718973/

Ledesma, K. J. L., Werner, F. A., Spotorno, A. E., & Albuja, L. H. 2009. A new species of Mountain Viscacha (Chinchillidae: *Lagidium Meyen*) from the Ecuadorean Andes. *Zootaxa* **2126**: 41–57.

Lewis, J., Farnsworth, M., Burdett, C., Theobald, D., Gray, M., & Miller, R. 2017. Biotic and abiotic factors predicting the global distribution and population density of an invasive large mammal. *Scientific Reports* **7(1)**. https://www.researchgate.net/publication/314394782/

Lopes, G. P., Rohe, F., Bertuol, F., Polo, E., Lima, I. J., Valsecchi, J., Santos, T. C. M., Nash, S. D., da Silva, M. N. F., Boubli, J. P., Farias, I. P., & Hrbek, T. 2023. Taxonomic review of *Saguinus mystax* (Spix, 1823) (Primates, Callitrichidae), and description of a new species. *PeerJ.* https://peerj.com/articles/14526/

Lynx Nature Books. 2023. *All the Mammals of the World*. Barcelona.

Marsh, L. K. 2014. A taxonomic revision of the saki monkeys, *Pithecia* Desmarest, 1804. *Neotropical Primates* **21(1)**: 1–163.

Martins-Junior, A. M. C., Carneiro, J., Sampaio, I., Ferrari, S. F., & Schneider, H. 2018. Phylogenetic relationships among Capuchin (Cebidae, Platyrrhini) lineages: An old event of sympatry explains the current distribution of *Cebus* and *Sapajus*. *Genetics and Molecular Biology* **41(3)**: 699–712.

Menezes, F. H., Feijó, A., Fernandes-Ferreira, H., da Costa, I. R., & Cordeiro-Estrela, P. 2021. Integrative systematics of Neotropical porcupines of *Coendou prehensilis* complex (Rodentia: Erethizontidae). *Journal of Zoological Systematics and Evolutionary Research*. https://doi.org/10.1111/jzs.12529

Menezes, F. H., Garbino, G. S. T, Semedo, T. B. F., Lima, M., Feijó, A., Cordeiro-Estrela, P., & da Costa, I. R. 2020. Major range extensions for three species of porcupines (Rodentia: Erethizontidae: Coendou) from the Brazilian Amazon. *Biota Neotropica* **20(2)**, e20201030. Epub June 19, 2020. https://dx.doi.org/10.1590/1676-0611-bn-2020-1030

Menon, V. 2014. *Indian Mammals A Field Guide*. Hachette Book Publishing India Pvt. Ltd. Gurugram.

Mercês, M. P., Lynch Alfaro, J. W., Ferreira, W. A., Harada, M. L., & Silva, J. S. Jr. 2015. Morphology and mitochondrial phylogenetics reveal that the Amazon River separates two eastern squirrel monkey species: *Saimiri sciureus* and *S. collinsi*. *Molecular Phylogenetics and Evolution* 426–435.

Miranda, F. R., Casali, D. M., Perini, F. A., Machado, F. R., & Santos, F. R. 2017. Taxonomic review of the genus *Cyclopes* Gray, 1821 (Xenarthra: Pilosa), with the revalidation and description of new species. *Zoological Journal of the Linnean Society*, zlx079. https://doi.org/10.1093/zoolinnean/zlx079

Miranda, F. R., Garbino, G. S., Machado, F. A., Perini, F. A., Santos, F. R., & Casali, D. M. 2022. Taxonomic revision of maned sloths, subgenus *Bradypus* (*Scaeopus*), Pilosa, Bradypodidae, with revalidation of *Bradypus crinitus* Gray, 1850. *Journal of Mammalogy* **104(1)**: 86–103. https://doi.org/10.1093/jmammal/gyac059

Mittermeier, R. A., Nash, S., & Rylands, A. B. 2022. *Monkeys of the Atlantic Forest of Eastern Brazil Pocket Identification Guide* (Second Edition). Re:wild.

Mittermeier, R. A., Rylands, A. B., & Wilson, D. E. (Eds). 2013. *Handbook of the Mammals of the World – Volume 3: Primates*. Lynx Edicions, Barcelona.

Nunes, A. V., & Serrano-Villavicencio, J. E. 2017. Rediscovery of Vanzolini's Bald-Faced Saki, *Pithecia vanzolinii* Hershkovitz, 1987 (Primates, *Pitheciidae*): first record since 1956. *Check List* **13(1)**: 2048, 16 February 2017. https://doi.org/10.15560/13.1.2048

Orsini, V. S., Nunes, A. V., & Marsh, L. K. 2017. New distribution records of *Pithecia rylandsi* and *Pithecia mittermeieri* (Primates, Pitheciidae) and an updated distribution map. *Check List* **13(3)**: 2123. https://doi.org/10.15560/13.3.2123. ISSN 1809-127X.

Rengifo, E. M., D'Elía, G., García, G., Charpentier, E., & Cornejo, F. M. 2022. A New Species of Titi Monkey, Genus *Cheracebus* Byrne et al., 2016 (Primates: Pitheciidae), from Peruvian Amazonia. *Mammal Study* **48(1)**: 3–18. https://doi.org/10.3106/ms2022-0019

Royle, M. 2015. Privately published Venezuela tour report. http://www.mammalwatching.com/wp-content/uploads/2016/08/Anacondas-of-Los-Llanos-Jan-2015-MW.pdf

Ruedas, L. A. 2017. A new species of cottontail rabbit (Lagomorpha: Leporidae: *Sylvilagus*) from Suriname, with comments on the taxonomy of allied taxa from northern South America. *Journal of Mammalogy* **98(4)**: 1042–1059. https://doi.org/10.1093/jmammal/gyx048

Ruedas, L. A., Silva, S. M., French, J. H., Platt II, R. N., Salazar-Bravo, J., Mora, J. M., & Thompson, C. W. 2017. A prolegomenon to the systematics of South American Cottontail Rabbits (Mammalia, Lagomorpha, Leporidae, *Sylvilagus*): Designation of a neotype for *S. brasiliensis* (Linnaeus, 1758), and restoration of *S. andinus* (Thomas, 1897) and *S. tapetillus* (Thomas, 1913). *Publications of the Museum of Zoology, University of Michigan.* No. 205.

Ruedas, L. A., Silva, S. M., French, J. H., Platt II, R. N., Salazar-Bravo, J., Mora, J. M., & Thompson, C. W. 2019. Taxonomy of the *Sylvilagus brasiliensis* complex in Central and South America (Lagomorpha: Leporidae). *Journal of Mammalogy* **100(5)**: 1599–1630. https://doi.org/10.1093/jmammal/gyz126

Ruiz-García, M., Jaramillo, M. F., Cáceres-Martínez, C. H., & Shostell, J. M. 2020. The phylogeographic structure of the mountain coati (*Nasuella olivacea*; Procyonidae, Carnivora), and its phylogenetic relationships with other coati species (*Nasua nasua* and *Nasua narica*) as inferred by mitochondrial DNA. *Mammalian Biology* **100**: 521–548. https://doi.org/10.1007/s42991-020-00050-w

Ruiz-García, M., Jaramillo, M. F., López, J. B., Rivillas, Y., Bello, A., Leguizamon, N., & Shostell, J. M. 2021. Mitochondrial and karyotypic evidence reveals a lack of support for the genus *Nasuella* (Procyonidae, Carnivora). *Journal of Vertebrate Biology* **71(21040)**: 21040.1–25. https://doi.org/10.25225/JVB.21040

Ruiz-García, M., Pinedo-Castro, M., & Shostell, J. M. 2023. Morphological and Genetics Support for a Hitherto Undescribed Spotted Cat Species (Genus *Leopardus*; Felidae, Carnivora) from the Southern Colombian Andes. *Genes* **14(6)**: 1266. https://doi.org/10.3390/genes14061266

Rylands, A. B., Heymann, E. W., Lynch, Alfaro J., Buckner, J. C., Roos, C., Matauschek, C., Boubli, J. P., Sampaio, R., & Mittermeier, R. A. 2016. Taxonomic review of the New World tamarins (Primates: Callitrichidae). *Zoological Journal of the Linnean Society* **177**: 1003–1028. https://doi.org/10.1111/zoj.12386

Rylands, A. B., Kierulff, M. C. M., & Mittermeier, R. A. 2005. Notes on the taxonomy and distributions of the tufted capuchin monkeys (Cebus, Cebidae) of South America. *Lundiana* 97–110.

Sampaio, R., Röhe, F., Pinho, G., Silva-Júnior, J. S., Farias, I. P., & Rylands, A. B. 2015. Re-description and assessment of the taxonomic status of *Saguinus fuscicollis cruzlimai* Hershkovitz, 1966 (Primates, Callitrichinae). Primates **56**: 131–144. https://doi.org/10.1007/s10329-015-0458-2

Sánchez, R. T. 2020. *Pequeños Mamíferos de La Rioja*. Ruben Barquez.

Schiaffini, M. I., Gabrielli, M., Prevosti, F., Cardoso, Y. P., Castillo, D., Bo, R., Casanave, E., & Lizarralde, M. 2013. Taxonomic status of southern South American *Conepatus* (Carnivora: Mephitidae). *Zoological Journal of the Linnean Society* **167**: 327–344.

Serrano-Villavicencio, J. E., Hurtado, C. M., Vendramel, R. L., & do Nascimento, F. O. 2019. Reconsidering the taxonomy of the *Pithecia irrorata* species group (Primates: Pitheciidae). *Journal of Mammalogy*, gyy167. https://doi.org/10.1093/jmammal/gyy167

Shirihai, H., & Jarrett, B. 2006. *Whales, Dolphins and Seals: A Field Guide to the Marine Mammals of the World.* A&C Black Publishers Ltd., London.

Smith, A. T. & Yan Xie (Eds). 2013. *Mammals of China. Princeton Pocket Guides*. Princeton University Press, Princeton USA and Woodstock, UK.

Silva, F. E., Valsecchi do Amaral, J., Roosf, C., Bowler, M., Rohe, F., Sampaio, R., Janiak, M. C., Bertuol, F., Santana, M. I., de Souza Silva Júnior, J., Rylands, A. B., Gubili, C., Hrbek, T., McDevitt, A. D., Boubli, J. P. 2022. Molecular phylogeny and systematics of bald uakaris, genus *Cacajao* Lesson, 1840 (Primates: Pitheciidae), with the description of a new species. *Molecular Phylogenetics and Evolution* **173**. https://doi.org/10.1016/j.ympev.2022.107509

Strier, K. B., & Boubli, J. P. 2006. A History of Long-term Research and Conservation of Northern Muriquis (*Brachyteles hypoxanthus*) at the Estação Biológica de Caratinga/RPPN-FMA. *Primate Conservation* **20**: 53–63.

Sunquist, M., & Sunquist, F. 2002. *Wild Cats of the World*. The Chicago University Press.

Teta, P., Jayat, J. P., & Ortix, P. E. 2022. A new species of the genus *Microcavia* (Rodentia, Caviidae). *Associación Mexicana de Mastozoología* **13(1)**. DOI: 10.12933/therya-22-1217. ISSN 2007-3364

Teta, P., Ojeda, R. A., Lucero, S. O., & D'Elía, G. 2017. Geographic variation in cranial morphology of the Southern Mountain Cavy *Microcavia australis* (Rodentia, Caviidae): taxonomic implications, with the description of a new species. *Zoological Studies* **56**. DOI: 10.6620/ZS.2017.56-29

Teta, P., & Reyes-Amaya, N. 2021. Uncovering species boundaries through qualitative and quantitative morphology in the genus *Dasyprocta* (Rodentia, Caviomorpha), with emphasis in *D. punctata* and *D. variegata*. *Journal of Mammalogy* **102(6)**: 1548–1563. https://doi.org/10.1093/jmammal/gyab101

Thorington, Jr, R. W., Koprowski, J. L., Steele, M. A., & Whatton, J. F. 2012. *Squirrels of the World*. The Johns Hopkins University Press, Baltimore.

Tirira, D. 2017. *A Field Guide to the Mammals of Ecuador*. University of Texas.

van Roosmalen, M. G. M., Frenz, L., van Hooft, W. F., de Iongh, H. H., & Leirs, H. 2007. A New Species of Living Peccary (Mammalia: Tayassuidae) from the Brazilian Amazon. *Bonner zoologische Beitrage* **55(2)**: 105–112.

van Roosmalen, M. G. M., & van Roosmalen, T. 2003. The Description of a New Marmoset Genus, Callibella (Callitrichinae, Primates), Including Its Molecular Phylogenetic Status. *Neotropical Primates* **11(1)**: 1–10.

van Roosmalen, M. G. M., van Roosmalen, T., & Mittermeier, R. A. 2002. A taxonomic review of the Titi Monkeys, genus *Callicebus* Thomas, 1903, with the description of two new species, *Callicebus bernhardi* and *Callicebus stephennashi*, from Brazilian Amazonia. *Neotropical Primates* **10** (Suppl.).

van Roosmalen, M. G. M., van Roosmalen, T., Mittermeier, R. A., and Rylands, A. B. 2000. Two new species of marmoset, genus *Callithrix* Erxleben, 1777 (Callitrichidae, Primates), from the Tapajos/Madeira interfluvium, south central Amazonia, Brazil. *Neotropical Primates* **8(1)**: 2–18.

Voss, R. S. 2011. Revisionary notes on Neotropical porcupines (Rodentia: Erethizontidae). 3. An annotated checklist of the species of *Coendou* Lacépède, 1799. *Am. Mus. Novit.* **3720**: 1–36.

Voss, R. S., Helgen, K. M., & Jansa, S. A. 2014. "Extraordinary claims require extraordinary evidence: a comment on Cozzuol *et al.* (2013)." *Journal of Mammalogy* **95(4)**: 893–898.

Voss, R. S., Hubbard, C., & Jansa, S. A. 2013. Phylogenetic relationships of New World porcupines (Rodentia, Erethizontidae): implications for taxonomy, morphological evolution, and biogeography. *Am. Mus. Novit.* **3769**: 1–36.

Wallace, R. B., Gómez, H., Felton, A., & Felton, A. M. 2006. On a New Species of Titi Monkey, Genus *Callicebus* Thomas (Primates, Pitheciidae), from Western Bolivia with Preliminary Notes on Distribution and Abundance. *Primate Conservation* **20(1)**: 29–39

Wilson, D. E., Lacher Jr, T. E., & Mittermeier, R. A. (Eds) 2016. *Handbook of the Mammals of the World – Volume 6: Lagomorphs and Rodents* I. Lynx Edicions, Barcelona.

Wilson, D. E., Lacher Jr, T. E., & Mittermeier, R. A. (Eds) 2017. *Handbook of the Mammals of the World – Volume 7: Rodents* II. Lynx Edicions, Barcelona.

Wilson, D. E., & Mittermeier, R. A. (Eds) 2009. *Handbook of the Mammals of the World – Volume 1: Carnivores*. Lynx Edicions, Barcelona.

Wilson, D. E., & Mittermeier, R. A. (Eds) 2011. *Handbook of the Mammals of the World – Volume 2: Hoofed Mammals*. Lynx Edicions, Barcelona.

Wilson, D. E., & Mittermeier, R. A. (Eds) 2014. *Handbook of the Mammals of the World – Volume 4: Sea Mammals*. Lynx Edicions, Barcelona.

Wilson, D. E., & Mittermeier, R. A. (Eds) 2015. *Handbook of the Mammals of the World – Volume 5: Monotremes and Marsupials*. Lynx Edicions, Barcelona.

Wilson, D. E., & Mittermeier, R. A. (Eds) 2018. *Handbook of the Mammals of the World – Volume 8: Insectivores, Sloths and Colugos*. Lynx Edicions, Barcelona.

Useful Websites

The following list includes a number of important websites for the South American mammal-watcher, most of which were important sources of information in the production of this book.

General

www.facebook.com/groups/192466010712 – Neotropical Mammalogy Facebook Group. A closed group but a useful source of up-to-date information on new papers and other information relating to South American mammals.

www.gbif.org/dataset/672aca30-f1b5-43d3-8a2b-c1606125fa1b – A checklist of the Mammal Species of the World produced by the National Museum of Natural History, Smithsonian Institution.

www.iucnredlist.org – The IUCN Red List of Threatened Species: separate pages for each species.

www.mammaldiversity.org – The Mammal Diversity Database of the American Society of Mammalogists (ASM) tracks the latest taxonomic changes to living and recently extinct species and higher taxa of mammals.

www.mammalwatching.com – An important source of information on where to see many of South America's most sought-after mammals. Provides access to reports of mammal-watching trips within South America.

www.robjansenphotography.com – A website featuring great mammal photos and trip reports that provide information on where to find some of South America's most sought-after mammals.

IUCN Specialist Group websites

www.canids.org – The website of the IUCN Canids Specialist Group, with downloadable copies of its journal *Canid Biology & Conservation*.

www.catsg.org – The website of the IUCN Cat Specialist Group.

www.smallcarnivore.org – The website of the IUCN Small Carnivores Specialist Group, with downloadable copies of its journal *Small Carnivore Conservation*.

www.deerspecialistgroup.org – The website of the IUCN Deer Specialist Group, including access to its downloadable newsletter.

www.primate-sg.org – The website of the IUCN Primate Specialist Group, including access to the downloadable journals *Neotropical Primates* and *Primate Conservation*.

www.tapirs.org – The website of the IUCN Tapir Specialist Group, with downloadable copies of its newsletter *Tapir Conservation*.

https://xenarthrans.org – The website of the IUCN Anteater, Sloth and Armadillo Specialist Group, a useful source of information on sloths, anteaters and armadillos, and contains downloadable copies of its newsletter *Edentata*.

Primate websites

www.alltheworldsprimates.org – An extremely valuable source of information on South American primates. Subscription required.

www.neoprimate.org – The website of Neotropical Primate Conservation.

www.facebook.com/neoprimate - Neotropical Primate Conservation Facebook Group. A useful source of up-to-date information on new papers and other information relating to South American primates.

https://primate.wisc.edu/primate-info-net – Maintained by the Wisconsin National Primate Research Center, Primate Info Net provides original content and links to resources about primates in research, education and conservation.

Recommended Mammal Watchers' Code of Conduct

Users of this book are encouraged to be conscious of the potential impacts of mammal watching on both the mammals they are watching and local communities. Responsible mammal watching can have a positive impact on local communities and encourage them to preserve habitats and their associated wildlife, particularly if they can see tangible benefits by doing so. This Code of Conduct was first developed in conjunction with Jon Hall of mammalwatching.com, although this version has evolved separately.

General
- Avoid deliberately disturbing mammals and/or their habitats. The welfare of the animal should always come first.
- Familiarize yourself with and obey the laws of the country in which you are operating.
- Stay on roads, trails and paths where they exist to keep habitat disturbance to a minimum, and never enter private land without the landowner's explicit permission.
- Stay a sensible distance away from the animal. This will depend on the species involved and whether you are on foot, in a vehicle or in a boat. Formal regulations exist for cetacean-watching in many areas.
- Never follow a clearly stressed animal for better views, approach nests or dens too closely or get between a mother and young.
- Using recordings, *e.g.* to call in primates, or lures to call in predators is discouraged.

Spotlighting and thermal imaging
- As thermal imagers become more affordable, wherever practical use these to locate nocturnal mammals and only switch to spotlights when you find something of interest.
- Use the lowest wattage/lumen light necessary to spotlight in the relevant habitat. Lower-power lights are generally adequate for picking up spotlight in forested habitats.
- If using brighter lights, use these only to search for eyeshine, and lower the power once you have found a mammal.
- Where possible, switch to green or red filters when watching an animal and ensure that the light is shone away from the eyes. The mammal should be towards the edge rather than in the center of the beam.
- Do not keep the light on an animal for longer than necessary, particularly at sites where spotlighting occurs regularly and mammals are frequently disturbed. The welfare of the animal should always come before a photograph.

Small-mammal trapping
- Trapping should be carried out only in countries where it is legal to do so. In many countries, licences or permits from the national or regional wildlife management authorities are required.
- Even if you are legally permitted to do so, think carefully before carrying out any small-mammal trapping as even the most careful trapping is invasive and creates a genuine risk of mammals being accidentally injured or even killed.
- Small-mammal trapping should be undertaken only with great care and should not be undertaken in extreme heat or cold, when the risks increase dramatically. If you do intend trapping, it is important to attend specialized training courses offered by naturalist groups and conservation organizations first in order to limit the risks to the animals.

- Traps should have suitable food for any species likely to be trapped, including insectivores, and should have adequate bedding to enable trapped animals to stay warm, particularly in cold conditions. Traps should also be positioned to avoid exposure to high early-morning temperatures.
- In particularly cold conditions, traps should be checked at gaps of no more than every 4–5 hours and should not be left unattended overnight.
- Traps should be thoroughly cleaned, particularly after visiting new areas, in order to reduce the risk of transmitting disease between different populations of mammals.

Baiting

- Baiting should be used only sparingly at all times. Repeated baiting at a site increases the risk of disease and predation and also creates a dependency.
- Live baiting to attract predators should be discouraged.

Bat roosts

- Avoid disturbing bat roosts, particularly at hibernation sites or when bats have young. Never handle bats without the required licences.
- Do not use bright lights in bat roosts and avoid using flash photography.
- You should also ensure that you are vaccinated against the risk of contracting rabies and ensure that you wash clothes thoroughly after visiting bat roosts, particularly large roosts in caves, to reduce the risk of spreading disease.

Reporting

- If possible, please submit data to the relevant local wildlife management agency, citizen science database or, where threatened species are involved, to the relevant IUCN Specialist Group.
- Please write a meaningful report at the end of the trip and post it on mammalwatching. com. Do not just use other people's reports or rely on others to do the hard work.
- Try to include details of the logistics on how to get to sites and not purely lists of species seen, as the latter are of limited value to future visitors. If using guides, include their contact details in reports.
- Be realistic about what you report having seen. Some online mammal-watching reporting sites contain records that seem to be overly ambitious, given that they include species that are not reliably identifiable in the field.

Organized tours

- Guides wanting to please clients may push the boundaries to unacceptable levels. Please encourage them (and anyone else you see behaving inappropriately) to adopt these guidelines for the benefit of the animals. If you feel that guides are getting too close, please say so.
- Tip guides for good practice and not just for finding you the species.

And finally...

Last, but not least, please help locals to appreciate the value of their wildlife and help protect the wildlife and their habitats. Try to be generous with your thanks and tips when you can see animals being protected by people with much less than you have. Respect local people, their land and their customs. Avoid upsetting locals or being disrespectful. The negative effects of such behavior can cause considerable problems in the long term.

Photographic Credits

The following is a complete list of all 563 photographs included in this book, with the name of the photographer and their website where requested. Images sourced via the photographic agencies Alamy (alamy.com), Shutterstock (shutterstock.com), Nature Picture Library (naturepl.com), Minden Pictures (mindenpictures.com) or Calphotos (calphotos.berkeley.edu) are indicated in after the photographer's name.

In total, ten images are published under the terms of one of the following Creative Commons licenses: Attribution 2.0 Generic license (CC BY 2.0), Creative Commons Attribution-ShareAlike 3.0 Unported license (CC BY 3.0), Creative Commons Attribution-ShareAlike 3.0 Unported license (CC BY-SA 3.0), or the Creative Commons Attribution-ShareAlike 4.0 International license (CC BY-SA 4.0). In addition. one image has No Rights Reserved (CC0), The relevant license is indicated using the appropriate code after the photographer's name in the list.

Cover – Jaguar: Andrew Griffin.
Frontispiece – Giant Anteater: Ondřej Prosický.
p. 4 – Puma: Andy and Gill Swash.
p. 12 – Satellite image of South America: Source https://www.worldmap1.com/south-america-map.asp). **p. 16 – Monito del Monte:** Rob Jansen.
p. 17 – Monito del Monte: Mark Chappell [image background replaced]. **p. 18 – Northern Black-eared Opossum:** Rob Jansen. **p. 19 – Big Lutrine Opossum:** Rodolfo Pani, Reserva Costanera.
p. 22 – Southern Four-eyed Opossum: (TOP) Paulo Tomasi, (BOTTOM) Roberto L. M. Novaes.
p. 24 – McIlhenny's Four-eyed Opossum: James L. Patton (Calphotos). **p. 25 – Mondolfi's Four-eyed Opossum:** André de Souza Pereira (CC BY 3.0). **p. 27 – Gray Four-eyed Opossum:** Jon Hall.
p. 28 – Brown Four-eyed Opossum: Ronald Bravo. **p. 29 – Water Opossum:** Eduardo Chacon. **p. 30 – Brazilian White-eared Opossum:** (LEFT) Rob Jansen, (RIGHT) Phil Telfer. **p. 32 – Andean White-eared Opossum:** (TOP) Rob Jansen, (BOTTOM) Victor Quiroz.
p. 33 – Southern Black-eared Opossum: Rob Jansen. **p. 34 – Northern Black-eared Opossum:** (LEFT) Ignacio Yufera, (RIGHT) Kenneth Ross.
p. 35 – Derby's Woolly Opossum: Gianfranco Gómez, Drake Bay (www.thenighttour.com).
p. 36 – Brown-eared Woolly Opossum: Rob Jansen. **p. 37 – Bare-tailed Woolly Opossum:** (LEFT) Antoine Baglan, (RIGHT) Mario Sacramento. **p. 38 – Black-shouldered Opossum:** Marcio Martins. **p. 39 – Bushy-tailed Opossum:** Thiago Oliveira. **p. 40 – Six-banded Armadillo:** Rob Jansen; **Southern Three-banded Armadillo:** Bill Bouton. **p. 41 – Southern Long-nosed Armadillo:** Fabrice Schmitt.

p. 45 – Nine-banded Armadillo: Jeff Blincow.
p. 46 – Hairy Long-nosed Armadillo: Andre Baertschi. **p. 47 – Northern Long-nosed Armadillo:** Diego Rodriguez.
p. 49 – Seven-banded Armadillo: Rodrigo Conte. **p. 50 – Screaming Hairy Armadillo:** Rob Jansen. **p. 51 – Large Hairy Armadillo:** Andrea Izzotti (Shutterstock). **p. 52 – Pichi:** Jeff Blincow.
p. 53 – Six-banded Armadillo: (TOP) Steve Davis, (BOTTOM) Jon Hall. **p. 54 – Greater Fairy Armadillo:** Thomas and Sabine Vinke.
p. 55 – Pink Fairy Armadillo: Pablo Rinaudo.
p. 56 – Southern Three-banded Armadillo: Rob Jansen. **p. 57 – Brazilian Three-banded Armadillo:** Jerome Boisard. **p. 58 – Greater Naked-tailed Armadillo:** Alex Meyer.
p. 59 – Chacoan Naked-tailed Armadillo: Pablo and Maria Zingerling. **p. 60 – Northern Naked-tailed Armadillo:** Cristian Castro Morales.
p. 61 – Southern Naked-tailed Armadillo: Alex Meyer. **p. 62 – Giant Armadillo:** Kevin Schafer.
p. 63 – Southern Tamandua: Andrew Griffin; **Giant Anteater:** Rob Jansen.
p. 64 – Northern Tamandua: (TOP) Rob Jansen, (BOTTOM) Miguel Siu. **p. 65 – Southern Tamandua:** (TOP) Jaime Valenzuela, (BOTTOM) Cheryl Antonucci.
p. 66 – Giant Anteater: Andy and Gill Swash.
p. 67 – Common Silky Anteater: Kevin Schafer.
p. 68 – Rio Negro Silky Anteater: Chris Collins.
p. 69 – Central American Silky Anteater: (LEFT) Jon Hall, (RIGHT) Miguel Siu. **p. 70 – Amboro Silky Anteater:** (BOTH) Andre Baertschi.
p. 71 – Thomas's Silky Anteater: Bernardo Roca-Rey (www.bernardorocarey.com).
p. 74 – Linnaeus's Two-toed Sloth: Rob Jansen.

p. 75 – Hoffmann's Two-toed Sloth: Milan Zygmunt (Shutterstock). **p. 76 – Maned Three-toed Sloth:** (LEFT) Kevin Schafer, (RIGHT) Eduardo Menezes (Shutterstock).

p. 77 – Pale-throated Three-toed Sloth: Fabrice Schmitt.

p. 78 – Brown-throated Three-toed Sloth: (LEFT) Stuart Elsom, (RIGHT) Rob Jansen.

p. 79 – Forest Rabbit: Andy Reago and Chrissy McClarren (CC BY 2.0).

p. 81 – Eastern Cottontail: Pedro Peloso.

p. 82 – Forest Rabbit: Rob Jansen.

p. 83 – Guianan Squirrel: Steve Davis.

p. 84 – Guianan Squirrel: Rob Jansen.

p. 85 – Yellow-throated Squirrel: Andy and Gill Swash. **p. 86 – Red-tailed Squirrel:** (LEFT) Rob Jansen, (RIGHT) Alejandra Rendón.

p. 87 – Bolivian Squirrel: (LEFT) Jon Irvine, (RIGHT) Andy and Gill Swash.

p. 88 – Andean Squirrel: Priscilla Burcher.

p. 89 – Sanborn's Squirrel: Martin Royle.

p. 90 – Guayaquil Squirrel: Andy and Gill Swash.

p. 92 – Northern Amazon Red Squirrel: Phillip Edwards. **p. 93 – Southern Amazon Red Squirrel:** Yasmy Medina Valdivia.

p. 94 – Junín Red Squirrel: Aniket Sardana.

p. 95 – Central American Dwarf Squirrel: Laval Roy. **p. 96 – Amazon Dwarf Squirrel:** Roger Ahlman. **p. 97 – Western Dwarf Squirrel:** (LEFT) Rob Jansen, (RIGHT) Victor Quiroz.

p. 99 – Neotropical Pygmy Squirrel: Fabrice Launay. **p. 100 – Bicolor-spined Porcupine:** Rob Jansen. **p. 101 – Quichua Porcupine:** (TOP) Jacobo Quero, (BOTTOM) Paul Carter.

p. 102 – Eastern Amazonian Dwarf Porcupine: Nereston J. Camargo [image background replaced]. **p. 103 – Yellow Quill-tipped Porcupine:** Fabiano Gumier Costa.

p. 104 – Amazonian Long-tailed Porcupine: Paul Carter. **p. 105 – Baturité Porcupine:** Hugo Fernandes-Ferreira. **p. 106 – Stump-tailed Porcupine:** (LEFT) Rob Jansen, (RIGHT) Jorge M. Brito. **p. 107 – Western Amazonian Dwarf Porcupine:** Paul Carter. **p. 108 – Bahian Hairy Dwarf Porcupine:** János Oláh. **p. 109 – Black-tailed Porcupine:** Antoine Baglan.

p. 110 – Paraguay Hairy Dwarf Porcupine: Alan Martin. **p. 111 – Frosted Porcupine:** Dr Mathias Scharmann, M. S. Silvia Alejandra Sotelo Lopez and Dr Ana Marcela Florez Rueda.

p. 113 – Blackish Hairy Dwarf Porcupine: Sergio Chaparro-Herrera. **p. 114 – Pernambuco Dwarf Porcupine:** János Oláh.

p. 115 – Broomstraw-spined Porcupine: Leonardo Mercon (Shutterstock). **p. 116 – Red Acouchi:** Pavel Filatov (Shutterstock). **p. 117 – Green Acouchi:** Stefan Köder. **p. 118 – Azara's Agouti:** (TOP) Andy and Gill Swash, (BOTTOM) Rob Jansen. **p. 119 – Black Agouti:** Roger Ahlman. **p. 121 – Kalinowski's Agouti:** Roland Seitre (Minden Pictures).

p. 122 – Common Red-rumped Agouti: (TOP) John Caddick, (BOTTOM) Leonardo Mercon (Shutterstock). **p. 123 – Orange Agouti:** Klaus Rudloff. **p. 124 – Jack's Red-rumped Agouti:** Hugo Fernandes-Ferreira. **p. 125 – Black-rumped Agouti:** Pablo Cerqueira.

p. 126 – Central American Agouti: Jeff Blincow. **p. 127 – Brown Agouti:** Phillip Edwards.

p. 128 – Lowland Paca: James Adams.

p. 129 – Mountain Paca: Rob Jansen.

p. 130 – Pacarana: Rob Jansen.

p. 131 – Patagonian Mara: (LEFT) Jesus Carbo, (RIGHT) Phillip Edwards. **p. 132 – Chacoan Mara:** Rob Jansen. **p. 133 – Brazilian Guinea Pig:** Andy and Gill Swash; **Montane Guinea Pig:** James Adams. **p. 135 – Greater Guinea Pig:** Arnaud Delberghe; **Southern Mountain Cavy:** Rob Jansen; **Common Yellow–toothed Cavy:** Keith Dover; **Rock Cavy:** Allan Hopkins.

p. 136 – Lesser Capybara: (TOP) Jon Hall, (BOTTOM) Miguel Siu.

p. 137 – Greater Capybara: (BOTH) Jeff Blincow.

p. 138 – Coypu: Rob Jansen.

p. 139 – Common Plains Viscacha: Jon Hall.

p. 140 – Ecuadorian Mountain Viscacha: Carlos Narváez Romero. **p. 141 – Common Mountain Viscacha:** Rob Jansen.

p. 142 – Wolffsohn's Viscacha: (TOP) Paul Jones, (BOTTOM) Jeff Foott (Minden Pictures / Alamy).

p. 143 – Short-tailed Chinchilla: Francisco Erize.

p. 144 – Chilean Chinchilla: Rob Jansen.

p. 145 – Northern Oncilla: Rob Jansen.

p. 146 – Puma: Janco van Gelderen.

p. 147 – Jaguarundi: (TOP) slowmotiongli (Shutterstock), (BOTTOM) Gabriel Arroyo.

p. 148 – Geoffroy's Cat: (TOP) Enrique Couve / Far South Expeditions, (BOTTOM) Rob Jansen. **p. 149 – Kodkod:** (TOP) Enrique Couve / Far South Expeditions, (BOTTOM) Eduardo Minte. **p. 151 – Southern Colocolo:** (TOP) Nicolás García Del Castello, (BOTTOM) Ashley Howe. **p. 152 – Pantanal Cat:** Marc Faucher. **p. 153 – Muñoa's Pampas Cat:** Carol Fontes. **p. 154 – Northern Colocolo:** (LEFT) Rob Jansen, (RIGHT) David Robichaud.

p. 155 – **Andean Cat:** Sebastian Kennerknecht (Minden Pictures / Alamy).
p. 156 – **Northern Oncilla:** (TOP) Ignacio Yufera, (BOTTOM) Phil Telfer.
p. 157 – **Southern Oncilla:** Groumfy69 (CC BY-SA 3.0). p. 158 – **Eastern Oncilla:** John Philip Medcraft. p. 159 – **Margay:** Luis Piovani.
p. 160 – **Ocelot:** Mike Watson. p. 161 – **Jaguar:** Mike Watson. p. 162 – **Maned Wolf:** Rob Jansen.
p. 163 – **Northern Gray Fox:** Bill Bouton.
p. 164 – **Culpeo:** Jeff Blincow. p. 165 – **Darwin's Fox:** Jono Dashper. p. 166 – **Pampas Fox:** Jeff Blincow. p. 167 – **Sechuran Fox:** Rob Jansen.
p. 168 – **South American Gray Fox:** Jeff Blincow.
p. 169 – **Hoary Fox:** Phil Telfer.
p. 170 – **Maned Wolf:** Andy and Gill Swash.
p. 171 – **Crab-eating Fox:** Rob Jansen.
p. 172 – **Short-eared Dog:** Galo Zapata Rios.
p. 173 – **Bush Dog:** Vladimir Wrangel (Shutterstock). p. 174 – **Andean Bear:** Jeff Blincow. p. 175 – **South American Fur Seal:** Otaria1 (CC BY-SA 4.0). p. 176 – **South American Fur Seal:** (TOP) Jonathan Chancasana (Shutterstock), (BOTTOM) Melody Lytle.
p. 177 – **Juan Fernández Fur Seal:** David Fisher; **Antarctic Fur Seal:** Phillip Edwards; **Galápagos Fur Seal:** Andy and Gill Swash.
p. 178 – **Galápagos Sea Lion:** Andy and Gill Swash. p. 179 – **South American Sea Lion:** (LEFT) Rob Jansen, (RIGHT) Andy and Gill Swash. p. 180 – **Southern Elephant Seal:** fieldwork (Shutterstock). p. 181 – **Leopard Seal:** (BOTH) Chris Collins; **Weddell Seal:** (BOTH) Chris Collins; **Crabeater Seal:** Chris Collins.
p. 182 – **Marine Otter:** Kevin Schafer.
p. 183 – **Southern River Otter:** Nicolas McPhee.
p. 184 – **Neotropical Otter:** John Richardson.
p. 185 – **Giant Otter:** Andy and Gill Swash.
p. 186 – **Tayra:** Andy and Gill Swash.
p. 187 – **Patagonian Weasel:** Dario Podesta.
p. 188 – **Lesser Grison:** Rob Jansen.
p. 189 – **Greater Grison:** Tony Hisgett.
p. 190 – **Tayra:** Phillip Edwards.
p. 191 – **Long-tailed Weasel:** (TOP) Ron Dudley, (BOTTOM) Keith and Kasia (CC BY 2.0).
p. 194 – **Humboldt's Hog-nosed Skunk:** Rob Jansen. p. 195 – **Striped Hog-nosed Skunk:** (LEFT) Rob Jansen, (RIGHT) Rafael Martos Martins. p. 196 – **Humboldt's Hog-nosed Skunk:** Jeff Blincow. p. 197 – **Molina's Hog-nosed Skunk:** Rob Jansen. p. 198 – **Western Mountain Coati:** Romain Bocquier.
p. 199 – **Kinkajou:** James Adams.

p. 200 – **Western Lowland Olingo:** Alex Meyer.
p. 201 – **Eastern Lowland Olingo:** (LEFT) Rod Williams (Naturepl.com), (RIGHT) Marc Faucher.
p. 202 – **Olinguito:** Carlos Bocos.
p. 203 – **Crab-eating Raccoon:** (BOTH) Rob Jansen. p. 204 – **White-nosed Coati:** Jeff Blincow.
p. 205 – **South American Coati:** (BOTH) Rob Jansen. p. 207 – **Western Mountain Coati:** (TOP) Ben Schweinhart, (BOTTOM) Romain Bocquier.
p. 208 – **Marmosets (Common Marmoset):** Andy and Gill Swash; **Tamarins (Weddell's Saddle-back Tamarin):** Angus McNab; **Lion Tamarins (Golden-headed Lion Tamarin):** Ciro Albano; **Squirrel Monkeys (Black-capped Squirrel Monkey):** Glenn Bartley; **Capuchins (Humboldt's White-fronted Capuchin):** Jon Hall; **Night Monkeys (Brumback's Night Monkey):** Jon Hall.
p. 209 – **Titi Monkeys (Toppin's Titi):** Rob Jansen; **Saki Monkeys (Pithecia) (Golden-faced Saki):** Andy and Gill Swash; **(Bearded) Saki Monkeys (Chiropotes) (Guianan Bearded Saki):** Andy and Gill Swash; **Uakaris (White Bald Uakari):** Dick Filby; **Howlers (Eastern Red-handed Howler):** Bruno Carvalho; **Spider Monkeys (Colombian Black Spider Monkey):** Ignacio Yufera; **Woolly Monkeys (Humboldt's Woolly Monkey):** Andy and Gill Swash; **Muriquis (Northern Muriqui):** Rob Jansen.
p. 210 – **Geoffroy's Tufted-ear Marmoset:** Gabriel Mello. p. 214 – **Northern Pygmy Marmoset:** Rob Jansen. p. 215 – **Southern Pygmy Marmoset:** Cheryl Antonucci.
p. 216 – **Goeldi's Monkey:** (BOTH) Rob Jansen.
p. 217 – **Black-crowned Dwarf Marmoset:** Jon Hall. p. 218 – **Silvery Marmoset:** (LEFT) Pablo Cerqueira, (RIGHT) Cheryl Antonucci.
p. 219 – **Golden-white Bare-ear Marmoset:** Roland Seitre (Naturepl.com). p. 220 – **Golden-white Tassel-ear Marmoset:** (BOTH) Jeff Blincow.
p. 221 – **Munduruku Marmoset:** Marlyson Costa. p. 222 – **Snethlage's Marmoset:** FLPA / Alamy. p. 223 – **Black-tailed Marmoset:** Rob Jansen. p. 224 – **Rondon's Marmoset:** (TOP) Caio Brito, (BOTTOM) Cheryl Antonucci.
p. 226 – **Schneider's Marmoset:** (LEFT) Diego Silva, (RIGHT) Felipe Arantes. p. 228 – **Marca's Marmoset:** Albert Burgas Riera.
p. 230 – **Maués Marmoset:** (LEFT) Jeff Blincow, (RIGHT) Ignacio Yufera. p. 231 – **Santarém Marmoset:** (LEFT) Andy and Gill Swash, (RIGHT) Pablo Cerqueira. p. 232 – **Sateré Marmoset:** (LEFT) Mark Sullivan, (RIGHT) Ignacio Yufera.

p. 233 – **Buffy-headed Marmoset:** (BOTH) Rob Jansen. p. 234 – **Buffy Tufted-ear Marmoset:** Rob Jansen. p. 235 – **Black Tufted-ear Marmoset:** (LEFT) Steve Davis, (RIGHT) Rob Jansen. p. 236 – **Geoffroy's Tufted-ear Marmoset:** Steve Davis. p. 237 – **Wied's Tufted-ear Marmoset:** Andy and Gill Swash. p. 238 – **Common Marmoset:** (BOTH) Andy and Gill Swash. p. 239 – **Cotton-top Tamarin:** Rob Jansen. p. 240 – **Black-mantled Tamarin:** Rob Jansen. p. 241 – **Lesson's Saddle-backed Tamarin:** (TOP) Victor Quiroz, (BOTTOM) Rob Jansen. p. 242 – **Golden-mantled Saddle-back Tamarin:** (LEFT) Rob Jansen, (RIGHT) Alan Dahl. p. 243 – **Red-mantled Saddle-back Tamarin:** (LEFT) Rob Jansen, (RIGHT) Marc Faucher. p. 244 – **Andean Saddle-back Tamarin:** Rob Jansen. p. 245 – **Illiger's Saddle-back Tamarin:** Paul Tavares. p. 247 – **Cruz-Lima's Saddle-back Tamarin:** Ricardo Sampaio. p. 248 – **Spix's Saddle-backed Tamarin:** Ben Schweinhart. p. 249 – **Weddell's Saddle-back Tamarin:** Rob Jansen. p. 250 – **Weddell's Saddle-back Tamarin:** Angus McNab. p. 251 – **Spix's Mustached Tamarin:** (LEFT) Rob Jansen, (RIGHT) Cheryl Antonucci. p. 254 – **Red-bellied Tamarin:** blickwinkel / Hummel (Alamy). p. 255 – **Mottle-face Tamarin:** (LEFT) Rob Jansen, (RIGHT) Luis G. Restrepo. p. 256 – **Black-chinned Emperor Tamarin:** blickwinkel / Hummel (Alamy). p. 257 – **Bearded Emperor Tamarin:** Rob Jansen. p. 258 – **Western Black-handed Tamarin:** Nereston J. Camargo. p. 259 – **Eastern Black-handed Tamarin:** (LEFT) Vincent Lo, (RIGHT) Nailson Júnior. p. 260 – **Midas Tamarin:** Ariadne Van Zandbergen (Alamy). p. 261 – **Martins's Bare-faced Tamarin:** (BOTH) Felipe Arantes. p. 262 – **Pied Tamarin:** (LEFT) Mark Sullivan, (RIGHT) guentermanaus (Shutterstock). p. 263 – **White-footed Tamarin:** (LEFT) Ignacio Yufera, (RIGHT) Victor Quiroz. p. 264 – **Cotton-top Tamarin:** (LEFT) NataliaVo (Shutterstock), (RIGHT) Ignacio Yufera. p. 265 – **Geoffroy's Tamarin:** (LEFT) Andrew Griffin, (RIGHT) Glenn Bartley. p. 266 – **Golden Lion Tamarin:** Mike Watson. p. 267 – **Golden-headed Lion Tamarin:** Ciro Albano. p. 268 – **Black Lion Tamarin:** Rob Jansen. p. 269 – **Black-faced Lion Tamarin:** Cheryl Antonucci. p. 270 – **Humboldt's Squirrel Monkey:** Jeff Blincow. p. 271 – **Colombian Squirrel Monkey:** Rob Jansen.

p. 272 – **Ecuadorian Squirrel Monkey:** Rob Jansen. p. 273 – **Golden-backed Squirrel Monkey:** Ignacio Yufera. p. 274 – **Guianan Squirrel Monkey:** (LEFT) Anthony Carole, (RIGHT) David Robichaud. p. 275 – **Collins's Squirrel Monkey:** (LEFT) Funtay (Shutterstock), (RIGHT) Ernst Vikne (CC BY-SA 3.0). p. 276 – **Black-capped Squirrel Monkey:** Rob Jansen. p. 277 – **Black-headed Squirrel Monkey:** Chris Collins. p. 278 – **Colombian White-fronted Capuchin:** Rob Jansen. p. 279 – **Northern Black-horned Capuchin:** Sjef Ollers. p. 280 – **Northern Black-horned Capuchin:** Andy and Gill Swash. p. 281 – **Southern Black-horned Capuchin:** guentermanaus (Shutterstock). p. 282 – **Hooded Capuchin:** Rob Jansen. p. 283 – **Crested Capuchin:** (BOTH) Rob Jansen. p. 284 – **Bearded Capuchin:** Tiago Falótico (CC BY-SA 3.0). p. 285 – **Yellow-breasted Capuchin:** (TOP) Andy and Gill Swash, (BOTTOM) Ciro Albano. p. 286 – **Guianan Brown Capuchin:** Rob Jansen. p. 287 – **Guianan Brown Capuchin:** (TOP) Chris Collins, (BOTTOM) Andy and Gill Swash. p. 288 – **Blond Capuchin:** (LEFT) Thiago Tolêdo, (RIGHT) Regina Ribeiro. p. 289 – **Shock-headed Capuchin:** Ivan Mlinaric (CC BY 2.0). p. 290 – **Marañón White-fronted Capuchin:** (LEFT) Rob Jansen, (RIGHT) Glenn Bartley. p. 291 – **Spix's White-fronted Capuchin:** Jeff Blincow. p. 292 – **Humboldt's White-fronted Capuchin:** Jon Hall. p. 293 – **Ka'apor Capuchin:** (BOTH) Pablo Cerqueira. p. 294 – **Weeper Capuchin:** Gabbro (Alamy). p. 296 – **Venezuelan Brown Capuchin:** (BOTH) Oswaldo Hernández. p. 298 – **Varied White-fronted Capuchin:** Rob Jansen. p. 299 – **Río Cesar White-fronted Capuchin:** Jaime Alejandro Montañez Mendez. p. 300 – **Santa Marta White-fronted Capuchin:** Cheryl Antonucci. p. 301 – **Trinidad White-fronted Capuchin:** Faraaz Abdool. p. 302 – **Ecuadorian White-fronted Capuchin:** Ivette Solis Ponce. p. 303 – **Colombian White-fronted Capuchin:** Jerome Boisard. p. 304 – **Brumback's Night Monkey:** Jon Hall; **Gray-legged Night Monkey:** Rob Jansen; **Black-headed Night Monkey:** Rob Jansen. p. 305 – **Lemurine Night Monkey:** Rob Jansen. p. 306 – **Gray-legged Night Monkey:** Pete Morris. p. 307 – **Panamanian Night Monkey:** Glenn Bartley. p. 308 – **Brumback's Night Monkey:** (TOP) Rob Jansen, (BOTTOM) Jon Hall.

p. 309 – **Humboldt's Night Monkey:** Blickwinkel / Layer (Alamy). **p. 310 – Spix's Night Monkey:** Rob Jansen.

p. 312 – Andean Night Monkey: (LEFT) Rob Jansen, (RIGHT) János Oláh. **p. 313 – Ma's Night Monkey:** (LEFT) Rob Jansen, (RIGHT) Jeff Blincow. **p. 314 – Black-headed Night Monkey:** (LEFT) Andy and Gill Swash, (RIGHT) James Adams. **p. 315 – Azara's Night Monkey:** Jéssica dos Anjos Oliveira. **p. 317 – Rio Beni Titi:** Jesus Martinez / WCS. **p. 319 – Rio Beni Titi:** Rob Jansen. **p. 320 – White-eared Titi:** Rob Jansen.

p. 321 – White-coated Titi: Rob Jansen.

p. 322 – Olalla Brothers' Titi: Rob Jansen.

p. 323 – Rio Mayo Titi: Michael Greenfelder (Alamy). **p. 324 – Coppery Titi:** Ernie Janes (Alamy). **p. 325 – Red-crowned Titi:** (BOTH) Rob Jansen. **p. 326 – Ornate Titi:** (LEFT) Marc Faucher, (RIGHT) Rob Jansen. **p. 327 – Chestnut-bellied Titi:** Arnaud Delberghe.

p. 328 – Doubtful Titi: Fabio Schunck.

p. 329 – Stephen Nash's Titi: Roland Wirth [image background replaced]. **p. 330 – Caquetá Titi:** Rob Jansen. **p. 331 – Ashy Titi:** Jeff Blincow.

p. 333 – Lake Baptista Titi: (LEFT) Chris Collins, (RIGHT) Cheryl Antonucci. **p. 334 – Red-bellied Titi:** (LEFT) Andy and Gill Swash, (RIGHT) Cheryl Antonucci. **p. 335 – Vieira's Titi:** Alex Meyer [image background replaced].

p. 336 – Brown Titi: blickwinkel / Hummel (Alamy). **p. 338 – Madidi Titi:** (LEFT) Rob Jansen, (RIGHT) Candy McNamee.

p. 339 – Milton's Titi: Adriano Gambarini.

p. 340 – Toppin's Titi: Rob Jansen.

p. 341 – Urubamba Brown Titi: (LEFT) Kevin Schafer, (RIGHT) Angus McNab.

p. 342 – Groves's Titi: (BOTH) Regina Ribeiro.

p. 344 – Medem's Titi: Jon Hall.

p. 345 – White-collared Titi: Gustavo Alarcón-Nieto. **p. 346 – White-chested Titi:** (LEFT) Trevor Ellery, (RIGHT) Rob Jansen. **p. 347 – Yellow-handed Titi:** (TOP) Jesus Carbo, (BOTTOM) LABETAA Andre (Shutterstock).

p. 351 – Black-fronted Titi: Daniel Mello.

p. 352 – Masked Titi: Rob Jansen.

p. 353 – Southern Bahian Titi: Jacek Kisielewski (CC BY-SA 3.0). **p. 354 – Blond Titi:** (LEFT) Cristine Prates, (RIGHT) Cheryl Antonucci.

p. 355 – Coimbra-Filho's Titi: (LEFT) Wylde Vieira, (RIGHT) Eduardo Menezes (Shutterstock).

p. 356 – Gray's Bald-faced Saki: Pablo Cerqueira; **Uta Hick's Bearded Saki:** Diego Grandi (Shutterstock) [image background replaced].

p. 357 – White-faced Saki: (LEFT) Pablo Cerqueira, (RIGHT) Danny Iacob (Shutterstock).

p. 358 – Golden-faced Saki: (LEFT) Andy and Gill Swash, (RIGHT) Pablo Cerqueira. **p. 359 – Hairy Saki:** Russell Mittermeier. **p. 360 – Miller's Saki:** (LEFT) Rob Jansen, (RIGHT) Laura Marsh.

p. 361 – Burnished Saki: (LEFT) Russell Mittermeier, (RIGHT) Laura Marsh [image background replaced]. **p. 362 – Geoffroy's Monk Saki:** Alex Meyer [image background replaced].

p. 363 – Geoffroy's Monk Saki: Mark Bowler.

p. 365 – Equatorial Saki: Jesus Carbo.

p. 366 – Napo Saki: Ben Queenborough (Alamy).

p. 367 – Isabel's Saki: Amazon-Images (Alamy).

p. 368 – Buffy Saki: Russell Mittermeier.

p. 369 – Vanzolini's Bald-faced Saki: (BOTH) Prof. Marcelo Ismar Silva Santana.

p. 371 – Gray's Bald-faced Saki: Andy and Gill Swash. **p. 372 – Mittermeier's Tapajós Saki:** Ignacio Yufera. **p. 373 – Rylands's Bald-faced Saki:** Rob Jansen. **p. 374 – Pissinatti's Bald-faced Saki:** Philip Stettler. **p. 375 – Red-nosed Bearded Saki:** Andy and Gill Swash. **p. 376 – Black Bearded Saki:** Andy and Gill Swash.

p. 377 – Rio Negro Bearded Saki: (BOTH) Thomas Fuhrmann. **p. 378 – Uta Hick's Bearded Saki:** (TOP) João Marcos Rosa, (BOTTOM) blickwinkel / Hummel (Alamy).

p. 379 – Guianan Bearded Saki: (BOTH) Andy and Gill Swash. **p. 380 – Ucayali Bald Uakari:** Mark Bowler (Naturepl.com).

p. 381 – Red Bald Uakari: GTW (Shutterstock).

p. 382 – Ucayali Bald Uakari: Jeff Blincow.

p. 384 – White Bald Uakari: Dick Filby.

p. 385 – Kanamari White Uakari: Edward Parker (Alamy). **p. 386 – Neblina Black-headed Uakari:** (BOTH) Radana Dungalova.

p. 387 – Humboldt's Black-headed Uakari: (TOP) Rob Jansen, (BOTTOM) Radana Dungalova.

p. 389 – Bolivian Red Howler: Rob Jansen.

p. 390 – Colombian Red Howler: (TOP) Josh Vandermeulen, (BOTTOM) Alejandra Rendón.

p. 391 – Juruá Red Howler: Jéssica dos Anjos Oliveira. **p. 392 – Purús Red Howler:** Rodrigo Costa Araujo. **p. 393 – Ursine Red Howler:** Oswaldo Hernández. **p. 394 – Guianan Red Howler:** (LEFT) Kester Clarke, (RIGHT) Andy and Gill Swash. **p. 395 – Bolivian Red Howler:** Ignacio Yufera.

p. 396 – Amazon Black Howler: Chris Collins.

p. 397 – Eastern Red-handed Howler: (LEFT) Pablo Cerqueira, (RIGHT) Bruno Carvalho.

p. 398 – Spix's Red-handed Howler: (TOP) Cheryl Antonucci, (BOTTOM) Rich Hoyer. **p. 399 – Maranhão Red-handed Howler:** (BOTH) Pablo Cerqueira. **p. 400 – Brown Howler:** (LEFT) Rob Jansen, (RIGHT) Steve Davis. **p. 401 – Black-and-gold Howler:** Lee Dingain. **p. 402 – Mantled Howler:** Rob Jansen. **p. 403 – Brown-headed Spider Monkey:** (TOP) John Rogers, (BOTTOM) Stephen John Davies. **p. 404 – Colombian Black Spider Monkey:** (BOTH) Ignacio Yufera. **p. 405 – Black Spider Monkey:** (BOTH) Rob Jansen. **p. 406 – Red-faced Black Spider Monkey:** (BOTH) Chris Collins. **p. 407 – White-whiskered Spider Monkey:** (LEFT) Minister2009 (CC BY 3.0), (RIGHT) Cheryl Antonucci. **p. 408 – White-bellied Spider Monkey:** Whaldener Endo. **p. 409 – Variegated Spider Monkey:** Joachim S. Müller. **p. 410 – Humboldt's Woolly Monkey:** Rob Jansen. **p. 411 – Humboldt's Woolly Monkey:** James Adams. **p. 412 – Humboldt's Woolly Monkey:** Jeff Blincow. **p. 413 – Yellow-tailed Woolly Monkey:** (LEFT) Bernardo Roca-Rey (www.bernardorocarey.com), (RIGHT) Mark Bowler (Alamy). **p. 414 – Southern Muriqui:** (LEFT) Cheryl Antonucci, (RIGHT) Rob Jansen. **p. 415 – Northern Muriqui:** Rob Jansen. **p. 416 – Franciscana:** (BOTH) Rob Jansen. **p. 417 – Guiana Dolphin:** Alex Meyer. **p. 418 – Tucuxi:** (LEFT) Jeff Blincow, (TOP RIGHT) Ben Schweinhart, (BOTTOM RIGHT) Fernando Trujillo (www.mamiferoscolombia.org) **p. 419 – Amazon River Dolphin:** (LEFT) Chris Collins, (TOP RIGHT & BOTTOM RIGHT) Rob Jansen. **p. 422 – Collared Peccary:** James Adams. **p. 423 – White-lipped Peccary:** Bruno Carvalho. **p. 424 – Chacoan Peccary:** Jon Hall. **p. 425 – Collared Peccary:** Dennis W Donohue (Shutterstock). **p. 426 – Guanaco:** Jon Hall. **p. 427 – Vicuña:** Rob Jansen. **p. 428 – Guanaco:** Jeff Blincow. **p. 429 – White-tailed Deer:** Jeff Blincow. **p. 430 – White-tailed Deer:** (LEFT) Andy and Gill Swash, (RIGHT) ecuadorplanet (Shutterstock). **p. 431 – Pampas Deer:** Angelo Gandolfi (NPL/Alamy). **p. 432 – Common Red Brocket:** (LEFT) Greg Dean, (RIGHT) Pablo Cerqueira. **p. 433 – Mexican Red Brocket:** Marc Faucher. **p. 434 – Small Red Brocket:** Alex Meyer. **p. 435 – Mérida Brocket:** T-34-85 (CC0, via Wikimedia Commons). **p. 436 – Common Dwarf Brocket:** Carlos Otávio Gussoni.

p. 437 – Common Brown Brocket: (BOTH) Rob Jansen. **p. 438 – Amazonian Brown Brocket:** Alex Meyer. **p. 439 – Lesser Brocket:** Alex Meyer. **p. 440 – Little Red Brocket:** Roger Ahlman. **p. 441 – Northern Pudu:** Roger Ahlman. **p. 442 – Southern Pudu:** Rob Jansen. **p. 443 – Marsh Deer:** (TOP) Steve Davis, (BOTTOM) Andy and Gill Swash. **p. 444 – North Andean Huemul:** Rob Jansen. **p. 445 – South Andean Huemul:** Farjana.rahman (Shutterstock). **p. 446 – Lowland Tapir:** Rob Jansen. **p. 447 – Central American Tapir:** Arco/G. Lacz (Imagebroker/Alamy). **p. 448 – Mountain Tapir:** Roger Ahlman. **p. 449 – Lowland Tapir:** (TOP) Andy and Gill Swash, (BOTTOM) James Adams. **p. 450 – Amazonian Manatee:** Tharuka Photographer (Shutterstock). **p. 451 – West Indian Manatee:** Ethan Daniels (Shutterstock). **p. 452 – European Rabbit:** David Kjaer; **European Hare:** Glyn Sellors. **p. 453 – Pallas's Squirrel:** Marut Sayarnnikroth (Shutterstock); **North American Beaver:** Christian Musat (Shutterstock). **p. 454 – Common Muskrat:** Claude Allaert; **American Mink:** David Kjaer. **p. 455 – Small Indian Mongoose:** Gillian Holliday; **Wild Boar:** David Kjaer; **Feral domestic pig:** Andy and Gill Swash. **p. 456 – Fallow Deer:** (TOP) SanderMeertinsPhotography (Shutterstock), (BOTTOM) Jeff Blincow; **Western Red Deer:** (TOP) Jeff Blincow, (BOTTOM) Jeff Blincow. **p. 457 – Chital:** (TOP) Karel Bartik (Shutterstock), (BOTTOM) Andy and Gill Swash; **Sambar:** (TOP) Signature Message (Shutterstock), (BOTTOM) Piotr Poznan (Shutterstock). **p. 458 – Pere David's Deer:** (BOTH) Mikhail Kolesnikov (Shutterstock); **Blackbuck:** (TOP) Erni (Shutterstock), (BOTTOM) Sourabh Bharti (Shutterstock). **p. 459 – Alpine Ibex:** (TOP) Dario Pautasso (Shutterstock), (BOTTOM) Dario Pautasso (Shutterstock); **Mouflon:** (TOP) Jordi Guerrero Morera (Shutterstock), (BOTTOM) Luciavonu (Shutterstock). **p. 461 – Llama:** buteo (Shutterstock); **Alpaca:** Bruno M Photographie (Shutterstock); **Domestic Water Buffalo:** Andy and Gill Swash.

Index

This index includes the English and (in *italics*) scientific names of all the mammals referred to in the book. **CAPITALIZED** text is used for 'types' of mammals; **bold text** highlights families or subfamilies and those species that are the subject of a main species account or are illustrated; regular text is used for alternative English names and other species that are mentioned but not subject to a full account. *Italicized* numbers indicate other pages where is a photograph or illustration.